Unsaturated Soils

Unsaturated Soils

A fundamental interpretation of soil behaviour

E. J. Murray

Director
Murray Rix Geotechnical
UK

V. Sivakumar

Senior Lecturer
Queen's University Belfast
UK

A John Wiley & Sons, Ltd., Publication

This edition first published 2010
© E. J. Murray

Blackwell Publishing was acquired by John Wiley & Sons in February 2007. Blackwell's publishing programme has been merged with Wiley's global Scientific, Technical, and Medical business to form Wiley-Blackwell.

Registered office
John Wiley & Sons Ltd. The Atrium, Southern Gate, Chichester, West Sussex, PO19 8SQ, United Kingdom

Editorial offices
9600 Garsington Road, Oxford, OX4 2DQ, United Kingdom
2121 State Avenue, Ames, Iowa 50014-8300, USA

For details of our global editorial offices, for customer services and for information about how to apply for permission to reuse the copyright material in this book please see our website at www.wiley.com/wiley-blackwell.

Library of Congress Cataloging-in-Publication Data

Murray, E. J. (Edward John), 1949-
    Unsaturated soils : a fundamental interpretation of soil behaviour / E.J. Murray, V. Sivakumar.
        p. cm.
    Includes bibliographical references and index.
        ISBN 978-1-4443-3212-4 (pbk. : alk. paper)   1. Soil mechanics.   2. Zone of aeration.   I. Sivakumar, V.
    II. Title.
        TA711.M87 2010
        624.1′5136–dc22   2010007734

A catalogue record for this book is available from the British Library.

Set in 9.5/11.5pt Sabon by Aptara® Inc., New Delhi, India
Printed and bound in Malaysia by Vivar Printing Sdn Bhd

1   2010

Contents

Preface

The book examines the mechanical properties of unsaturated soils. The aim is to provide students and researchers in geotechnical engineering with a framework for understanding soil behaviour at a fundamental level and a model for interpreting experimental data, as well as providing practitioners with a deeper appreciation of the important characteristics of unsaturated soils. A notable theme of the book is the interpretation of experimental strength and compression data from stress path tests in the triaxial cell.

The three phases making up an unsaturated soil are the soil particles, water and air; the physical and chemical properties of which vary widely. The behaviour of soils is influenced by many factors but must be compliant with the general laws of thermodynamics that provide broad principles to which material behaviour must adhere. The theoretical analyses are based on sound thermodynamic principles and provide a rigorous methodology, justified by comparison with test data, to predict and investigate the mechanical behaviour of unsaturated soils. We have endeavoured to keep the theoretical part of the book interesting but thorough, and have drawn on analogous behaviour in physics and chemistry to explain important phenomena.

Chapter 1 provides a basic introduction to soil variables, the phases, the phase interactions and the relevance of soil structure to subsequent discussions. Particular reference is made to mercury intrusion porosimetry (MIP) testing in describing the aggregated, bi-modal structure of unsaturated soils.

An up-to-date review of laboratory testing techniques is presented in Chapters 2 and 3, including suction measurement and control techniques in laboratory triaxial cell testing. The basis of the testing is important if the ideas developed from thermodynamics are to be properly applied.

Chapter 4 introduces the stress state variables, critical state and theoretical models in unsaturated soils. This review of current ideas provides a background to subsequent analyses, which differ principally in the description of unsaturated soils as controlled by a dual stress regime with the relative volumes of the phases playing an essential role in defining the volumes through which the stresses act.

Chapters 5 and 6 dive into a detailed description of thermodynamic principles as applied to multiphase materials under equilibrium conditions. In particular, the significance of the thermodynamic potentials, including enthalpy, is described. It is shown that it is possible to deal with anisotropic stress conditions as the thermodynamic potentials are extensive variables. The significance of the minimisation of the thermodynamic potentials at equilibrium and the meanings of equilibrium and meta-stable equilibrium are outlined.

Chapters 7 and 8 use the thermodynamic principles established in Chapters 5 and 6 to develop an alternative theoretical basis for analysing unsaturated soils. Soil structure is broken down into its component parts to develop an equation(s) describing the dual stress regime. The critical state strength and compression characteristics of unsaturated soils are examined and it is shown how the behaviour may be viewed as a three-dimensional model in dimensionless stress–volume space.

The analysis is extended in Chapter 9 to the work input into unsaturated soils and the development of conjugate stress, volumetric and strain-increment variables. These are used to examine experimental triaxial shear strength data on kaolin. The formulation for energy dispersion allows not only the anisotropic stress–strain behaviour on a macro-mechanical level to be appraised but also the stress–strain behaviour of the aggregates of soil particles and water on a micro-mechanical level to be examined. Analysis of the experimental data highlights phenomena that cannot be explained by other means.

E. J. Murray
V. Sivakumar

Acknowledgements

We thank the reviewers who commented on the text and suggested amendments that we have endeavoured to accommodate. We also extend our thanks to researchers at Queen's University Belfast, as much of the laboratory test data analysed and contributing to the book is based on their experimental work. This includes laboratory test data obtained by Dr Rajaratnam Sivakumar, Dr Rachel Thom and, in particular, Dr Jane Boyd (nee Brown) who also assisted with some of the figures and the text. Rachel Jones of Murray Rix Geotechnical also deserves our gratitude for contributions to discussions.

We are as always indebted to our respective wives Rhianydd Murray and Shirleen Sivakumar for their encouragement and for putting up with our moods when work got us down.

Gratitude is also expressed to staff at the library of the Institution of Civil Engineers for their assistance with references and to Wiley-Blackwell, and in particular Madeleine Metcalfe (Senior Commissioning Editor), for encouragement and advice in the preparation of the book.

E. J. Murray
V. Sivakumar

Introduction

Civil engineering structures interact with the ground and some structures are composed entirely of ground-derived materials. In this context, the ground is as much an engineering material as concrete and steel. We will be concerned principally with those fine-grained clays and silts, or soils containing coarser sand and gravel particles but with a significant percentage of fines, which constitute the ground. Soil–structure interaction must be taken into account for structures such as foundations, roads and tunnels; and the behaviour of earth structures such as earth dams and slopes requires the development of sound principles on which to base analyses. There is a fundamental need to interpret the ground conditions and geology in an engineering context, to provide warning of natural disasters such as landslides, to deal with environmental issues such as landfills or contaminant migration in the ground and groundwater, amongst other equally important issues. Soil mechanics is a specialist branch of civil engineering that addresses the investigation, analysis and prediction of soil behaviour. Yet, even among practising engineers there is often lack of awareness of the intricacies of the subject, with solutions to problems appearing to emanate from a 'black box'. Unsaturated soil, which is the focus of this book, is an important, complex and not well-understood component of soil mechanics that presents geotechnical engineers with many challenges. The study of unsaturated soils is an absorbing, practical subject linking fundamental science to nature. While soils, in general, are inherently variable and not readily amenable to analysis, unsaturated soils raise the complexity to a higher level.

This book provides a perspective of unsaturated soils based on recent research, and shows how this dovetails with the general discipline of soil mechanics. Thermodynamic principles underpin the analyses and allow the development of a rigorous theoretical model. In developing the model, reference is made to branches of physics and chemistry and analogous behavioural trends. In this respect, imagination is not considered a vice but must be tempered with realism and experimental evidence. Attempt is made to clearly distinguish between verifiable conclusions and imagination while pushing ideas as far as we dare. Where assumptions are made and where potential errors arise in testing, these are highlighted, but we do not apologise for raising people's blood pressure if it generates meaningful discussion.

Geotechnical engineers have a different perspective of the ground than other professionals involved in associated fields. However, there is a great deal of overlap in the disciplines of geology, environmental science and soil science; and professionals in these fields have a lot of knowledge and expertise to share. Nevertheless, each has a different view of the ground and the most important problems to be solved. It is only in recent times that the significance of suction in unsaturated soils has transcended from soil science applications, involving plant water deficiency, into geotechnical engineering applications.

There are also different approaches to problems within geotechnical engineering. A practising geotechnical engineer dealing with day-to-day problems is likely to be more interested in a prescriptive, quick solution than the niceties of a detailed analysis. The use of California Bearing Ratio (CBR) tests for the design of road pavements exemplifies this approach; though highly empirical, such tests nevertheless provide a basis for achieving a satisfactory design. Researchers in the field of soil mechanics sometimes adopt a similar approach to a problem, arguing that it is not possible to fully analyse soil behaviour as the materials are not manufactured under controlled conditions. Further, soils do not obey simple stress–strain rules and, importantly, exhibit time-dependent behaviour. Thus, unless proven otherwise, a pragmatist would argue that tried and tested methods be adopted in solving simple problems. However, other researchers strive for a better understanding of the underlying principles behind soil behaviour and try to answer the question: 'what is actually happening?' This more realistic approach attempts to discover what the world is actually like with protagonists arguing that a better understanding should lead to better predictive models and ultimately to cost savings. The development of predictive models based on a sound theoretical footing is a necessity in allaying the perception that geotechnical engineering comprises empiricism and guesswork.

Arguably, geotechnical engineers, possibly more than most other associated professionals, must keep up with current research. However, research, by definition, is at the forefront of knowledge and opinions will vary on the methods and products of endeavours to reach solutions to problems. It is often only with years of verification testing and experience that a consensus is achieved and research is accepted as valid, though subject to review and improvement with time. The accumulation of research promotes and progresses concepts and established theories may turn out to be special cases of more far-reaching ones. The prime scientific example is Newton's laws, which Einstein showed to be a limited case of relativity. On a less grand scale, Terzaghi's well-established effective stress equation for saturated soils is shown in Chapter 8 to be a special case of a more general equation for unsaturated soils relating the three-phase relationship of stress and pressure to the volumes of the phases. In this context research is a self-developing system of investigation in which errors and idealisations are, sooner or later, detected by experiment and more realistic analysis. Findings are customarily put forward fraught with errors and unperceived assumptions that are gradually worked out before the underlying theory is accepted. We hope this is not the case here, but if we could predict the future we might not have written the book. It is imperative, however, that theory fits well with sound, repeatable experimental data. This is a major thrust of this book.

We hope you find the book interesting and readable.

Symbols

Symbol	Definition	Symbol	Definition
A	Helmholz potential	G	Gibb's potential
A_a	area of air voids cut by plane	G_s	specific gravity of the soil particles
A_c	shape factor for the contractile skin of an aggregate of soil particles and water	H	enthalpy (suffixes 'm' and 'r' are used in Chapter 6 to differentiate between the enthalpy in a soil specimen and the enthalpy in the triaxial cell water reservoir)
A_p	area of flat plane (cross-sectional area of soil specimen)		
A_s	area of solids cut by flat plane		
$A_{(s)}$	parameter dependent on suction	H_A	enthalpy of adsorbed water
A_w	area of water voids cut by flat plane	H_a	enthalpy of the air phase and interactions
a	dimensionless variable with minimum value of 1	H_C	enthalpy of contractile skin
		H_{Cf}	component of enthalpy of the contractile skin in an aggregate of water and soil particles
$B_{(s)}$	parameter dependent on suction		
b	dimensionless variable		
C	intercept q/s when $p'_c/s = 0$	H_{Ci}	component of enthalpy of the contractile skin of a spherical water droplet
C_m	compression index with respect to s		
C_t	compression index with respect to $\overline{\sigma}$	H_D	enthalpy of dissolved air
		H_i	enthalpy of the individual phases or interactions between phases
D_m	water content index with respect to s		
		H_s	enthalpy of solid phase and interactions
D_t	water content index with respect to $\overline{\sigma}$		
		H_v	enthalpy of water vapour
c'	intercept of the failure envelope with the shear stress axis (apparent cohesion)	H_w	enthalpy of water phase and interaction
		H_{wi}	initial enthalpy of water phase
d	pore diameter	H_{si}	initial enthalpy of solid phase
e	voids ratio	H_{sf}	final enthalpy of solid phase
e_0	initial voids ratio	H_{wf}	final enthalpy of water phase
g	acceleration due to gravity	h_c	capillary water rise

Symbol	Definition
h	specimen height (suffixes 'm' and 'r' are used in Chapter 6 to allow virtual infinitesimal changes in axial displacement to be analysed)
I_P	plasticity index
K_0	lateral earth pressure at rest
k	negative constant
L	characteristic length
M	critical state stress ratio parameter for a saturated soil
M_a	total stress ratio parameter
M_b	suction stress ratio parameter
M_s	mass of solid phase
$M_{(s)}$	stress ratio parameter dependent on suction
M_t	total mass
M_w	mass of water phase
M_i	mass of soil phase or interaction
m_p	slope of psychrometer correction factor
N	specific volume of *iso-ncl* at $p' = 1.0$ kPa for a saturated soil
N_a	number of aggregates per unit volume of soil
$N_m \, N_{(s)}$	specific volume of *iso-ncl* at $\overline{p}/\overline{p}^c = 1.0$ kPa or $\overline{p}/p_{atm} = 1.0$ kPa
N_t	specific volume of compression lines at $p_c' = 1.0$ kPa
n	porosity
n_a	porosity term for air phase
n_s	porosity term for solid phase
n_w	porosity term for water phase
n^*	porosity for dry aggregated soil
p	pressure and mean total stress (suffixes 'm' and 'r' are used in Chapter 6 to differentiate between the pressure in a soil specimen and the applied triaxial cell pressure respectively)
p_{atm}	atmospheric pressure
p'	Terzaghi's mean effective stress $(p - u_w)$
$\overline{p}$	mean net stress $(p - u_a)$
p_B'	Bishop's mean stress

Symbol	Definition
$\overline{p}^c$	net stress at a reference stress state for which $v = N(s)$
p_{cs}'	mean effective stress at the critical state
p_c'	average volumetric 'coupling' stress
p_d	differential pressure in mercury intrusion porosimetry
p_f'	inter-particle stress within an aggregate
p_i	pressure (or spherical stress component) arising from a phase or interaction between phases
$\overline{p}_o$	pre-consolidation stress
$\overline{p}_o^*$	pre-consolidation stress for saturated conditions
p_p	partial pressures for miscible interaction obeying Dalton's divisional law
$\overline{p}_s$	intercept on $\overline{p}$ axis
Q	heat (suffixes 'm' and 'r' are used in Chapter 6 to differentiate between the heat in a soil specimen and the heat in the triaxial cell water reservoir)
q	deviator stress (suffixes 'm' and 'r' are used in Chapter 6 to differentiate between the deviator stress in a soil specimen and the applied deviator stress respectively in the triaxial cell)
q_{a-f}	components of the deviator stress for an unsaturated soil
q_{os}	intercept of the *csl* with the q-axis
R	radius of curvature of air–water meniscus and radius of spherical air bubble
R_e	air bubble equilibrium radius
R_h	relative humidity
R_u	universal gas constant
R^*	radius of spherical water droplet
R_1 and R_2	radius of curvature of contractile skin on two perpendicular planes

Symbol	Definition
r	specimen radius (suffixes 'm' and 'r' are used in Chapter 6 to allow virtual infinitesimal changes in radial displacement to be analysed)
r_c	constant related to the maximum stiffness of the soil (for infinite suction) in Equation 4.35
r_t	radius of capillary tube
S	entropy (suffixes 'm' and 'r' are used in Chapter 6 to differentiate between the entropy in a soil specimen and the entropy in the triaxial cell water reservoir)
S_i	entropy of a phase or interaction
S_r	degree of saturation
s	matric suction $(u_a - u_w)$
s^*	modified suction component
T	absolute temperature (suffixes 'm' and 'r' are used in Chapter 6 to differentiate between the temperature in a soil specimen and the temperature in the triaxial cell water reservoir)
T_c	surface tension of air–water interface (contractile skin)
T_m	surface tension of mercury
t	temperature in °C
U	internal energy (suffixes 'm' and 'r' are used in Chapter 6 to differentiate between the internal energy in a soil specimen and the internal energy in the triaxial cell water reservoir)
U_a	internal energy of air phase
U_i	internal energy of a phase or interaction
U_s	internal energy of solid phase
U_w	internal energy of water phase
u_a	pore air pressure
u_s	imposed stress in the soil particles
u_{sf}	final fluid pressure acting through solid particles
u_{si}	initial fluid pressure acting through solid particles
u_v	partial pressure of pore water vapour

Symbol	Definition
u_{vo}	saturation pressure of pore water vapour over a flat surface
u_w	pore water pressure
u_w^*	water pressure in a droplet of radius R^*
V	total volume (suffixes 'm' and 'r' are used in Chapter 6 to differentiate between the volume of a soil specimen and the volume of the the triaxial cell water reservoir)
V_a	volume of air phase
V_i	volume of phase or interaction between phases
V_p	volume of aggregates
V_s	volume of solid phase
V_v	volume of voids
V_w	volume of water phase
v	specific volume
v^*	specific volume for a dry aggregated soil
v_a	specific air volume
v_a^*	specific air volume for a dry aggregated soil
v_{os}	specific volume of the soil at critical state when $\bar{p} = 1$ kPa
v_w	specific water volume
v_{wos}	specific water volume of the soil at critical state when $\bar{p} = 1$ kPa
v_k	specific volume of url at $p' = 1.0$ kPa for a saturated soil
v_λ	specific volume of $1d$-ncl at $p' = 1.0$ kPa for a saturated soil
W	total work (suffixes 'm' and 'r' are used in Chapter 6 to differentiate between the work in a soil specimen and the work in the triaxial cell water reservoir)
W'	work input in saturated or perfectly dry soil per unit volume
W_v'	volumetric work input in a saturated or perfectly dry soil per unit volume
W_q'	deviatoric work input in a saturated or perfectly dry soil per unit volume
W_u'	total work input in an unsaturated soil per unit volume

Symbol	Definition	Symbol	Definition
W'_{vu}	volumetric work input due to the stress state variables in an unsaturated soil per unit volume	ε_{qa}	deviator strain of air voids between aggregates
W'_{qu}	deviatoric work input due to the components of the deviator stress in an unsaturated soil per unit volume	ε_{qw}	deviator strain of aggregates
		ε_v	volumetric strain (suffixes 'm' and 'r' are used in Chapter 6 to allow virtual infinitesimal changes in volumetric strain to be analysed)
W^e	applied elastic work		
W^p	applied plastic work	ε_w	volumetric strain of aggregates
$\overline{W}'$	work input per unit volume due the stress differences	$\varepsilon_{w,11}$	axial strain of aggregates
		$\varepsilon_{w,33}$	radial strain of aggregates
w	gravimetric water content	$\varepsilon_{11}, \varepsilon_{22},$	compressive strains in
w_0	initial gravimetric water content	ε_{33}	directions 1, 2 and 3 respectively
w_{fp}	gravimetric water content of filter paper in Figure 2.17		
w_L	liquid limit	$\varepsilon_{12}, \varepsilon_{21}, \varepsilon_{13},$	shear strains on planes 1, 2
w_P	plastic limit	$\varepsilon_{31}, \varepsilon_{23}, \varepsilon_{32}$	and 3
x	length in x-direction	η	mobilised stress ratio M
y	length in y-direction	η_a	mobilised stress ratio M_a
z	length in z-direction	η_b	mobilised stress ratio M_b
α	dimensionless variable with value between 0 and 1	Θ	normalised volumetric water content
β	parameter that controls the rate of increase of soil stiffness with suction	θ	contact angle of air–water interface with wall of capillary
		θ_m	contact angle of air–mercury interface with wall of capillary
Γ	specific volume intercept of the *csl* at $p' = 1.0$ kPa for a saturated soil	θ_r	residual volumetric water content
Γ_{ab}	specific volume intercepts of the csl at $p' = 1.0$ kPa	θ_s	saturated volumetric water content
Γ_c	specific volume intercept of *iso-ncl* at $p'_c/s = 1$	θ_w	volumetric water content
$\Gamma_{(s)}$	parameter dependent on suction	κ	elastic stiffness parameter for swelling and recompression line (slope of *url*) for a saturated soil
γ_{ij}	shear strains (tensor) often taught to engineers where the strains are twice the shear strains ε_{ij} and of opposite sign: $\varepsilon_{ij} = -\gamma_{ij}/2$ where $i \neq j$		
		K	fitting parameter
		$\kappa_{(s)}$	elastic stiffness parameter for changes in suction
δ_{ij}	Kronecker delta	λ	stiffness parameter of *iso-ncl* and *csl* for a saturated soil
ε_a	volumetric strain of air voids		
$\varepsilon_{a,11}$	axial strain of air voids	λ_a	stiffness parameter as a result of the influence of $\overline{p}$
$\varepsilon_{a,33}$	radial strain of air voids		
ε_{ij}	strain tensor	λ_b	stiffness parameter as a result of the influence of s
ε_q	deviator strain (suffixes 'm' and 'r' are used in Chapter 6 to allow virtual infinitesimal changes in shear strain to be analysed)	λ_c	stiffness parameter of *iso-ncl*
		$\lambda_{(0)}$	stiffness parameter for *iso-ncl* for $s = 0$
		$\lambda_{(s)}$	stiffness parameter for *iso-ncl* for $s = $ constant

Symbol	Definition
λ_t	stiffness parameter of compression line
λ_w	stiffness parameter of csl in the $v_w : \bar{p}$ plane
μ_i	chemical potential of soil phase or interaction
$\mu_{(s)}$	strength parameter dependent on suction
μ_w	chemical potential of water phase
ρ_b	bulk density
ρ_d	dry density
ρ_s	density of soil particles
ρ_w	density of water
ρ_d^*	density of dry aggregated soil
σ	total stress
σ_{ij}	total stress tensor
σ'	Terzaghi's effective stress $(\sigma - u_w)$
σ'_{ij}	Terzaghi's effective stress tensor
$\bar{\sigma}$	net stress $(\sigma - u_a)$
$\bar{\sigma}_{ij}$	net stress tensor
$\sigma'_{A,ij}$	Aitchison's stress tensor
σ'_B	Bishop's stress
$\sigma'_{B,ij}$	Bishop's stress tensor
σ'_c	coupling stress
$\sigma'_{c,ij}$	'coupling stress' tensor
$\sigma'_{c,11}$	axial coupling stress in triaxial test
$\sigma'_{c,33}$	radial coupling stress in triaxial test
σ''_{ij}	average soil skeleton stress tensor
σ_{11}	total principal axial stress (suffixes 'm' and 'r' are used in Chapter 6 to allow virtual infinitesimal strains to be analysed)
σ'_{11}	axial effective stress in triaxial cell

Symbol	Definition
σ_{33}	radial total stress or cell pressure in triaxial cell (suffixes 'm' and 'r' are used in Chapter 6 to allow virtual infinitesimal strains to be analysed)
σ'_{33}	radial effective stress in triaxial cell
$\bar{\bar{\sigma}}_{11}$	total axial stress internal to a soil specimen
$\bar{\bar{\sigma}}_{33}$	total radial stress internal to a soil specimen
τ	shear stress
Φ	dissipated plastic work
ϕ	psychrometer correction factor
ϕ'	effective angle of friction
ϕ^b	friction angle associated with changes in $(u_a - u_w)$
ϕ_s	osmotic suction
φ_i	gravitational potential of phase or interaction
χ	empirical variable in Bishop's equation
χ_m	empirical matric suction parameter
χ_s	empirical osmotic suction parameter
Ψ	total suction
Ψ^e	elastic work not frozen
Ψ^p	elastic work frozen as applied plastic work
Ψ_p	water potential or total suction in psychrometer
$\Psi_{(s)}$	parameter dependent on suction
Ω	intercept on the q/s axis at $p'_c/s = 1$
ω	mobilised intercept Ω
ω	angle in Figure 4.1
ω_v	molecular mass of water vapour

Prefix	Usage
d	small increment of change where the quantity is considered independent of path and an exact differential

Prefix	Usage
δ	small increment of change where the quantity is path dependent and not an exact differential
∂	partial derivative
Δ	large increment of change

Abbreviation	Meaning	Abbreviation	Meaning
General			
BBM	Barcelona Basic Model	SD	suction decrease yield locus
CSSM	Critical state soil mechanics	SI	suction increase yield locus
HAE	high air entry (in reference to HAE discs)	OWC	optimum water content
		SWCC	soil–water characteristic curve
LC	load-collapse yield locus		

Triaxial test			
iso-cs	constant suction shearing test on true isotropically prepared specimen	1d-cs	constant suction shearing test on initially one-dimensionally compressed specimen
iso-cwm	constant water mass shearing test on true isotropically prepared specimen	1d-cwm	constant water mass shearing test on initially one-dimensionally compressed specimen
Critical state construct			
csl	critical state line	*url*	un-loading and re-loading line
iso-ncl	isotropic normal compression line	*1d-ncl*	one-dimensional normal compression line

Chapter 1
Properties of Unsaturated Soils

1.1 Nature and genesis of unsaturated soils

The term *soil* as used in geotechnical engineering encompasses a wide spectrum of particulate materials. In the saturated state all the void spaces between the particles are filled with water, but in the unsaturated state a proportion of the void spaces is filled with air. We will primarily be concerned with the analysis of fine-grained soils comprising clays and silts, or soils containing coarser sand and gravel that have a significant percentage of fines. As shown in Figure 1.1, which illustrates the gradings of a number of soils of different lithologies, the finer clay soils have particle sizes less than 2 μm that are not visible to the naked eye, while the coarser gravels range up to 60 mm. The following can serve as a yardstick to the range of particle sizes: if a clay platelet was expanded to the size of a saucer, a coarse gravel particle would have a proportional dimension of around 10 km. This wide range of particle sizes and the inherent variability of soils give rise to behavioural characteristics not readily amenable to rigorous analysis. Stress history, particle shape and time-dependent characteristics also influence their multi-faceted behaviour, generally requiring simplifications and generalisations in formulating solutions to geotechnical problems.

The solid particles, water and air are the phases making up a soil mass. Classically, researchers have achieved recognisable success in developing an understanding of the behaviour of saturated, fine-grained soils and the behaviour of dry, coarse materials, such as sand and gravel. Extending our understanding to the behaviour of unsaturated soils, particularly those with a significant percentage of fines, has proved problematic. This is principally because of the additional fluid phase of air or other gases in the void spaces, which complicates the thorny issue of the controlling stress regime. Interpretation of the behaviour of unsaturated soils requires the differences in the air and water pressures, the phase compressibilities and their interactions, as well as chemical effects, to be taken into account. The interactions include the contractile skin between the fluid phases, which gives rise to a surface tension effect that is particularly influential in creating the characteristic aggregated structure of fine-grained soils. The significances of the phases and their interactions in influencing soil behaviour are investigated in later chapters.

Both natural soils and engineered ground are liable to be in an unsaturated condition. Natural soils in an unsaturated state are common in arid or semi-arid areas where the groundwater table is often many metres deep. Around one-third of the earth's surface is

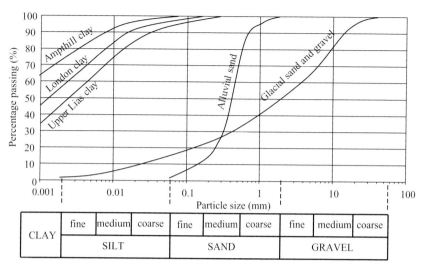

Figure 1.1 Particle size distribution.

situated in arid or semi-arid regions where the potential evaporation exceeds the precip-itation (Barbour, 1999). However, any soil near the ground surface in a relatively dry environment is liable to have a negative pore water pressure (water pressure relative to a datum of atmospheric air pressure) and could experience de-saturation or air entry into the pore spaces. Though the soil may be saturated for some height above the water table, if the pore water pressure drops sufficiently, air will enter the pore spaces. Figure 1.2 illustrates the change from a positive pore water pressure below the water table to a negative pore wa-ter pressure above the water table. While the plot indicates a reduction in negative pore wa-ter pressure close to the ground surface, where precipitation would increase the degree of saturation, increased desiccation due to evaporation can be expected to occur in a hot en-vironment. Compacted fills such as in earth dams, road subgrades and embankments are

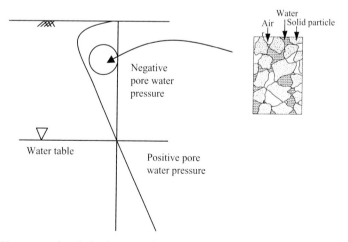

Figure 1.2 Unsaturated soils in the ground.

also usually placed in an unsaturated state as complete compaction and closure of all air voids is generally impractical. Negative pore water pressure[1] is the key to understanding unsaturated soil behaviour and in interpreting the significance to engineering structures.

Climate plays an important role in the formation of unsaturated soils (Lu and Likos, 2004), with evaporation in hot weather drying out the ground leading to shrinkage of fine-grained soils and, ultimately, to shrinkage cracking. Subsequent wetting, following rain, leads to swelling and closure of cracks, but not necessarily to total eradication of suction-induced soil structure. Aggregation of fine particles following drying affects subsequent behaviour characteristics. Future climate changes due to global warming could potentially cause significant changes in the soil moisture regime, and thus in soil conditions over large areas of the world. Uptake of water by vegetation can also lead to significant ground de-saturation due to evapotranspiration, while removal of vegetation can lead to subsequent re-saturation leading to potential stability problems, notably instability of hillsides due to deforestation. Where unsaturated soils comprise *highly* plastic clays, large swelling and shrinkage phenomenon can occur due to uptake of water or reduction in water content, resulting in ground movements capable of causing severe structural damage. The primary cause of expansive clays is the presence of swelling clay minerals such as montmorillonite. In temperate zones such as Britain, seasonal volume change in clay is generally restricted to the upper 1.0–1.5 m (Bell and Culshaw, 2001). Nevertheless, in the South and South Midlands regions of England, the presence of clay formations such as the London clay, Oxford clay, Kimmeridge clay and Lias clay has, in dry summers, particularly in conjunction with tree root action, led to large numbers of insurance claims for cracking of domestic properties as a result of excessive ground shrinkage.

Collapsible soils represent an important phenomenon. Loess or loosely compacted fills in an unsaturated state can experience large collapse settlements if they are wetted, bringing about damage to overlying structures. In developing areas of the world, notably East Asia, residual soils from decomposed rock exhibit complex behaviour characteristics, attributable in part to partial saturation (e.g. Lee and Coop, 1995; Ng and Chiu, 2003). The rapid structure and infrastructure development in areas such as the tropics is liable to lead to enhanced awareness of potential problems with unsaturated soils. Further reading on the wide range of geotechnical problems, where an understanding of unsaturated soil behaviour is a prerequisite to determining engineering solutions, is given by Alonso and Olivella (2006).

Within the context of the foregoing discussion, the remainder of this chapter deals with the phases and their interactions and addresses some of the basic principles governing the behaviour of soils, particularly those soils in an unsaturated state. First, however, it is necessary to define some of the variables used to describe soils.

1.2 Soil variables

Subsequent sections of this book employ the following variables. It is these tools that are used to form the basic equations governing the behaviour of unsaturated soils.

1.2.1 Volume relationships

An important theme within this book is the thermodynamic link between the pressures (and stresses) and the volumes of the phases in a soil at equilibrium, and the link between

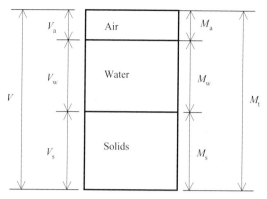

Figure 1.3 Phase diagram.

the pressures (and stresses) and specific strain increments in analysing changes between equilibrium states. These 'conjugate' variables are particularly relevant in unsaturated soils. However, we are racing ahead of ourselves and it is first necessary to introduce a number of simple variables for the relative volumes of the phases that are illustrated in Figure 1.3. The first are the porosity terms that relate the volume of the voids (V_v) and the volumes of the phases V_s, V_w and V_a for the solid particles, the water and the air respectively to the total volume of the soil V:

$$n = \frac{V_v}{V} \qquad n_s = \frac{V_s}{V} \qquad n_w = \frac{V_w}{V} \qquad n_a = \frac{V_a}{V} \qquad [1.1]$$

where n is porosity and the suffices 's', 'w' and 'a' attached to the volume and porosity terms relate to the solid particles, the water and the air phases respectively.

The porosity term for the solid particles does not fit well with the idea of solid particles but reflects the relative volume filled by the solid particles and is convenient as it allows us to write the following equation:

$$n_s + n = n_s + n_w + n_a = 1 \qquad [1.2]$$

The porosity terms are expressed here as ratios but are often expressed as a percentage by multiplying by 100. The air porosity n_a is more frequently referred to as the air voids content and is particularly useful in expressing the adequacy of compaction in earthworks operations where effort is usually made to reduce the air voids content to a low level.

Voids ratio e is an alternative representation of the volumetric variables and is conveniently used to express volumetric changes in consolidation or compression tests, particularly for saturated soils. The voids ratio expresses the ratio of the volume of the voids to the volume of the solids:

$$e = \frac{V_v}{V_s} \qquad [1.3]$$

The relationship between porosity and voids ratio is given by:

$$n = \frac{e}{1+e} \qquad [1.4]$$

The degree of saturation S_r is also a commonly used variable in unsaturated soils and is expressed in percentage terms as:

$$S_r = \frac{V_w}{V_v} \times 100\%$$ [1.5]

For dry soils $S_r = 0$ and for saturated soils $S_r = 100\%$. Between these two extremes the degree of saturation has an intermediate value. The degree of saturation is an important term where there is fluid movement as it expresses the relative volumes of the water and air in the pore spaces. However, the controlling stresses and the strength of unsaturated soils are often correlated with the degree of saturation. The use of the degree of saturation, on its own, is considered a poor choice of variable to which to relate stresses and strength. There is no term for the volume of the solids in the definition of S_r, yet it is the solid particles that give a soil its strength. Far better variables and ones that emerge naturally from thermodynamic analysis later in this book are the specific volume v, the specific water volume v_w and the specific air volume v_a defined as:

$$v = 1 + e = \frac{V}{V_s} \quad v_w = 1 + eS_r = \frac{V_s + V_w}{V_s} \quad v_a = e\,(1 - S_r) = \frac{V_a}{V_s}$$ [1.6]

The specific volume terms are additive in a similar manner to the porosity terms other than the term for the volume of solids which is included within the specific water volume; thus:

$$v = v_w + v_a$$ [1.7]

The specific volume v is defined as the total volume of soil per unit volume of solids, and the difference $(v - v_w)$ is the volume of air voids per unit volume of solids v_a, frequently defined in the literature as the air voids ratio e_a. In fine-grained soils in an unsaturated state, the soil particles have a tendency to collect together as *aggregates* of particles, typically of silt or sand size, with relatively large inter-aggregate void spaces. The water phase and the soil particles have a close affinity, and where the aggregates of soil particles in an unsaturated soil contain all the water, v_w can be defined as the volume of the aggregates per unit volume of solids. Accordingly, the ratio of the specific water volume to the specific volume (Equation 1.8) expresses the volume of the aggregates per unit volume of soil:

$$\frac{v_w}{v} = \frac{V_s + V_w}{V}$$ [1.8]

This equation contains a term for the volume of the solid particles and a term for the volume of water phase, and because there is a term for the total volume, there is inbuilt into the formulation a term for the air phase. The equation presents the volumetric terms for all the phases in an unsaturated soil in a single formulation. The ratio v_w/v is shown in the following chapters to be a far better variable than S_r to which to relate the controlling stresses and the strength of unsaturated soils. While there would be some merit in replacing v_w/v with a unifying symbol, this is resisted as it would mask the importance of the volumetric terms in the subsequent analysis.

1.2.2 Water content relationships

The amount of water in a soil is generally expressed in one of two ways. The water content w (frequently termed the gravimetric water content) is expressed as:

$$w = \frac{M_w}{M_s} \times 100\%$$ [1.9]

where M_w is the mass of water and M_s is the mass of solid particles.

An alternative definition is the volumetric water content θ_w:

$$\theta_w = \frac{V_w}{V}$$ [1.10]

This is synonymous with the definition of n_w in Equation 1.1.

1.2.3 Density relationships

There are two important density variables for a soil mass: the bulk density and the dry density. The bulk density ρ_b is the ratio of the total mass M_t to the total volume, where the total mass and the total volume are the sums of the masses and volumes of the phases respectively, as illustrated in Figure 1.3:

$$\rho_b = \frac{M_t}{V}$$ [1.11]

The dry density ρ_d is the ratio of the mass of the solid particles M_s to the total volume:

$$\rho_d = \frac{M_s}{V}$$ [1.12]

The density of the solid particles ρ_s is often related to the density of water ρ_w in a dimensionless term variously referred to as the relative density, specific weight or specific gravity G_s. The relative density is the density of the soil particles divided by the density of water at a temperature of $4°C$ under normal atmospheric pressure:

$$G_s = \frac{\rho_s}{\rho_w}$$ [1.13]

In earthworks operations such as in road embankments and earth dams, soils are compacted to improve their strength and compression properties. For a given compaction effort, there is a water content, termed the optimum water content (OWC), at which a soil achieves a maximum dry density. A typical laboratory dry density – water content relationship is illustrated in Figure 1.4. Greater compactive effort leads to an increase in the maximum dry density and a reduction in the OWC. It is important to carefully select the appropriate laboratory compaction test to replicate as closely as possible the compaction conditions achievable on site. Standard Proctor compaction is normally considered achievable under reasonably controlled site operations. Modified Proctor compaction, using greater compactive energy, can be achieved under carefully controlled site conditions using modern compaction plant[2]. Wet of optimum, compaction plots lie close to the zero air voids line, where the air voids content is defined as n_a in Equation 1.1. The zero air voids line represents the maximum achievable dry density for given water content, and on this line all the air voids are eradicated. Dry of optimum, the soil becomes progressively less saturated, suction increases and there is increased evidence of a fissured soil structure.

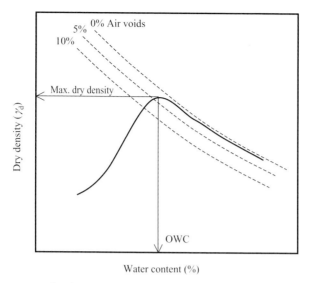

Figure 1.4   Compaction dry density – water content relationship.

1.3 Particle properties

Destructive processes such as chemical and physical weathering lead to the formation of soils from rock. Clay soils are generally the products of chemical weathering. The chemical processes due to the actions of water (especially if it is acidic or alkaline), oxygen and carbon dioxide (Craig, 1997) change the mineral form of the rock, leading to the formation of crystalline particles of colloidal sizes known as clay minerals. Natural clays generally contain a range of clay minerals, the most widely encountered minerals being kaolinite, calcite, illite, dolomite and smectite (the last clay mineral species including montmorillonite).

The types and proportions of minerals present in clay significantly influence its index properties. McLean and Gribble (1988) reported a range of liquid limits (w_L) and plastic limits (w_P) for the clay minerals kaolinite, illite and montmorillonite, as shown in Table 1.1. Also included are the limits for sodium-rich bentonite (impure clay consisting mostly of montmorillonite) and some common British soils.

We will not describe the tests to determine the limits (w_L) and (w_P) for fine-grained soils as these are provided in many textbooks and standards, but we will describe their significance in general terms based on Figure 1.5. The liquid limit is the water content above which a disturbed soil starts to flow and to behave as a liquid, and the plastic limit is the water content below which a disturbed unconfined soil fissures and behaves as a brittle solid. Between these limits the soil is said to behave as a plastic material.

Activity, defined as the ratio of plasticity index I_P (where $I_P = w_L - w_P$) to the percentage clay content, describes the sensitivity of clay (Skempton, 1953). The activity of London clay is typically within the range 0.5–0.83 (Gasparre *et al.*, 2007; Monroy *et al.*, 2010) while for Gault clay it is approximately 0.6 and for Belfast Upper Boulder clay approximately 0.5 (Sivakumar, 2005). Much of the experimental data reported later in this book uses kaolin with an activity of approximately 0.4. Skempton (1953) quoted an activity for sodium-rich montmorillonite of 7.2, while Blatz and Graham (2003) reported an activity

Table 1.1   Typical clay mineral and clay soil properties.

	(w_L) (%)	(w_P) (%)
Kaolinite	60–73	26–37
Illite (or hydrous mica)	63–120	34–60
Montmorillonite	108–700	51–60
Sodium-rich bentonite[a]	354	27
Ampthill clay	66–89	23–30
Mercia mudstone[b]	25–60	17–33
Etruria marl[c]	35–52	14–21
Belfast Upper Boulder Clay	59–65	27–31
Upper Lias clay[d]	57–65	24–31
London clay[e]	60–83	21–32

[a] After Stewart *et al.* (2001).
[b] After Chandler and Davis (1973).
[c] After Hutchison *et al.* (1973).
[d] After Chandler (1974).
[e] After Ward *et al.* (1959), Gasparre *et al.* (2007) and Monroy *et al.* (2010).

for sodium-rich bentonite of around 2. Clays are generally categorised into three groups in terms of their activity:

Inactive clay: activity <0.75
Normal clay: activity 0.75–1.25
Active clay: activity >1.25

In a relatively dry, fine-grained soil the particles are held together by a relict water phase, under a tensile suction force, in the fine intra-aggregate pore spaces. On wetting, the suction is reduced and the soil expands. This inter-particle swelling takes place in all clay soils. However, some clay minerals also exhibit significant intra-crystalline swelling. This is a particular characteristic of smectitic clay minerals such as montmorillonite. The

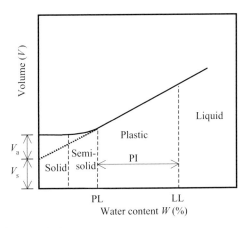

Figure 1.5   Plasticity characteristics.

individual molecular layers that make up the crystal structure are weakly bonded so that on wetting water enters not only between crystals but also between the layers that comprise the crystals. The adsorption of water into clay particle surfaces and the crystalline swelling of some clay minerals are discussed further in Section 1.4.5.

Generally kaolinite has the smallest swelling capacity of the clay minerals with nearly all the swelling occurring as inter-particle swelling. Larger swelling potential, in order of increasing magnitude, is exhibited by illite, calcium-montmorillonite and sodium-montmorillonite. Natural soils comprise mixtures of clay minerals and exhibit the net effect of the mineral composition. Increasing plasticity index and activity give a general guide to increasing swelling potential and the significance of intra-crystalline swelling (Grim, 1962; Bell and Culshaw, 2001).

1.4 Phase properties and interactions

The proportions of the three phases making up unsaturated soils have a significant influence on their behaviour as do the interactions within and between the phases. The interactions are particularly influential in the perception of equilibrium and the rates of mass change and redistribution. The interactions need careful appraisal to appreciate their significance.

At any point in a fluid under equilibrium conditions, the pressure is equal in all directions. However, gravitational effects give rise to greater pressures with depth and closeness to the earth's centre. In a soil the particles interact and the stresses at any point are, in general, different in different directions. Thus, soils exhibit a heterogeneous stress regime with stress levels generally increasing with depth. In laboratory soil tests the size of the specimen[3] tested is usually sufficiently small that the effect of the change in gravitational effects between the top and bottom of the specimen can be ignored without any significant loss of accuracy. This is the case for specimens tested in triaxial cells, the results from which are usually considered to represent the conditions at a point, or at least a localised volume.

For a soil system in equilibrium, an applied external pressure (or stress regime) is balanced by the components of pressure and stress arising from the phases and their interactions, and given by:

- The water pressure acting through the volume of the water;
- The air pressure acting through the volume of air;
- Any chemical imbalance leading to osmotic suction effects
- The water vapour in air;
- The dissolved air in water;
- The contractile skin formed at the water interface where it meets the air;
- The adsorbed double layer of water on the soil particle surfaces and absorbed crystalline water;
- The pressure acting through the solid phase because of the surrounding fluid(s);
- The interaction between the soil particles, which can be viewed on a micro-mechanical scale as the interaction stresses between the soil particles, or on the macro-mechanical scale as the stresses acting over a total planar area.

While some of the interactions between the phases contribute very little to resisting applied forces, at least at equilibrium, and can be ignored, it is important to understand the contributions of the phases and the interactions to overall soil behaviour.

1.4.1 Soil suction

Soil suction is a major factor affecting the behaviour of unsaturated soils. Total suction Ψ has two components, namely matric suction and osmotic suction. This can be expressed as:

$$\Psi = (u_a - u_w) + \phi_s \qquad [1.14]$$

where $(u_a - u_w)$ is the matric suction, being the difference between the pore air pressure u_a and the pore water pressure u_w, and ϕ_s is the osmotic suction, being the result of chemical imbalance between the pore water in the soil volume under consideration and an external source of water.

The significance of suction in soil has been recognised for many decades. The ability of a soil to absorb additional water, whether it is fully saturated or unsaturated, is termed soil suction and can be defined as the free energy state of soil water (Edlefsen and Anderson, 1943). The partial vapour pressure of the soil water can be used to measure the free energy state of the soil water (Richards, 1965), and the following thermodynamic relationship exists between soil suction and the partial vapour pressure:

$$\Psi = -\frac{R_u T \rho_w}{\omega_v} \ln \left(\frac{u_v}{u_{vo}} \right) \qquad [1.15]$$

where R_u is the universal gas constant, T is absolute temperature, u_v is the partial pressure of pore water vapour, u_{vo} is the saturation pressure of pore water over a flat surface of pure water at the same temperature (u_v/u_{vo} is the relative humidity R_h) and ω_v is the molecular mass of water vapour.

While matric suction will be of prime interest in this book, it is important to recognise the role of osmotic suction. This results from retention energy due to the presence of salts in the pore water, or more precisely a difference in salt concentrations in the pore water in the system being analysed, and the surrounding water. Osmotic suction can attract or remove water from a system. The osmotic suction represents the ionic potential of the pore fluid in a soil (Fredlund and Rahardjo, 1993). Osmotic suction can be altered by either changing the mass of water or the amount of ions in solution. However, the strength of an unsaturated soil is principally controlled by the matric suction, even though the presence of salts within the soil water can give rise to some fundamental changes in mechanical behaviour (Alonso et al., 1987).

Matric suction is a result primarily of the phenomenon of capillarity, but is also influenced by surface adsorption effects. The capillarity phenomenon is directly related to the surface tension of water and results, for example, in water rising up thin capillary tubes, as illustrated in Figure 1.6, and forming a curved surface between the water and air known as a *meniscus*. In unsaturated soil mechanics, the water–air interface is often referred to as the *contractile skin*. For equilibrium at the air–water interface in the capillary tube, the pressure difference across the meniscus $(u_a - u_w)$ is given by:

$$(u_a - u_w) = \frac{2 T_c}{R} \qquad [1.16]$$

where $(u_a - u_w) = \rho_w g h_c$, h_c is the capillary rise, R is the radius of curvature of the meniscus (where $R = r_t / \cos\theta$), r_t is the radius of the capillary tube, T_c is the surface tension of the air–water interface (contractile skin) with units of force per unit length or energy per unit area and θ is the contact angle of the air–water interface with the wall of the capillary tube.

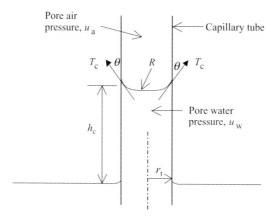

Figure 1.6 Capillary model.

In soils, the pores act as tortuous capillary tubes and result in the soil water rising above the water table. The finer the pores, the greater the meniscus curvature and the higher the water is elevated. The capillary water has a negative water pressure with respect to the air pressure and its magnitude is inversely proportional to the radius of curvature of the meniscus. In other words, negative pore water pressure, or matric suction, increases as the radius of the meniscus decreases. Accordingly, fine-grained soils normally experience greater capillary rise than coarse-grained soils where the pore spaces are larger, though there are recognisable effects on the overall capillary rise from changes in pore diameter within soils, as soils comprise discrete particles not uniform tubes.

Surface adsorption phenomena also influence the matric suction and are particularly important to clay minerals. Surface adsorption results from the negatively charged surfaces of clay particles (Figure 1.7) (Mitchell, 1993). The relatively powerful electrical force

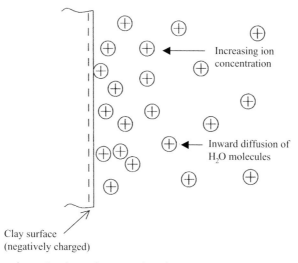

Figure 1.7   Water adsorption by a clay particle (after Mitchell, 1993).

around clay particles strongly attracts water molecules leading to an adsorbed water layer that effectively reduces the inter-particle pore spaces. The clay mineralogy plays an important role in the thickness of the adsorption layer and those soils with high-activity clay minerals, such as montmorillonite, are able to retain a greater amount of adsorbed water around their particles.

In unsaturated soils, the pore water has a pressure less than the air pressure and the contractile skin that forms at the interface between the phases is concave to the water phase, as illustrated in Figure 1.6. The water pressure deficiency in the pore spaces may have a magnitude many hundreds of times the atmospheric pressure, and the volume changes associated with the development or equalisation of the pressure difference can prove detrimental to engineering structures. The matric suction, that is the difference between the air and water pressures, is often abbreviated as s and is given by:

$$s = (u_a - u_w) \qquad [1.17]$$

As described in Section 1.1, the term negative pore water pressure $-u_w$ is equivalent to s when u_a in Equation 1.17 is atmospheric pressure and is adopted as the datum for pressure measurement.

Suction effects are generally associated with unsaturated soils and it is less well appreciated that the effect is also important for saturated soils. This has reduced the emphasis on the study and measurement of soil suction, and in design situations the soil is often assumed to be saturated with zero suction, though this may not necessarily be the case. Neglecting negative pore water pressure in assessing in situ soil strength usually represents a worst-case scenario for soils and allows for the possibilities of infiltration after rain and flooding. However, in many situations this is a simplification not reflective of site conditions.

The soil–water characteristic curve (SWCC) for a soil is the relationship between the water content and the soil suction (or negative pore water pressure). The water content variable can be defined in a number of ways: volumetric water content, gravimetric water content, degree of saturation or specific water volume, and is a measure of the amount of water in the soil pores. The amount of water can also be referenced to the residual water content as in Equation 1.18:

$$\Theta = \frac{\theta_w - \theta_r}{\theta_s - \theta_r} \qquad [1.18]$$

where Θ is the normalised volumetric water content, θ_r is the residual volumetric water content and θ_s is the saturation volumetric water content.

When the reference volumetric water content θ_r is taken as zero, $\Theta = \theta_w/\theta_s$. The normalised volumetric water content is used in Equations 4.16 and 4.17 to relate the strength of unsaturated soils to the SWCC.

In plotting the SWCC, the suction can be defined as either the matric suction $(u_a - u_w)$ or the total suction Ψ. At elevated suctions (e.g. >3000 kPa), the matric suction is, in practice, often significantly larger than the osmotic suction and assumed to be broadly equal to the total suction. The total suction corresponding to zero water content (close to 10^6 kPa) appears to be essentially the same for all soil types (Vanapelli et al., 1996).

Figure 1.8 shows typical SWCC plots of volumetric water content θ_w against suction $s = (u_a - u_w)$ for a range of soils from Holland (Koorevaar et al., 1983).

The shrinkage and swelling behaviour for London clay from the data of Croney and Coleman (1960) is presented in Figure 1.9. The plots are presented as negative pore water pressure to a log scale against specific volume v and specific water volume v_w (Murray

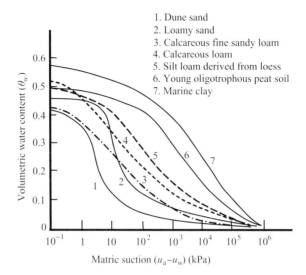

Figure 1.8 Soil–water characteristic curves (SWCC) for some Dutch soils (after Koorevaar *et al.*, 1983).

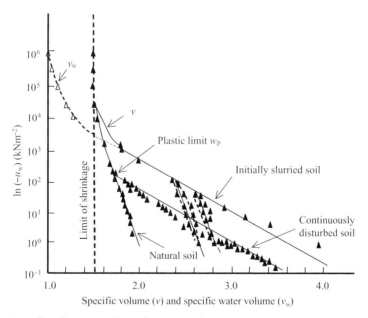

Figure 1.9 Specific volume *v* and specific water volume v_w against negative pore water pressure $-u_w$ to a log scale for London clay from the data of Croney and Coleman (1960) (after Murray and Geddes, 1995). © 2008 NRC Canadaor its licensors. Reproduced with permission.

and Geddes, 1995). Figure 1.9 shows plots for the soil dried from slurry. While saturated, the plots of v_w and v are the same and can be reasonably represented by a straight line up to a critical value of suction, known as the air entry value, where the soil ceases to be saturated. The data indicate that this change point is close to the plastic limit. The plot for v_w approaches a value $v_w = 1.0$ and v approaches the shrinkage limit of the soil as suction increases. Also shown in Figure 1.9 is a line for 'continuous disturbed soil' which represents the final suction on being sheared using a miniature vane apparatus. This again can be reasonably represented by a straight line parallel to the 'initially slurried line' for most of its length. Murray and Geddes (1995) argued that this represents the critical state strength line (*csl*) which forms an integral part of discussions in later chapters.

Mass constitutes 'energy' and the phases and their interactions in an unsaturated soil can be viewed as components of the soil energy, with changing conditions synonymous with exchange of energy. It is important to realise also that pressure is a manifestation of molecular activity, and thus the pressure of a liquid or gas in a closed system increases with temperature because of increase in molecular energy. Solids also generate large expansive or contractile forces due to temperature change though the shape of the solid mass is more rigidly defined. Consistent with this are the definitions by Aitchison (1965) of total suction as the free energy of the soil water, matric suction as the capillary component of free energy and osmotic suction as the solute component of free energy.

1.4.2 Water vapour in air

Vapour movement, in addition to mass water movement, is an important means of transport of water within an unsaturated soil and assists in the equalisation of water pressure. In laboratory tests in a closed system, vapour exchange assists in the unification of water pressure throughout a soil mass. Under equilibrium conditions the vapour pressure is defined as the saturated vapour pressure above the contractile skin. This is a special case not readily justifiable under normal conditions. At low degrees of saturation where continuous air passages in a soil are connected to the atmosphere, the vapour pressure will be less than the saturated vapour pressure and true equilibrium may not be established. There is always likely to be an imbalance and therefore ongoing mass transfer. Vapour pressure increases with temperature increase, decreases with increasing negative pore water pressure and is influenced by the solutes and the soil particles.

1.4.3 Contractile skin and air phase

A manifestation of the energy within the contractile skin is a surface tension effect that acts within a layer a few molecules thick. As illustrated in Figure 1.10, in three dimensions, equilibrium of the contractile skin is given by the following equation which includes a term for the vapour pressure u_v:

$$u_a + u_v - u_w = T_c \left[\frac{1}{R_1} + \frac{1}{R_2} \right] \qquad [1.19]$$

where R_1 and R_2 are the radii of curvature of the contractile skin for two perpendicular planes at right angles to the curvature of the interface.

As discussed by Childs (1969), the contractile skin may be wholly concave or a combination of concave and convex (in two orthogonal directions) to the higher pressure air.

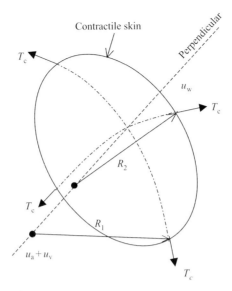

Figure 1.10 Contractile skin.

It is only necessary that the relative magnitudes of R_1 and R_2 are such so as to balance Equation 1.19. Indeed, where the contractile skin spans between collections of soil particles, dependent on the geometry, it appears necessary for continuity that the interface has a curvature both concave and convex to the air phase.

In idealised liquids containing no dissolved gases or impurities, it is the vapour pressure in relation to the liquid pressure that controls the development or the degradation of vapour bubbles. In natural systems, such as soils, there are dissolved gases and impurities in the liquid phase as well as the presence of the soil particles that influence the nucleation and bubble dynamics. Schuurman (1966) showed that for air–water mixtures the influence of the vapour pressure term u_v in Equation 1.15 is small and can be omitted for general purposes.

The idealised case of spherical occluded air bubbles ($R = R_1 = R_2$) does not represent a stable equilibrium condition and careful consideration needs to be given to the growth and decay of gas bubbles to appreciate their significance in the development of unsaturated conditions in soils. From Equation 1.19, for the case of a spherical bubble, ignoring the effect of the water vapour, the matric suction $s = (u_a - u_w)$ can be shown to be given by Equation 1.16, which is variously referred to as the Kelvin or Laplace equation. There is likely to be a relatively abrupt energy change in the formation of bubbles and the development of a contractile skin as the water ruptures. The growth and decay of gas bubbles in liquids is a dynamic process which can be described in terms of Kelvin's equation and the Rayleigh–Plesset equation (e.g. Keller, 1964; Brennen, 1995). A full treatment of this complex subject matter is beyond the scope of this text and we will deal with it in a descriptive manner in bringing out the significance to soils.

Nucleation is the seed to the development of bubbles. This can be divided into cavitation and boiling (Brennen, 1995). Cavitation is the process of nucleation in a liquid when the pressure falls below a critical value, and boiling is the process of nucleation that occurs when the temperature is raised above a critical value. Physically there is little difference between the two nucleation processes though the thermodynamic path followed is different. In both cases dissolved gases and/or vapour form bubbles at nucleation points. The nucleation points may arise due to thermal motions within a liquid forming temporary microscopic voids that act as nuclei for the growth of bubbles; this is termed homogeneous nucleation. Heterogeneous nucleation is more common in practice where bubbles form spontaneously around nuclei such as a solid particle or an ion, or form on the surface of a containing vessel.

A bubble in the pore water in a soil can decay in size as the gas dissolves or grow if gas is liberated from the solution into the bubble. In discussing the general framework of bubble growth and decay it is important to distinguish between the different degrees of air saturation in water. In a saturated condition the free air pressure is taken as the pressure of the dissolved air, in a supersaturated condition the dissolved air pressure exceeds the free air pressure and in an undersaturated condition the dissolved air pressure is less than the free air pressure.

Consider the situation under confined conditions of water initially in an air-saturated condition. Decreasing the confining pressure will result in the water becoming supersaturated and the dissolved air will start to be more readily liberated, forming bubbles around nuclei. There must be a reduction in air pressure in the bubbles and of the dissolved air as air is liberated from the solution but the pressure inside the bubbles must exceed the water pressure outside the bubbles. For spherical bubbles, there is an equilibrium radius R_e based on Equation 1.16 which will be dictated by the equilibrium of the pressure of the dissolved air and the air in the bubbles. This represents an unstable equilibrium and any bubbles smaller than R_e will tend to decay, as the pressure inside the bubbles will be greater than the dissolved air pressure. Conversely, any bubbles with radius greater than R_e will tend to expand, as the pressure will be less than the dissolved air pressure.

In a supersaturated closed water system with no soil particles, it is necessary for continuity to conjecture the tendency for the free air to form in a single bubble[4] that gradually approaches equilibrium. This bubble is found at the top of a container of gassy liquid. In support of this, Keller (1964) points to the observation of the Russian astronaut Colonel Nikolayev, who while orbiting around the world in a satellite reported that in a closed bottle containing liquid and gas 'the gas formed a single spherical bubble near the centre of the bottle'. Another example is the cavitation in the water back-pressure leads of triaxial testing equipment. Bubbles form if the water pressure is too low and the bubbles gradually combine to form large isolated bubbles[5]. For a water–air solution under reduced confining pressure, where there is no influence from the sides of the container the dissolved air pressure reduces more rapidly than the free air pressure and equilibrium (an air-saturated condition) is gradually approached with a single bubble present. This is consistent with the minimisation of the thermodynamic potential as discussed later in the book.

In soils the situation is more complicated because of the soil particles. Consider again the case of decreasing external pressure under confined conditions. Under decreasing external pressure, the bubbles will expand until they achieve apparent equilibrium within the confines of the pore spaces. Within these spaces the bubbles will interact with the soil particles, with which they are likely to have a close affinity, and will be influenced by the adsorbed double layer and gravity, and are unlikely to be spherical. Under decreasing confining pressure these air 'pockets' will have a tendency to expand and link with others to form larger pockets until eventually continuous air passages are established.

Under the alternative of increasing confining pressure from an equilibrium condition, the water will be undersaturated, as the free air will have a pressure greater than the dissolved air. The free air will tend to dissolve to equalise conditions, although again this will be influenced by the presence of the soil particles.

The rate of expansion and decay of the volume of free air is likely to influence the perceived equilibrium conditions. Barden and Sides (1967) concluded that in unsaturated soils there is evidence that equilibrium in terms of Henry's law[6] may require a considerable time interval, far greater than in the absence of soil particles.

During wetting and drying of a soil, hysteresis can be expected: when a soil is wetted, surface tension forces resist the expansion of a soil; when a soil is dried surface tension forces attempt to keep as many voids filled with water as possible. During wetting of a soil, partly because of the variation in pore dimensions, the water reluctantly fills the void spaces. Conversely, during drying of a soil the water resists being drawn into the smaller pore spaces. This hydraulic hysteresis is discussed further in Sections 4.7.4 and 7.10.

1.4.4 Air phase and dissolved air in water

In most analyses and tests, other than where fines are removed during fluid flow, the volume of solid particles remains unchanged. It is the volume of water and air in the voids that changes and gives rise to overall volume change. Water is usually considered incompressible and the volume of water change is due to exchange with the surrounding environment, although a small volume change may result from changes in water vapour in the air.

Air cannot be said to be incompressible and the change in the volume of air within a soil can be a result of the following effects:

■ Interchange of air with the external surroundings;
■ Compression or expansion of the free air in the pores;
■ Air being dissolved or liberated from the pore water.

Water molecules form a lattice-like structure with the dissolved gases occupying the spaces within the lattice. The volume available for dissolved air is approximately 2% of the total volume. Water is of very low compressibility and at equilibrium, when in contact with free air, the volume of air dissolved in water is essentially independent of air and water pressures. This can be demonstrated by an examination of the ideal gas laws and Henry's law (Fredlund and Rahardjo, 1993). It is the liberation of the dissolved air under decreasing external pressure and corresponding increasing suction that leads to the formation of air bubbles. As shown by Dorsey (1940), however, dissolved air produces an insignificant difference between the compressibility of de-aired water and air-saturated water.

Strictly speaking, the dissolved and undissolved gases in a soil may comprise mixtures of gases at odds with the proportions of gases normally found in free air. In fact, monitoring of ground gases rarely indicates gas concentrations in the soil, precisely those in free air. Even well away from sites influenced by factors such as landfills, elevated carbon dioxide and depleted oxygen are frequently detected. The ground gases are in a more restrained state than free air and segregation of gases and the influence of soil chemical reactions can give rise to gas concentrations far removed from a breathable environment. We will assume, however, that this does not influence the general conclusions in the following and when reference is made to dissolved and liberated gases that this is air and the terms gases and air are interchangeable.

1.4.5 Adsorbed double layer, crystalline swelling and soil particle interactions

Clay particles less than around 1 μm are often referred to as colloids. Colloids are small particles with irregular plate-like geometry, the behaviour of which is influenced greatly by surface forces. The surface area of clay particles for a given mass is large and the smaller the particles, the larger the *specific surface* (surface area per unit mass) and the influence of surface forces. Kaolinite has a specific surface of around 10–20 m^2/g and the influence of surface forces is significantly less than that of montmorillonite with a specific surface of around 800 m^2/g (Lambe and Whitman, 1969). Consequently, kaolinite particles are far less water-sensitive than the finer montmorillonite particles. The surface of clay particles carries a small electrical charge, generally considered to be negative, which depends on the soil mineral and may be affected by an electrolyte in the pore water. The charge gives rise to forces between fine soil particles in addition to self-weight effects. The charge is balanced by exchangeable ions on the particle surface, which along with the soil particle surface attracts water into the so-called adsorbed double layer. Under equilibrium conditions, isolated particles under these conditions have no net charge. However, the surface charge comes into play when:

- The double layers of particles come into contact;
- There is a reduction of the cations within the adsorbed layer, as would be expected when air comes into contact with the double layer or there is a change in pore fluid chemistry.

In these cases a net repulsive force exists. The particles also experience attractive forces, notably van der Waals force, which affects all adjacent pieces of matter and is essentially independent of the characteristics of the fluid between the particles (Sposito, 1989). The repulsive and attractive forces are of significance and play a role in the strength of soils. The physical interpretation of the stresses in fine-grained particulate soils is closely related to the stresses transmitted through the particles, but is not merely the contact stresses between the particles. Indeed in highly plastic, saturated, dispersed clay there may be little or no physical contact between the particles (Lambe and Whitman, 1969).

As a soil compresses or expands, the particles move closer together or further apart and this influences the 'effective stress[7]', which is a function of the resulting interactions. It is this stress that controls the strength and compressibility of a soil. The 'effective' stress is a net stress that includes the influence of the contact stresses and other interactions between the particles and the adsorbed water. In an unsaturated soil, where there are menisci between particles, there must also be interaction between the contractile skin and the adsorbed double layer and a reduction in the thickness of the double layer. This is unlikely to be a static environment, though a quasi-static equilibrium between surface tension forces, water vapour and the adsorbed water is likely to be established with time.

With a decrease in moisture content, however, there will be an increase in the repulsive forces between closely spaced particles, which along with the influence of the contractile skin can be perceived as helping to propagate air-filled void spaces and the creation of aggregated packets containing the soil particles and water.

While all clay soils can be expected to change volume on change of water content, and thus suction, the term *swelling clay* is often used to describe those clays containing a relatively high proportion of clay minerals with the potential for significant crystalline swelling. This is particularly evident with the colloidal clay minerals of the smectite group, which includes montmorillonite, that exhibit high activity (see Section 1.3). Increasing

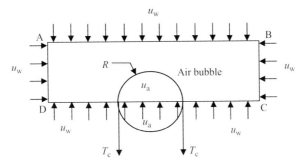

Figure 1.11 Equilibrium of element ABCD with air bubble.

suction draws water from the hydration layers of the clay minerals and the interlayer spaces contract resulting in particle shrinkage in addition to reduction in the void spaces between particles. Reducing suction reverses the phenomenon but hysteresis in volume change can be expected to occur. The change in particle size due to variation in water content obviously influences the mechanical response of swelling clays to external loading. Fityus and Buzzi (2009) have reviewed the place of expansive clays in the framework of unsaturated soil mechanics.

1.4.6 Soil particles and surrounding fluid pressure

All soil particles are surrounded by fluid(s), usually water and air. In a saturated soil, water is the only fluid and the water pressure acts not just through the volume of the water, but also through the volume of the soil particles. This important point is often missed in an analysis and we make no excuse for explaining this simple fact in detail. First, however, consider the spherical air bubble of radius R in Figure 1.11 as it is useful in understanding the significance of fluid pressure on a soil particle. The air bubble is intersected by the small element ABCD with horizontal plane area A_p at right angles to the plane of the paper. The element is considered to be sufficiently small that the influence of the change of gravitational field from top to bottom can be ignored, which is a reasonable assumption for small bubbles. For vertical equilibrium of the element, equating those forces acting downwards to those acting upwards:

$$u_w A_p + 2\pi R T_c = u_w(A_p - \pi R^2) + u_a \pi R^2 \qquad [1.20]$$

This reduces to Equation 1.16 for a spherical air bubble as A_p tends to πR^2.

Now consider the case of a soil particle of cross-sectional area A_s intersected by the small element ABCD in Figure 1.12. For vertical equilibrium of the element:

$$u_w A_p = u_w(A_p - A_s) + u_s A_s \qquad [1.21]$$

where u_s is the pressure in the soil particle. As A_p tends to A_s, u_w tends to u_s, indicating that the water pressure acts through the soil particle. This is taken into account in Section 1.4.7 to establish Terzaghi's equation of effective stress.

Gravity dictates that the air bubble in Figure 1.11, being less dense than water, rises through the water unless acted upon by other forces. In a soil, the confining spaces between particles and the interaction of air bubble with the particles inhibit bubble migration. For

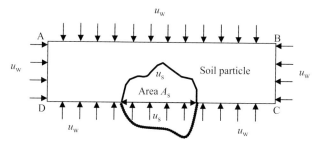

Figure 1.12   Equilibrium of element ABCD with soil particle.

the soil particle in Figure 1.12, gravity dictates that the particle will sink if unrestrained and denser than water. This is the converse of the behaviour of the air bubble. Both the air bubble and solid particle would move to minimise of the thermodynamic potential if free to do so.

1.4.7 'Effective' stress and stress state variables

In any analysis on a flat plane cut through a soil, the plane must cut through soil particles. In a saturated soil, the water pressure acting through the soil particles must be taken into account. Consider Figure 1.13 where the plane AB of area A_p cuts through soil particles and water.

The total upwards force is given by the total stress σ acting over the area A_p. This is balanced by the water pressure, acting through the volume of water and soil particles, and the effective stress. The effective stress is a consequence of the soil particle interactions and acts over the total planar area consistent with the definition of effective stress. Thus, the balance of forces gives $\sigma A_p = u_w A_w + u_w A_s + \sigma' A_p$. Since $A_p = A_w + A_s$, this reduces to Terzaghi's effective stress equation for saturated soil:

$$\sigma' = \sigma - u_w \qquad [1.22]$$

Equation 1.22 does not emerge from the analysis unless the water pressure acting through the volume of the soil particles is taken into account. An alternative presentation can be found in some textbooks and assumes that points of contact between particles are small and analysis is made of equilibrium on a 'wavy' surface passing through only the points of contact. In this case the water pressure acts over the total area of the wavy plane and

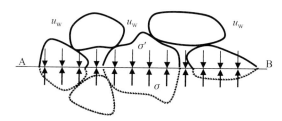

Figure 1.13   Equilibrium on plane AB cut through a saturated soil.

$\sigma A_p = u_w A_p + \sigma' A_p$. This again reduces to Equation 1.17, but the analysis masks the influence of the fluid pressure (in this case water) acting through the soil particles, which is important in the analysis of unsaturated soils presented in Chapter 8.

In a perfectly dry soil the surrounding fluid is air and it is the air pressure that acts through the soil particles. The equivalent effective stress (net stress) for a perfectly dry soil is given by:

$$\bar{\sigma} = \sigma - u_a \qquad [1.23]$$

This assumes no 'bonding' between soil grains and no influence from soil suction.

In unsaturated soils the situation is more complex because two fluids are present. In Chapter 5 current thinking around the use of independent stress state variables for unsaturated soils is discussed. The stress state variables are defined as the two, effective, stresses $(\sigma - u_a)$ and $(\sigma - u_w)$, along with the matric suction $(u_a - u_w)$. The suction variable is the difference between the other two stress state variables and thus only two of the variables can be considered independent.

We will deal with the state of the soil under triaxial stress conditions in later chapters but it is appropriate here to define the mean stresses. A schematic of triaxial stress conditions is presented in Figure 1.14. In the apparatus the total axial stress is given by the principal stress σ_{11} and the total radial or lateral stress (the cell pressure acting all around the soil

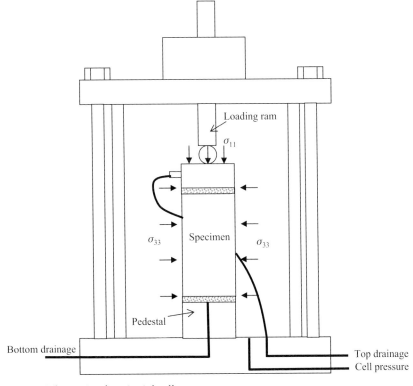

Figure 1.14 Schematic of a triaxial cell.

specimen) is given by the principal stress σ_{33}. Thus, for a saturated soil:

$$p' = p - u_w \qquad [1.24]$$

and for a perfectly dry soil:

$$\overline{p} = p - u_a \qquad [1.25]$$

where $p = \frac{1}{3}(\sigma_{11} + 2\sigma_{33})$ is the mean total stress, $p' = \frac{1}{3}(\sigma'_{11} + 2\sigma'_{33})$ is the mean effective stress, $\overline{p} = \frac{1}{3}(\overline{O}_{11} + 2\overline{O}_{33})$ is the mean net stress, $\overline{O}_{11}$ is the axial net principal stress and $\overline{O}_{33}$ is the radial net principal stress. σ'_{11} is the axial effective principal stress, σ'_{33} is the radial effective principal stress.

1.5 Soil structure

The fabric or structure of soil is derived from the arrangement of the particles and the presence of mass structural effects. Soil fabric significantly influences the mechanical and hydromechanical behaviour (Casagrande, 1932; Lambe, 1951; Leroueil and Vaughan, 1990; Mitchell, 1993). Structure is particularly important in fine-grained clayey soils, which comprise an abundance of fine, plate-like particles. Though also influential in coarser-grained sands and gravels, the effects are less pronounced. The presence of mass discontinuities such as fissures, laminations and bedding planes form part of the macro-fabric of a soil and are a consequence of deposition and past stress history. Such characteristics significantly influence behaviour in the field, not least the permeability (Hossain, 1992; Leroueil et al., 1992; Little et al., 1992). Small specimens of either natural soil or compacted soil may not contain a representative distribution of the macro-features, and the structural hierarchy in situ is rarely as uniform as in laboratory-prepared specimens. It is important to recognise this in transferring laboratory results to the prediction of behaviour in the field. However, the micro-fabric, or distribution of the soil particles, also plays a significant part in soil behaviour and is usually considered adequately represented in small-scale laboratory specimens.

The early descriptions of the structure of compacted clays was based on individual particles with the structure dry of OWC described as flocculated and the structure wet of OWC described as more dispersed with a face-to-face, parallel particle arrangement (Lambe, 1951). The generalised horizontal alignment of clay platelets in soil compacted wet of optimum is consistent with the description given by Lambe (1951) of a dispersed structure. However, an ordered soil structure of this kind can also result from natural deposition or be a consequence of particle reorientation on deformation planes. In fine-grained cohesive soil, preferential orientation of clay platelets leads to pronounced anisotropic behaviour characteristics. At right angles to the direction of particle alignment the permeability can be expected to be several orders of magnitude less than the permeability in the direction of particle alignment because of the greater tortuosity of fluid flow (Lambe, 1955; Witt and Brauns, 1983; Arch et al., 1993). Strength, deformation and consolidation characteristics can also be expected to be influenced by particle alignment. It may be that research into fractal geometry is able to shed further light on the distribution of different levels of soil structure.

The importance of soil structure and its dependence, amongst other factors, on water content, stress history and method of compaction have generated much research, with progress through the years relying on advances in technology. While there are several methods available to examine the pore size distribution of soils, the most productive have been the scanning electron microscope (SEM) and mercury intrusion porosimetry (MIP). These techniques are outlined in Section 1.6. Interpretation of the structure in unsaturated soils dry of OWC is now more refined and takes account of the aggregation of soil particles. Aggregation leads to a soil with a bi-modal pore size distribution comprising relatively small intra-aggregate void spaces within the particle aggregations and larger inter-aggregate void spaces between the aggregates (Barden and Sides, 1970; Ahmed *et al.*, 1974; Wan *et al.*, 1990; Alonso *et al.*, 1995; Gens *et al.*, 1995; Delage *et al.*, 1996; Lloret *et al.*, 2003; Romero *et al.*, 2003; Thom *et al.*, 2007; Sivakumar *et al.*, 2010b).

1.6 Experimental techniques for examining pore size distribution

The importance of soil fabric to behaviour cannot be overemphasised. Consequently, techniques that allow an examination of the microstructural arrangement of soil particles, soil structural features such as strain localisation and a determination of the variation of pore sizes within a soil are fundamental to characterising soil behaviour. The SEM allows a visual image of the structure of fine-grained soils to be obtained and is the most widely used surface diagnostic tool. Much research has also centred on MIP, which allows measurement of the pore size distribution and clearly illustrates a bi-modal distribution of pore sizes in most fine-grained unsaturated soils. There continue to be advancements in these approaches as well as the introduction of newer techniques such as the tomography methods: X-ray computed tomography, neutron tomography and electrical impedance tomography. These approaches are briefly discussed in the following sections. Other techniques referenced in the literature – time-domain reflectometry, near-infrared spectroscopy, the dual-energy gamma-ray technique, along with more powerful X-ray techniques provided by synchrotron radiation and micro-focus X-ray systems – are also finding increasing usage (see Romero and Simms, 2008).

1.6.1 SEM method

This technique employs a narrowly focused beam of high-energy electrons to scan across the surface of a specimen. The beam interaction with the surface generates a shower of secondary and backscattered electrons which are collected by a detector. The intensity of the emitted electrons varies according to differences in surface topography and/or material composition. A flat specimen is usually needed in order to discriminate between the topographic information and the compositional information, as well as phase identification. The specimens are carefully glued into plugs using epoxy or silpaste and are dried overnight in a low-temperature oven. The specimens are then coated with a conductive metal, normally gold, and placed inside the SEM unit from which images are obtained at the required magnifications.

Romero and Simms (2008) described the environmental scanning electron microscope (ESEM) technique, which is a promising advancement from SEM testing. The approach requires no conductive coating on the specimen, making it possible to examine wet soil specimens and to preserve the natural specimen characteristics. Amongst others who have reported testing using ESEM are Komine and Ogata (1999) and Villar and Lloret (2001), who have reported on the swelling characteristics of bentonite, and Monroy *et al.* (2010), who have reported on the structure of London clay.

1.6.2 MIP method

MIP is perhaps the most widely used method for measuring pore sizes of a soil or rock fragment. The general procedure for MIP was first suggested by Washburn (1921). Normally, mercury does not intrude pores in a specimen because the free energy of the soil–mercury interface is greater than the free energy of the soil–gas interface. However, the application of pressure can force mercury into the pores of a soil. The differential pressure required to intrude the pores is given by:

$$p_d = \frac{-4T_m \cos\theta_m}{d} \qquad [1.26]$$

where p_d is the differential pressure, T_m is the surface tension of mercury (0.484 N/m at 25°C), d is the pore diameter, θ_m is the contact angle of the air-mercury with the walls of the capillary (usually taken as between 139° and 147° (Diamond, 1970), although Penumandu and Dean (2000) suggest higher values of 162° for the advancing angle and 158° for the receding angle for kaolin using the sessile drop technique).

The use of mercury porosimetry requires that all of the pores be free of interfering liquids at the start of the intrusion measurement (Sridharan et al., 1971). The interfering liquids can be removed by either air drying or oven drying. However, in the case of soils, these approaches can lead to soil shrinkage (Diamond, 1970). In order to overcome the problem, freeze drying is used to remove the water by sublimation and desorption (Ahmed et al., 1974; Delage et al., 1996). Romero and Simms (2008) provide further discussion on the methodology and the use of MIP.

1.6.3 Tomography methods

Sun et al. (2004) described the development and use of real-time, X-ray computed tomography to the observation of density changes in both soils and rocks, and reported results from a conventional triaxial compression test on silty clay. This non-destructive method was used to provide cross-sectional images of the attenuation of an X-ray beam through the soil specimen, as the specimen was sheared under increasing deviator stress up to and beyond peak strength. The images provided evidence of the stages in development of defects from initial specimen conditions through to post-failure. The cross-sectional images appear to indicate the development of, or tendency towards, a complex pattern of fissuring, interpreted as radial, axial fissures, though more laterally inclined fissures are unlikely to be adequately represented on the images.

Koliji et al. (2006) described the use of neutron tomography followed by image processing to measure the macro-pore evolution during compaction. Neutron tomography is a non-destructive technique for investigating the distribution of neutron attenuating materials (Degueldre et al., 1996). Using the principles of computer tomography, a 3-D reconstruction of a soil specimen from a set of radiographies can be obtained by rotating the specimen.

Electrical impedance tomography can be used in the laboratory as well as in geophysical investigations in the field. The technique allows the estimation of the spatial distribution of the electrical conductivity within a soil to be determined from impedance measurements. Electrodes are applied to the surface of the soil and electric currents applied to selected electrodes while the resulting voltage is measured at the remaining electrodes. The distribution of conductivity within the specimen can then be estimated. Abu-Hassanein et al. (1996) and Borsic et al. (2005) described the use of electrical conductivity measurements in the laboratory to determine heterogeneities in soil properties such as porosity, degree of saturation and hydraulic conductivity.

1.7 Pore size distribution

There is a considerable body of research identifying the bi-modal pore size distribution of unsaturated fine-grained soils based primarily on SEM and MIP analyses of laboratory-prepared specimens. It is now widely accepted that the simplest representation of unsaturated soils is given by two levels of structure (Diamond, 1970; Ahmed *et al.*, 1974; Gens and Alonso, 1992; Al-Mukhtar, 1995; Al-Mukhtar *et al.*, 1996; Delage *et al.*, 1996; Cuisinier and Laloui, 2004; Koliji *et al.*, 2006). Figure 1.15 illustrates a compacted soil that possesses a bi-modal pore size distribution as a result of aggregation of the soil particles. Within the aggregates are the fine intra-aggregate pores, while between the aggregates are the larger inter-aggregate pores. The general conclusion from analysis of specimens compacted wet and dry of Proctor OWC is of an aggregated soil structure dry of optimum and a more dispersed, unimodal pore size distribution wet of optimum.

The dual-porosity model is a result of the micro-structural arrangement of the soil particles within the aggregations, and the macro-structural arrangement of the aggregates and their relation to mass characteristics. Within both natural and compacted clayey soils dry of optimum, the aggregation of soil particles results in what is often referred to as an 'open' soil structure. The drier a soil, the more likely the inter-aggregate pore spaces will develop into open fissures. Griffiths and Joshi (1989) found that volume change of unsaturated soils under increasing consolidation stress was due to deformation of inter-aggregate pore spaces and that intra-aggregate pore spaces remained largely unchanged. Gens and Alonso (1992) reached a similar conclusion from an examination of the deformation characteristics of unsaturated expansive clay, while Toll (1990) recognised the significance of an aggregated structure to the strength of unsaturated soils.

As a soil dries, the water phase is drawn into the smaller intra-aggregate pores and the suction increases, while the air phase fills the surrounding larger inter-aggregate void spaces (Alonso *et al.*, 1990; Delage and Graham, 1995). The concept of a bi-modal structure and the dual stress regime implied by this model is used in examining the strength and compression behaviour of unsaturated soils in later chapters. Consistent with this, in describing unsaturated soils, Barden and Sides (1970) carried out a series of SEM tests and referred to the creation of saturated packets, comprising the soil particles and water, surrounded by air-filled voids. While this represents an idealised view of unsaturated soils, theoretical justification for the close association of the water and soil particles is presented in Chapters 7 and 8. There is also increasing experimental evidence to support the view that saturated aggregates of particles provide a reasonable basis for

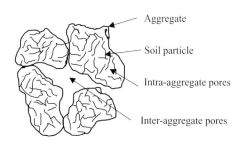

Figure 1.15   Illustration of aggregated soil structure.

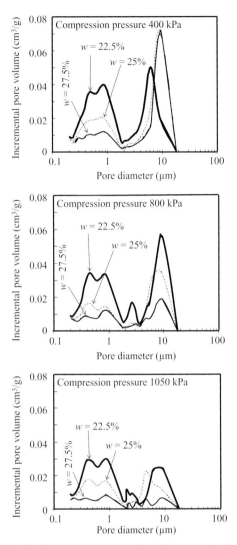

Figure 1.16 Pore size distribution of specimens of kaolin isotropically compressed to various pressures at water contents of 22.5, 25 and 27.5% (after Thom *et al.*, 2007).

the interpretation of the structure of unsaturated soils over a wide range of suction. Tang and Cui (2009) made this assumption in their analysis, arguing that it was supported by the experimental observations of Saiyouri *et al.* (2000). Monroy *et al.* (2010) described a bi-modal structure for unsaturated London clay in which measurements indicated the aggregates were saturated. Further justification is given by the MIP analyses discussed later in the chapter.

Thom *et al.* (2007) reported on the results of an investigation using MIP into the effect of post-compaction wetting, compaction water content, method of compaction and compactive effort on the existence of a bi-modal pore distribution in unsaturated specimens of speswhite kaolin. Figures 1.16–1.18 illustrate the pore size distributions based on the

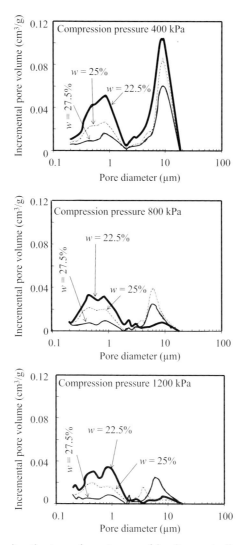

Figure 1.17   Pore size distribution of specimens of kaolin statically compressed to various pressures at water contents of 22.5, 25 and 27.5% (after Thom *et al.*, 2007).

MIP tests on specimens of kaolin isotropically compressed, statically compressed and dynamically compacted at water contents of 22.5, 25 and 27.5%. The plots indicate the incremental pore volume for 1 g of dry kaolin plotted against the pore diameter. Incremental pore volume refers to the volume of mercury that intruded into the pores between pressure increments. Irrespective of the method of specimen preparation, there is clear indication that the kaolin specimens possessed a bi-modal pore size distribution. The SEM image in Figure 1.19, taken of one of the specimens isotropically compressed, provides a pictorial illustration.

The plots of Figures 1.16–1.18 indicate a distinct division between those smaller voids, interpreted as comprising the intra-aggregate pore spaces, and the larger inter-aggregate voids. The intra-aggregate voids are shown as typically between 0.3 and 2 μm, with the

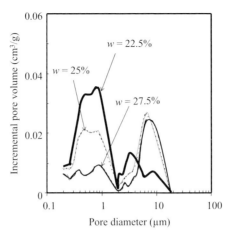

Figure 1.18   Pore size distribution of specimens of kaolin dynamically compacted at water contents of 22.5, 25 and 27.5% (after Thom *et al.*, 2007).

larger inter-aggregate voids >3 μm and typically around 10 μm. The difference between the two sets of void spaces of 2–3 μm corresponds approximately to the particle size of kaolinite. The inter-aggregate voids are shown as larger than those voids within the aggregations by around 10–20 times. In general, the plots indicate that the lower the compaction water content, the greater the percentage of the intra-aggregate pores and the lower the percentage of the inter-aggregate pores. However, a word of caution is necessary as the relative volume of inter-aggregate pore spaces for very dry soils may be influenced in the field by macro-fissures not being represented in the small specimens tested in the laboratory.

The plots clearly indicate that the form of compression and the water content at the time of compression influenced the determined pore size distribution of the unsaturated kaolin. For given water content, the increase in compression pressure in Figures 1.16 and 1.17 is shown to have had a far more pronounced influence on the inter-aggregate pores than on

Figure 1.19   SEM image of specimen of compacted kaolin (static compression to 800 kPa at water content of 25%) (after Thom *et al.*, 2007).

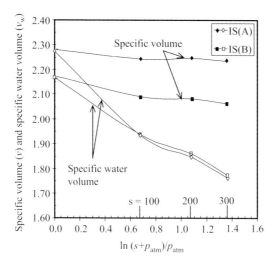

Figure 1.20 Suction-volume and suction-water volume relationship (after Sivakumar *et al.*, 2010b).

the intra-aggregate pores. This is consistent with the findings of Griffiths and Joshi (1989) and Gens and Alonso (1992).

It is also important to note that the bi-modal structure was not broken down by compaction even at water contents close to the optimum of 29% (based on standard Proctor compaction).

Figures 1.20 and 1.21 after Sivakumar *et al.* (2010b) and Sivakumar (2005) respectively present the results of swelling tests on specimens of kaolin. The specimens, IS(A) and IS(B), were prepared isotropically to two different initial specific volumes (IS(A) at

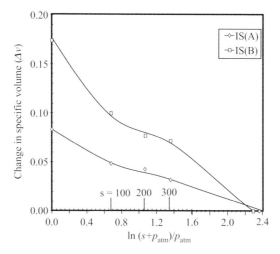

Figure 1.21 Wetting-induced volume change of specimens IS(A) and IS(B) (after Sivakumar, 2005).

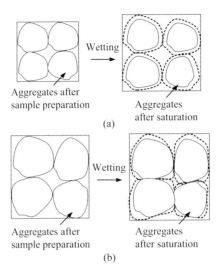

Figure 1.22   (a) Simple model and (b) actual model for aggregates expanding into macro-voids (after Sivakumar *et al.*, 2010b).

2.19 and IS(B) more compact at 1.99), but to similar specific water volumes corresponding to a water content of 25%. The method of isotropic soil preparation is detailed in Chapter 3. Figure 1.20 shows the specific volume and specific water volume against suction, and Figure 1.21 shows the change in specific volume against suction of the specimens from the initially prepared conditions. The inclusion of p_{atm} (atmospheric pressure) in the suction term avoids the natural logarithm of suction becoming indeterminate when the suction falls to zero. The results for specimens of both IS(A) and IS(B) show that significant swelling occurred during the wetting process, but the more compact specimens of IS(B) swelled twice as much as those of IS(A). The simple scenario of swelling of the clay specimens generated solely by the mechanism of expansion of the aggregates is illustrated in Figure 1.22(a). If expansion of the aggregates was the only mechanism for volume change, the volume change for specimens IS(A) and IS(B) would have been expected to be broadly similar. However, this was not the case and the wetting tests on kaolin confirm expansion of the individual aggregates into the inter-aggregate void spaces on uptake of water by the aggregates, reducing the overall potential expansion of a specimen (Thom *et al.*, 2007; Sivakumar *et al.*, 2010b). The conclusion is that the greater the initial compaction, the denser the specimen, the less the closure of the inter-aggregate voids and the greater the overall specimen expansion for a given increase in water content. The MIP results of Figure 1.24 indicate a significant reduction in the inter-aggregate pore volume. This can be explained only by inferred distortion and expansion of the aggregates into the inter-aggregate pore spaces during wetting, as illustrated in Figure 1.22(b). The aggregates in the tests thus acted as deformable structures in which particle rearrangement took place under changing conditions.

According to Alonso *et al.* (1995) and Sivakumar *et al.* (2006b), there are three mechanisms involved in the volume change characteristics of unsaturated soils: (a) swelling of individual aggregates due to water uptake; (b) aggregate slippage at the inter-aggregate contacts due to lack of strength to support the externally applied load, leading to collapse settlement; and (c) distortion of aggregates into the inter-aggregate pore spaces. The

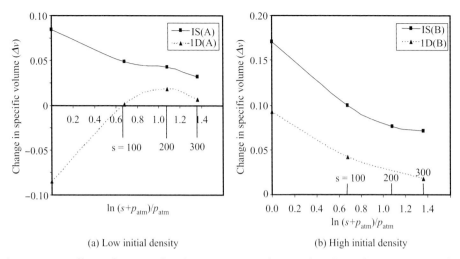

Figure 1.23 Effects of stress-induced anisotropy on the specific volume during wetting (after Sivakumar *et al.*, 2010b).

relative influence of these controlling mechanism determines the overall response of clays during wetting.

Further important insight can be gained into the aggregated structure of kaolin from the wetting results of Figure 1.23(a) and (b) and the corresponding MIP results of Figure 1.24(a) and (b). Cumulative mercury intrusion measurements for the specimens before wetting indicated that the volumes of mercury necessary to fill the inter-aggregate voids were approximately 0.18 cm³/g and 0.125 cm³/g for specimens IS(A) and IS(B) respectively. These volumes correspond to a macro-voids ratio of approximately 0.493 and 0.331 for IS(A) and IS(B). Since the respective overall voids

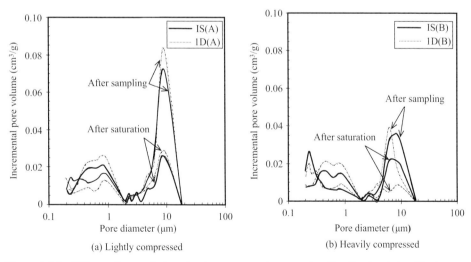

Figure 1.24 Effects of stress-induced anisotropy on the pore size distribution (after Sivakumar *et al.*, 2010b).

ratios of these specimens were 1.193 and 0.989, the magnitude of intra-aggregate voids ratios of the specimens of IS(A) and IS(B) were 0.700 and 0.658 respectively. The kaolin aggregates were prepared at a water content of 25%. If the aggregates were saturated and the inter-aggregates pores free of water, then, taking the specific gravity of kaolin as 2.65, the intra-aggregate voids ratios of the specimens would be approximately 0.663. This value agrees favourably with the measured intra-aggregate voids ratio from MIP results of IS(A) and IS(B), confirming the existence of saturated aggregates for the specimens with suctions up to at least 1000 kPa.

Figure 1.23(a) and (b) compares the changes in specific volumes from initial conditions of isotropically compressed and one-dimensionally compressed specimens subject to reductions in suction on wetting. The IS(A) and 1D(A) specimens of Figure 1.23(a) had an initial specific volume of 2.19 and the IS(B) and 1D(B) specimens of Figure 1.23(b) an initial specific volume of 1.99. Although the paired specimens had identical initial specific volumes and water volumes, the differences in the responses of the specimens to wetting (i.e. the reductions in suction) are shown as significant. In particular, the results for the specimen with isotropic previous stress history and relatively low initial density IS(A) show significant swelling during the wetting process, whereas the paired specimen with one-dimensional previous stress history 1D(A) exhibits significant collapse compression on reduction in suction from 200 kPa to zero. The collapse compression was a result of reduction of the larger inter-aggregate voids associated with a reduction in the resistance to compression instigated by the wetting process. For the denser one-dimensionally prepared specimen 1D(B) of Figure 1.23(b), no collapse is evident as the resistance to collapse was greater. Both IS(B) and 1D(B) experienced swelling throughout the equalisation process though the plots differ significantly. The magnitude of swelling from the initial conditions is shown as 75% greater for the isotropically compressed specimen than the one-dimensionally compressed specimen.

The MIP results of Figure 1.24(a) and (b) indicate the general trend of reduction in the larger inter-aggregate voids and increase in the smaller intra-aggregate voids following saturation from the as-prepared conditions. Despite the identical initial specific volumes and specific water volumes, the specimens with isotropic and one-dimensional previous stress history exhibit slightly different initial bi-modal pore size distributions, represented in the figure as 'after sampling'. The specimens prepared using one-dimensional (1D) compression are shown to exhibit a slightly more 'open' macro-structure at both levels of initial compression effort (i.e. light 1D(A) and heavy 1D(B) compression). The open structure exhibited by the statically compressed specimen of 1D(A) may have contributed to unstable inter-aggregate contacts susceptible to collapse during wetting. However, the observed small differences in the pore size distributions do not explain satisfactorily the disparity in the volume change behaviour reported in Figure 1.23(a) and (b), other than to suggest that small differences in structure may result in significantly different behaviour characteristics consistent with unstable or meta-stable conditions and a chaotic response to wetting.

1.8 Conclusions

The following outlines the main conclusions:

- Soils encompass a wide spectrum of particulate materials, which in the saturated state have all the voids between the particles filled with water and in the unsaturated state have a percentage of the void spaces filled with air. The wide range of particle sizes and the inherent variability of soils give rise to behavioural characteristics not readily amenable to analysis. Stress history and time also influence the soil's multi-faceted behaviour requiring simplifications and generalisations in formulating solutions to problems.

■ For a soil system in equilibrium, an applied external pressure (or stress regime) is balanced by the components of pressure and stress associated with the phases and their interactions.
■ The fabric or structure of soil is derived from the arrangement of the particles and can significantly affect the overall mechanical and hydromechanical properties.
■ Unsaturated fine-grained soils can be viewed as having a bi-modal pore size distribution due to aggregation of particles, with a clear division between those smaller voids comprising the intra-aggregate pore spaces and the larger inter-aggregate voids.
■ The bi-modal structure does not appear to be readily broken down by compaction even at water contents close to the optimum.
■ The wetting of unsaturated specimens results in individual aggregates expanding into the inter-aggregate void spaces, reducing the overall potential expansion of the specimen. The aggregates act as deformable structures in which particle rearrangement takes place under changing conditions.
■ Over a wide range of suctions, there is experimental evidence to suggest that the aggregates remain saturated with water, with air restricted to the larger inter-aggregate voids.

Notes

1. In later sections we will more frequently refer to matric suction of which negative pore water pressure may be considered a special case. Suction is the water pressure relative to an elevated datum air pressure and is frequently used in overcoming cavitation in laboratory testing.
2. B.S. 1377 (1990) 2.5 kg rammer compaction is frequently referred to as Standard Proctor compaction and 4.5 kg rammer compaction as the heavy or Modified Proctor compaction. The equivalent ASTM compaction tests are the Standard AASHTO and Modified AASHTO.
3. Throughout the book, the term *specimen* is used to describe soils carefully prepared for testing, while the term *sample* is used to describe volumes of material not yet prepared as specimens.
4. In some disciplines a bubble is defined as having two liquid surfaces, as for a soap bubble, one on the inside and one on the outside of the bubble. Where there is one surface, such as a gas bubble in a liquid, this is referred to as a cavity.
5. This is an example of Ostwald's ripening. Bubbles or cavities with relatively large curvature are less stable and larger bubbles will grow at the expense of smaller ones.
6. Henry's law is a physical principle which states that at a constant temperature, the solubility of a gas in a liquid varies directly with the partial pressure of the gas. Accordingly, as the partial pressure increases, the solubility also increases. The constant of proportionality is called the Henry's law constant and provides a measure of the *solubility* of a compound.
7. While the term effective stress is strictly applicable only to saturated soils or perfectly dry soils with no suction, we will from time to time refer to 'effective' stress as a generic term for the net inter-particle and interactive stresses internal to both saturated and unsaturated soils.

Chapter 2
Suction Measurement and Control

2.1 Introduction

Many engineering-related problems are associated with soils in an unsaturated state where the void spaces between particles are partly filled with air and partly with water. This leads to negative pore water pressures (or suctions), which greatly influences the controlling stress regime. The accurate measurement and interpretation of soil suction is thus vital to understanding the behaviour of unsaturated soils. However, magnitudes of suction can vary enormously (between 0 and 1 GPa) and the instruments and measurement techniques are usable over only specific suction ranges. Table 2.1 illustrates the general suction ranges and equalisation times for a number of methods of suction measurement and control.

Some instruments and techniques are suitable for laboratory testing only, while others can be used in the field. Some instruments record total suction, while others are used to measure matric or osmotic suction. The different methods can be broadly divided into direct and indirect techniques. The direct approach measures the equilibrium of a soil-water system without involving any external medium, while the indirect approach involves the use of an external medium that achieves moisture equilibrium with the soil. Further reading on suction measurement and control is provided by Fredlund and Rahardjo (1993), Guan (1996), Lu and Likos (2004), Rahardjo and Leong (2006), Ng and Menzies (2007), Bulut and Leong (2008), Delage *et al.* (2008) and Tarantino *et al.* (2009).

In this chapter a discussion is presented on the applicability and limitations of the hanging column technique, the pressure plate technique, psychrometers, chilled-mirror dew point device and tensiometers, along with the more indirect methods of total and matric suction measurements comprising the thermal and electric conductivity techniques and filter paper methods. The discussion is then extended to osmotic suction measurements, the axis translation technique, the osmotic control technique and the vapour equilibrium technique of suction measurement and control in laboratory tests to investigate the volume change, straining and strength of unsaturated soils.

Table 2.1 Approximate measurement ranges and times for equilibration in measurement and control of soil suction.

Instrument	Suction component measured	Typical measurement range (kPa)	Equilibration time
Suction measurement			
Pressure plate	Matric	0–1,500	Several hours to days
Tensiometers and suction probes	Matric	0–1,500	Several minutes
Thermal conductivity sensors	Matric	1–1,500	Several hours to days
Electrical conductivity sensors	Matric	50–1,500	Several hours to weeks
Filter paper contact	Matric	0–10,000 or greater	2–57 days
Thermocouple psychrometers	Total	100–8,000	Several minutes to several hours
Transistor psychrometers	Total	100–70,000	About 1 hour
Chilled mirror psychrometer	Total	1–60,000	3–10 minutes
Filter paper non-contact	Total	1,000–10,000 or greater	2–14 days
Electrical conductivity of pore water extracted using pore fluid squeezer	Osmotic	entire range	—
Suction control			
Negative (or Hanging) water column technique	Matric	0–30 or greater with multiple columns or vacuum control	Several hours to days
Axis translation technique	Matric	0–1,500	Several hours to days
Osmotic technique	Matric	0–10,000	up to 2 months
Vapour equilibrium technique	Total	4,000–600,000	1–2 months

2.2 Techniques for measurement of suction

2.2.1 Hanging water column

Vanapalli *et al.* (2008) provide a detailed discussion on the history and development of the hanging water column approach to matric suction measurement. This approach has been used mainly to examine the relationship between capillary potential and water content

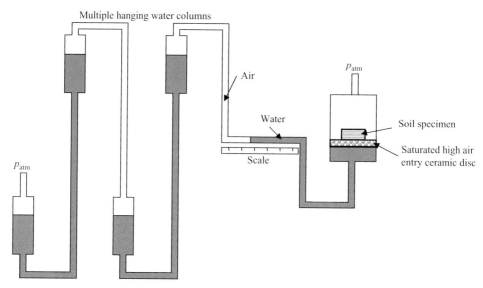

Figure 2.1   Multiple hanging water column apparatus.

in establishing the SWCC for a soil. The maximum suction that can be achieved by the conventional hanging column technique using a single burette to apply the suction is 20–30 kPa. Higher suctions can be achieved by 'multiple columns', as illustrated in Figure 2.1, or using the 'vacuum control technique', which permits matric suction measurements in the range of 0–80 kPa.

2.2.2 Pressure plate

Richards (1941) and Gardner (1956) appear to be amongst the first references to the pressure plate technique, which works on the assumption that the negative pore water pressure in the soil can be brought to a positive value by using the axis translation technique. This was a significant improvement on the original so-called suction plate approach, in which the air pressure was atmospheric and there was a lack of control over the equilibrium water content of the specimen under test. In the pressure plate apparatus, elevated air pressure is applied to an unconfined specimen in order to bring the pore water pressure to a positive value. The basis of the approach is that the increase in the air pressure is equal to the increase in the water pressure, and the matric suction $(u_a - u_w)$, which the apparatus measures, remains unchanged (Hilf, 1956). It is assumed that the water and soil particles are incompressible and the artificial increase in the datum atmospheric pressure does not influence the curvature of the meniscus between the water and air phases.

The apparatus consists of a saturated ceramic disc with a known high air entry (HAE) value and pressure transducers for measuring both the air pressure in the chamber and the water pressure in the drainage line (Figure 2.2). The ceramic disc acts as a membrane between the air and water and resists the flow of air into the pore water measuring system. The basic principle of the ceramic disc is illustrated schematically in Figure 2.3.

Even under careful control there may be an element of unavoidable liquid phase transfer through the disc, which is permeable to dissolved salts, thus influencing the measured

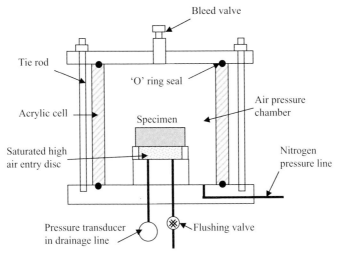

Figure 2.2 Pressure plate apparatus.

suction. If the air entry value of the ceramic disc is less than the soil matric suction, air will also pass through the disc. The presence of air bubbles in the measuring system must be avoided as this gives rise to erroneous pore water pressure measurements. The pressure plate apparatus shown in Figure 2.2 includes the important feature of a flushing valve to allow any air bubbles below the ceramic disc to be removed. Additionally, while the presence of a thin film of water on the high air entry filter before placement of the specimen helps to ensure continuity of the water phase between the specimen and ceramic disc, it can lead to an underestimate of the soil suction measured.

Two types of tests are normally carried out: the null-type test and the pore water extraction tests. For the so-called null-type measurements of matric suction, the air pressure is raised but the specimen is restrained to constant water mass conditions. Matric suction is the difference between the applied air pressure and the pore water pressure at equilibrium. In a pore water extraction test, the applied air pressure is increased and drainage from the specimen is allowed to occur through the high air entry porous disc. Drainage continues until the water content of the specimen achieves equilibrium with the applied matric

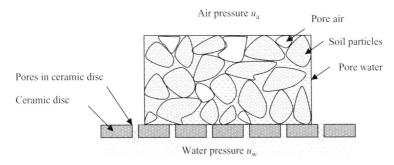

Figure 2.3 Schematic cross section of the interface between the soil specimen and HAE disc.

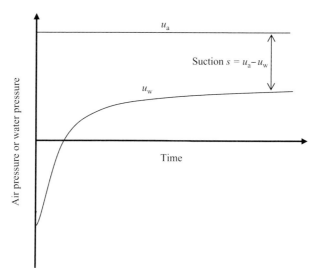

Figure 2.4 Pressure–time response in pressure plate apparatus.

suction. The matric suction is the difference between the water pressure on one side of the disc, which is frequently atmospheric, and the air pressure on the other side of the disc.

When placing the soil specimen on top of the saturated ceramic disc, it is important to ensure that there is good specimen to disc contact and no break in continuity between the water in the specimen and water in the drainage line. A weight of typically 0.1 to 0.2 kg on top of the specimen helps ensure good contact between the specimen and the disc. The placement of the soil specimen on the ceramic disc and assembly of the cell must be performed as quickly as possible, as any delay in action can lead to de-saturation of the previously saturated ceramic filter. Air pressure in the chamber is increased to the desired elevated value to bring the water pressure to a positive value, while drainage of pore water is prevented. Figure 2.4 shows a typical response of the water pressure transducer over time. The water pressure in the drainage line is shown to increase gradually and reach an equilibrium value. The suction is the difference between the pore air pressure applied in the chamber and the pore water pressure measured in the drainage line.

The maximum suction measurable by the pressure plate technique is dependent on the air entry value of the ceramic disc, and these are normally available up to 1500 kPa. Increased rate of air diffusion at higher air pressure and prolonged moisture loss through the vapour phase, due to long equalization times, can affect the accuracy of the suction measurements.

The so-called Tempe pressure cells for measuring matric suction are based on a concept similar to the pressure plate approach described here.

2.2.3 Thermocouple psychrometers

Thermocouple psychrometers allow a determination of the total suction by measuring the relative humidity of soil. The relative humidity R_h is the ratio u_v/u_{vo} of the partial pressure of pore water vapour u_v to the saturation pressure of water vapour over a flat surface of pure water at the same temperature u_{vo}. The thermodynamic relationship of

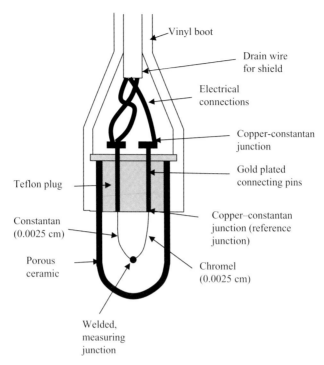

Vinyl boot

Drain wire
for shield

Electrical
connections

Copper-constantan
junction

Gold plated
connecting pins

Teflon plug

Constantan
(0.0025 cm)

Copper–constantan
junction (reference
junction)

Porous
ceramic

Chromel
(0.0025 cm)

Welded,
measuring
junction

Figure 2.5   Peltier thermocouple psychrometer (after Lu and Likos, 2004).

Equation 1.15 allows determination of the total suction from measurements of R_h. Thermocouple psychrometers operate on the basis of the temperature difference between a measuring or sensing junction (i.e. wet junction) and a reference junction (i.e. dry junction) of the thermocouple. Two types of thermocouple psychrometer are available: the Peltier type (Spanner, 1951) and the wet-loop type (Richards and Ogata, 1958). The instruments differ only in the way the measuring junction is wetted. In the Peltier type, evaporation is induced by passing a thermoelectric, cooling current through the measuring junction. In the wet-loop type, the measuring junction is wetted by placing a drop of water in a small silver ring.

A Peltier thermocouple psychrometer is illustrated in Figure 2.5. The thermocouple consists of 0.0025 cm diameter wires of constantan (a copper–nickel alloy) and chromel (a chrome–nickel alloy). The difference in the temperature is related to the relative humidity. It makes use of the cooling produced by the 'Peltier' effect (Figure 2.6) to condense a small amount of moisture (once the dew point is reached) inside the shield onto the measuring junction. The Peltier effect is the phenomenon that allows a thermocouple junction to be cooled by passing an electric current through the junction. As the condensed water evaporates, a potential difference is set up between the measuring junction and the reference junction. This creates an induced electric current in the thermocouple loop due to the 'Seebeck' effect. The Seebeck effect is the phenomenon resulting from the electromotive force generated in a closed circuit of two dissimilar metals when the two junctions of the circuit have different temperatures. The magnitude of the current depends primarily on the rate of evaporation of the condensed water, which in turn depends on the suction and the temperature of the soil (Bulut and Leong, 2008).

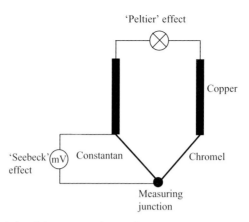

'Peltier' effect

Copper

'Seebeck' (mV)   Constantan   Chromel
effect

Measuring
junction

Figure 2.6   Basic principle of thermocouple psychrometer.

The response of a psychrometer depends on the protective covering to the thermocouple wires and the magnitude of total suction being measured. The protective covers take the form of a ceramic cup, a stainless steel screen or stainless steel (or Teflon) tubing with a screen end window. The type of protective cover provided depends on the application of the psychrometer and the response time required, which varies from a few hours to possibly 2 weeks. Brown (1970) suggested that a psychrometer with the ceramic cup protective cover required an equilibration periods of about three times that of a psychrometer with a stainless steel mesh cover. No protective cover reduced the equilibration period to about half that for a stainless steel mesh. The long equilibration time required by a ceramic cup–type psychrometer renders it impractical in many situations. In addition, the single-junction thermocouple shown in Figure 2.5 has been shown to be extremely sensitive to temperature changes and cannot measure the ambient temperature around the measuring junction. Double-junction psychrometers overcome these shortcomings (Meeuwig, 1972; Campbell, 1979).

Psychrometers can be used to measure the total suction in a soil specimen by comparing results with calibration curves. The calibration of a psychrometer consists of determining the relationship between microvolt output from the thermocouple and known total suctions. The calibration can be carried out by suspending the psychrometer over a salt solution with a known osmotic suction in a sealed chamber under isothermal conditions. The water vapour pressure or the relative humidity in the calibration chamber corresponds to the osmotic suction of the salt solution. A range of NaCl or KCl solutions of known osmotic potential are typically used to establish the relation between suction and micro-volt output. Tables 2.2 and 2.3 show the suction values for a range of temperatures and concentrations of NaCl and KCl solutions, respectively. Detailed calibration procedures have been provided by Wiebe $et\ al.$ (1970) and Sivakumar (2005).

The output readings of the psychrometer are highly dependent on the temperature and the standard procedure is to adjust all calibrations and measurements to a temperature of 25°C, using the following formula:

$$\text{Corrected reading} = \frac{(\text{measured reading})}{(0.325 + 0.027t)} \qquad [2.1]$$

where t is the temperature (in °C) at which the reading was made. However, Comstock (2000) showed that the correction factor $\phi = 0.325 + 0.027t$ depends on both

Table 2.2   Total suction (kPa) of the sodium chloride (NaCl) solution at temperatures from 0 to 40°C. Reproduced by permission of CSIRO Publishing / © CSIRO 1967

NaCl molality	Temperature								
	0°C	5°C	10°C	15°C	20°C	25°C	30°C	35°C	40°C
0.05	214	218	222	226	230	234	238	242	245
0.10	423	431	439	447	454	462	470	477	485
0.20	836	852	868	884	900	915	930	946	961
0.30	1247	1272	1297	1321	1344	1368	1391	1415	1437
0.40	1658	1693	1727	1759	1791	1823	1855	1886	1917
0.50	2070	2115	2158	2200	2241	2281	2322	2362	2402
0.70	2901	2967	3030	3091	3151	3210	3270	3328	3385
1.0	4169	4270	4366	4459	4550	4640	4729	4815	4901
1.2	5032	5160	5278	5394	5507	5620	5730	5835	5941
1.5	6359	6529	6684	6837	6986	7134	7276	7411	7548
1.7	7260	7460	7640	7820	8000	8170	8330	8490	8650
2.0	8670	8920	9130	9360	9570	9780	9980	10 160	10 350

Adapted from Lang (1967).

the temperature and the suction. It was suggested that the slope of the correction factor m_p can be calculated using the following formula:

$$m_p = -0.000210\psi_p^2 - 0.002606\psi_p + 0.023332 \qquad [2.2]$$

where m_p is the variable slope given as a constant (0.027) in Equation 2.1 and Ψ_p is the water potential or total suction. Therefore the corrected psychrometer readings can be

Table 2.3   Total suction (kPa) of the potassium chloride (KCl) solution at temperatures from 0 to 35°C. Reproduced by permission of SSSA

KCl molality	Temperature						
	0°C	10°C	15°C	20°C	25°C	30°C	35°C
0.0	0.0	0.0	0.0	0.0	0.0	0.0	0.0
0.10	421	436	444	452	459	467	474
0.20	827	859	874	890	905	920	935
0.30	1229	1277	1300	1324	1347	1370	1392
0.40	1628	1693	1724	1757	1788	1819	1849
0.50	2025	2108	2148	2190	2230	2268	2306
0.60	2420	2523	2572	2623	2672	2719	2765
0.70	2814	2938	2996	3057	3116	3171	3226
0.80	3208	3353	3421	3492	3561	3625	3688
0.90	3601	3769	3846	3928	4007	4080	4153
1.0	3993	4185	4272	4366	4455	4538	4620

After Campbell and Gardner (1971).

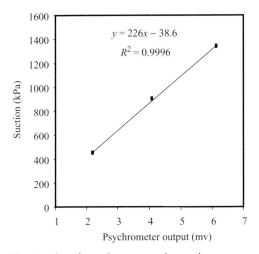

Figure 2.7 Typical calibration line for a thermocouple psychrometer at 20°C.

adjusted using the following formula (with temperature at 20°C):

$$\text{Corrected reading} = \frac{(\text{measured reading})}{(0.486 + 0.0257t)} \qquad [2.3]$$

At 20°C, the corrected reading and measured reading are the same. If the temperature is greater than 20°C, the corrected reading will be less than the measured reading, whereas if the temperature is less than 20°C, the corrected reading will be greater than the measured reading. Figure 2.7 shows a typical calibration curve for a thermocouple psychrometer at 20°C.

Spanner (1951) suggested that the apparatus must be capable of distinguishing dew points to an order of 0.001°C in order to measure suctions to an accuracy of about 10 kPa. This puts a severe limitation on the accuracy of the thermocouple psychrometer technique. The conclusions of tests reported by Krahn and Fredlund (1972) were that thermocouple psychrometers are unreliable for measuring total suctions less than 100 kPa and that psychrometer measurements between 100 and 300 kPa should be very carefully evaluated for validity. For these reasons, the range of suctions measurable using thermocouple psychrometers is usually limited to 100–8000 kPa. The lower limit corresponds to an approximate humidity of around 100% and the upper limit is restricted by the maximum degree of cooling produced by the chromel–constantan thermocouple and corresponds to a relative humidity of around 94%. The susceptibility of the device to errors as a result of temperature fluctuations renders it unsuitable for measuring suctions in situ where significant temperature changes can occur.

2.2.4 Transistor (or thermister) psychrometer

The transistor psychrometer (Dimos, 1991; Woodburn *et al.*, 1993; Truong and Holdon, 1995) has extended the range of psychrometer total suction measurement to 15 MPa. Woodburn and Lucas (1995) extended the measuring range further, to 70 MPa, by using a millivoltmeter to read outputs rather than the standard data logger system of measurement. Delage *et al.* (2008) discussed the instrument and its use, which consists of two bulbs that

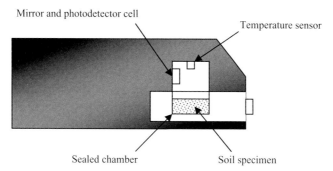

Mirror and photodetector cell

Temperature sensor

Sealed chamber   Soil specimen

Figure 2.8   Schematic of chilled-mirror dew point hygrometer (after Leong *et al.*, 2003). Reproduced by permission of the authors and the Institution of Civil Engineers

act as 'wet' and 'dry' thermometers. The instrument is placed in a sealed, insulated chamber in equilibrium with a soil specimen. A standard drop of distilled water is used in the wet thermometer. Evaporation from the wet bulb lowers its temperature and the difference in temperature between the wet and dry bulbs after a standard period of 1 hour gives a measure of the relative humidity and thus of the soil suction.

2.2.5 Chilled-mirror dew point hygrometer

The relative humidity of the air in the pore spaces of a soil can be determined by the chilled-mirror dew point technique (Leong *et al.*, 2003). This approach allows the total suction to be determined in a similar manner to a thermocouple psychrometer but with an extended upper limit of 60 MPa and a reduced equilibrium time of around 5 minutes. The apparatus is illustrated in Figure 2.8. A small specimen is placed in a sealed chamber containing a mirror and a photodetector cell. The soil specimen is allowed to reach equilibrium, which occurs when the relative humidity of the air in the chamber is the same as the relative humidity of the air spaces in the soil specimen. The mirror is cooled and the temperature carefully controlled by a Peltier current. The photoconductor records the exact point at which water condenses from the air in the chamber onto the mirror. This is the dew point and is the temperature at which the water vapour in the air is just sufficient to saturate the air. Relative humidity R_h is the ratio of the saturated vapour pressure of water at the dew point to the saturated vapour pressure of water above a flat surface at the air temperature. The device has a control to set the temperature of the soil sample to that at which the relative humidity is to be measured.

2.2.6 Tensiometers

Tensiometers can be used in the laboratory and under site conditions and are inserted into the soil via a pre-drilled hole. A number of different types are available and discussed in the literature (Delage *et al.*, 2008). A discussion on the background to the development and applicability of the tensiometer is given next. High-capacity tensiometers or suction probes developed principally for laboratory investigations are discussed in Section 2.2.6.

Tensiometers are used to measure the matric suction or negative pore water pressure where the pore air pressure is not raised as in the axis translation technique. Osmotic suction is not measured in the instrument as soluble salts can move freely through the porous filter.

Tensiometers rely on an exchange of water between the instrument and the soil to equalise the water pressure in the instrument with the pore water pressure in the soil. The instruments consist of a high air entry ceramic cup or filter connected to a pressure measuring device (usually a vacuum gauge or a pressure transducer) through a small pore tube. The pore tube and the cup are filled with de-aired water via the top of the tensiometer. Achieving good contact between the instrument tip and the soil is very important. When the tensiometer is inserted, the suction in the soil draws water from the tensiometer via the porous filter until the stress holding the water in the tensiometer is equal to the suction in the soil. The response time depends on the permeability of the porous filter, the sensitivity of the measuring gauge, the volume of water in the system and the amount of undissolved air in the system, but the instruments usually exhibit a relatively rapid response to pore water pressure changes. A miniature tensiometer is used in a laboratory environment, whereas larger probes are required for site application. Gosmo *et al.* (2000) reported the use of tensiometers in the field up to 3.0 m below ground level. In such cases, it is necessity to correct the measured negative pore water pressure to account for the difference in elevation of the pressure transducer and the measuring point. While the suction values recordable by a miniature tensiometer can be typically as high as 1–1.5 MPa, for larger scale tensiometers, due to the higher risk of cavitation of the water in the instrument, the suction values are limited to about 85–90 kPa (Stannard, 1990).

Watson (1967) introduced the electronic pressure transducer tensiometer for field measurement. An electronic pressure transducer tensiometer is shown in Figure 2.9. This type of tensiometer is suitable for automatic monitoring of negative pore water pressure. However, the thermal expansion/contraction of the water in the system can affect the measurements and result in errors thus continuous monitoring is necessary.

A mercury manometer–type tensiometer to measure the tension in the soil water was proposed by Mottes (1975). Alternative measuring systems include a Bourdon vacuum gauge or a pressure transducer connected to the tensiometer for measuring the equilibrium of water pressure (Rahardjo and Leong, 2006).

While other commercially available tensiometers are available, they are generally based on similar principles to the foregoing. Ridley *et al.* (2003) described a vacuum gauge tensiometer with a 'jet fill' reservoir. The porous cup or filter is connected to a vacuum gauge via a rigid pipe (Figure 2.10). The advantage of this type of tensiometer is that the vacuum gauge directly reads the negative pore water pressure in the porous cup.

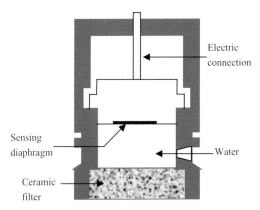

Figure 2.9 Electronic pressure transducer tensiometer (after Watson, 1967). Copyright Elsevier. Reproduced with Permission.

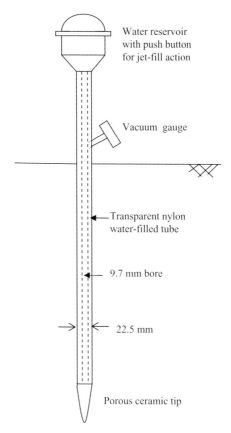

Water reservoir
with push button
for jet-fill action

Vacuum gauge

Transparent nylon
water-filled tube

9.7 mm bore

22.5 mm

Porous ceramic tip

Figure 2.10 Vacuum gauge tensiometer with a 'jet fill' reservoir (after Ridley *et al.*, 2003). Reproduced by permission of the authors and the Institution of Civil Engineers

The practice of applying tensile stress to the water can cause air bubbles to form within the tensiometer. Richards *et al.* (1937) proposed removal of the air bubbles using a double-tubing tensiometer, allowing flushing of the air from the interior of the cup. Figure 2.11 provides an example of a double-tubing tensiometer. Using a similar principle, the incorporation of a jet fill system at the top of the tensiometer tube allows flushing of air bubbles from the tube. During flushing, air bubbles rise to the water reservoir as water from the reservoir fills the tensiometer tube. However, the presence of air in the reservoir influences the readings of negative pore water pressure and it is necessary to remove the collected air (Rahardjo and Leong, 2006).

Peck and Rebbidge (1966) introduced an osmotic tensiometer to overcome the problem of cavitation in conventional tensiometers. Water in this tensiometer is replaced with polyethylene glycol (PEG), and the instrument uses a semi-permeable membrane that is permeable to water but impermeable to the larger PEG molecules. A confined PEG solution in contact with the pore water via the semi-permeable membrane reaches equilibrium when the hydrostatic pressure is equal to the negative pore water pressure. When the tensiometer is placed in soil and equilibrium is achieved, the pressure in the solution decreases by an amount equal to the negative pore water pressure. An advantage of this instrument over conventional tensiometers is that direct measurement of negative pore water pressure can

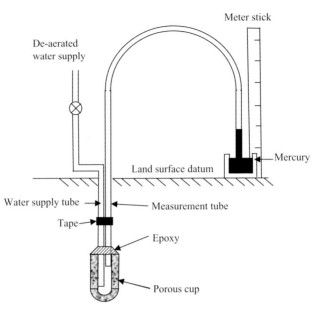

Figure 2.11   Double-tubing tensiometer (after Stannard, 1986). Reprinted from Ground Water Monitoring Review with permission of the National Ground Water Association. © 1986

be extended up to 1500 kPa, provided osmotic potential in the soil is not significant. The instrument does not exhibit hysteresis effects, as water in the tensiometer is always subjected to positive pressure and the formation of air bubbles is generally inhibited. The range of measurement is determined by the reference hydrostatic pressure when the confined solution is in contact with free water. Peck and Rebbidge (1969) found that the reference pressure could deteriorate with time as a result of leakage or a decrease in concentration of the PEG solution and that the instrument was temperature dependent. Dineen and Burland (1995) suggested that the suction determined from vapour pressure of a PEG solution is greater than that measured via a semi-permeable membrane as a result of a small amount of the PEG passing through the membrane.

2.2.7 Suction probes

Ridley and Burland (1993, 1995 and 1996) introduced and described improvements to the suction probe or Imperial College tensiometer (Figure 2.12) designed to overcome the problem of cavitation associated with conventional tensiometers. The suction probe can measure matric suctions in excess of 100 kPa and is usually used in laboratory applications. The suction probe consists of a ceramic porous filter with a high air entry value (1500 kPa), a diaphragm to which a miniature strain gauge rosette is attached and a very small quantity of de-aired water (3 mm^3) contained in a reservoir between the porous filter and the diaphragm. The very small volume of water in the reservoir inhibits cavitation. The probe can be inserted into a pre-drilled hole but its close contact with the soil must be ensured. When the probe is inserted into a soil specimen, the soil extracts water from the reservoir due to the effect of suction. As a result there is a drop in the reservoir pressure, which is measured using a standard pressure transducer. The preparation procedure for

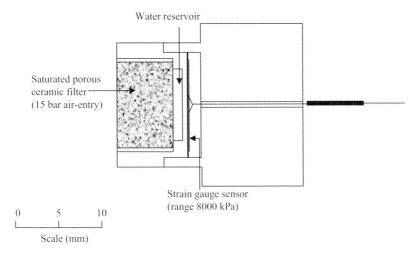

Water reservoir

Saturated porous
ceramic filter
(15 bar air-entry)

Strain gauge sensor
(range 8000 kPa)

0   5   10

Scale (mm)

Figure 2.12   Imperial College suction probe (after Ridley and Burland, 1995). Reproduced by permission of ASME

the suction probe has been described by Ridley and Burland (1999). Proper preconditioning allows the maximum suction of 1500 kPa, the air entry value of the filter, to be measured.

Descriptions of an increasing number of tensiometers or suction probes can be found in the literature. The Trento high-capacity tensiometer was described by Tarantino and Mongiovi (2002) and is similar in design to the Imperial College tensiometer but with modifications. This instrument was used by Tarantino and De Col (2008) to investigate the behaviour of non-active clay subject to one-dimensional static compression in an oedometer and by Tarantino (2009) to study the water retention behaviour of reconstituted and compacted soils. Toker *et al.* (2004) introduced the MIT tensiometer with a face diameter of 38 mm developed for use in a triaxial cell. They described its use in determining matric suction and a continuous SWCC. Meilani *et al.* (2002) and Rahardjo and Leong (2006) discussed the NTU mini suction probe for direct measurement of matric suction in the range of 100–500 kPa, although the range can be extended up to 1500 kPa by replacing the 500 kPa high air entry discs with a 1500 kP high air entry disc. Rahardjo and Leong (2006) discussed tensiometers with ceramic tips of 6 mm diameter and 25 mm length used for measuring negative pore water in laboratory tests. The water reservoir of the jet fill tensiometer can be installed on top of the tensiometer tube for flushing air bubbles. Lourenço *et al.* (2006) described the development of a commercial tensiometer, the WF-DU, which is a miniature suction probe having an approximate diameter of 14 mm and length of 35 mm.

2.2.8 Thermal conductivity sensors

According to Fredlund and Rahardjo (1993), Shaw and Baver (1939a, 1939b) were the first to propose using thermal conductivity to measure the water content in a soil. They employed heating and temperature-sensing elements in direct contact with the soil. Subsequently, Johnston (1942) used plaster of Paris to encase the heating and temperature-sensing elements. Various porous materials have since been tested as potential encasement material including gypsum, carstone (a stone powder), polymer synthetics, sintered metal or glass as well as mixtures of these materials,but fired clay (or ceramic) is the most common material used in thermal conductivity sensors (Bloodworth and Page, 1957; Phene *et al.*, 1971 amongst others). When a porous block is embedded in soil and allowed to reach equilibrium, any change in subsequent suction of the soil results in a change in water content of the block. Since the thermal conductivity of water is around 25 times that of air, the thermal conductivity of the block varies as the water content of the block varies. The water content of the block is measured by supplying a pulse of heat from a heater embedded in the centre of the block and measuring the temperature rise as a result of the heating. The temperature rise is directly related to the thermal conductivity of the porous material and the water content in the block and indirectly to the suction of the soil.

Fredlund *et al.* (2000) and Shuai and Fredlund (2000) described the development of a ceramic thermal conductivity sensor, the UOS (University of Saskatchewan) sensor. This was subject to both laboratory and field trials. The ceramic block was 28 mm in diameter and 38 mm in length. A drawback was the failure of the sensor in moist environments due to cracking of the ceramic, resulting in moisture penetration into the electronics. An improved version of the instrument with a moisture barrier around the electronics was described by Padilla *et al.* (2004). Calibration curves for the instrument for suction up to 1000 kPa were presented.

Thermal conductivity sensors can be used to measure suctions in the laboratory as well as negative pore water pressures on site. One of the advantages is that the readings are not affected significantly by soil salinity although the life of a block can be reduced in some soil types. As long as the block is touching the surrounding soil, it will work. However, porous blocks are susceptible to hysteresis and their response to change in suction can take a relatively long time. Calibration of individual blocks is required because of variations in properties of the porous blocks. This can be achieved by burying the block in a soil sample that is subsequently subjected to known suction values.

2.2.9 Electrical conductivity sensors

The electrical conductivity sensors comprise a porous block and two concentric electrodes embedded in the block. The porous block serves the same purpose as the porous block in the thermal conductivity sensor. However, instead of measuring thermal conductivity, the electric conductivity sensor measures the electric conductivity of the porous block. In wet conditions, the electrical resistance is small. As a soil dries, water is drawn from the block, the larger pores emptying of water first, followed gradually by emptying of the smaller pores. Electric current travelling between electrodes in the block must travel a longer path through smaller pores, and experiences greater resistance as a result of the decrease in water content. Calibration allows electrical resistance of the block to be related to water content of the block and indirectly to suction in the soil.

Gypsum has been found to be a suitable medium for the measurement of electrical conductivity as it takes a relatively short time to saturate, it is relatively quick to respond when placed in the ground and has stable electrical properties. Gypsum blocks are commercially available from a number of sources. However, gypsum has the distinct disadvantage of softening when saturated. Water in the gypsum blocks dissolves the gypsum, giving a solution rich in calcium sulphate. The electrical conductivity is influenced by dissolved solutes and influences the applicability of electrical conductivity sensors. The suction measurement range of gypsum blocks is typically taken as between 50 and 1500 kPa.

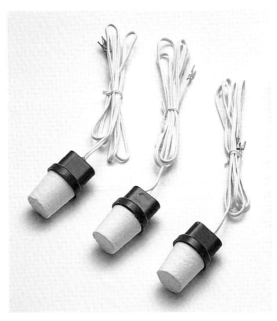

Figure 2.13   Gypsum blocks of Soilmoisture Equipment Corporation. Courtesy Soilmoisture Equipment Corp

Typical gypsum blocks are illustrated in Figure 2.13. He (1999) evaluated the performance of the Soilmoisture Equipment Corporation model 5201 gypsum block of 32 mm length and 35 mm diameter. The gypsum blocks were used in conjunction with a Soilmoisture Equipment Corporation model 5901-A meter to give a digital readout of soil matric suction. Delmhorst gypsum blocks GB-1 measure 22 mm diameter and 29 mm in length with the gypsum cast around two concentric, stainless steel electrodes. Dela (2001) used gypsum blocks with two electrodes comprising concentric rings of wire mesh embedded in a gypsum block of 25 mm diameter and 35 mm in length.

Calibration may be achieved in a similar manner to that for thermal conductivity sensors. Temperature correction is necessary because of the general reduction in electrical resistance with increasing temperature. Electrical conductivity sensors also experience hysteresis on wetting and drying.

2.2.10 Filter paper method

The filter paper method was initially developed in the soil sciences and was later introduced into geotechnical engineering. This is an indirect method of measuring soil suction. The measurement of suction is based on the moisture equilibrium achieved between the paper and the soil specimen in a closed system. The measurements are either of total suction or matric suction depending on how the filter paper interacts with the soil specimen. Total suction can be measured when the filter paper is not in contact with soil specimen (Figure 2.14(a)), whereas matric suction can be measured when the filter paper is in contact with soil specimen (Figure 2.14(b)). The upper range of suction measured by the filter

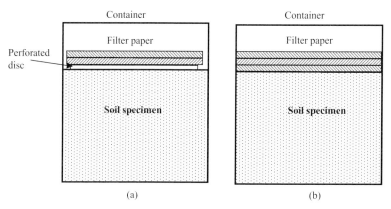

Figure 2.14 (a) Non-contact filter paper method for total suction measurement. (b) Contact filter paper method for matric suction measurement.

paper approach is often taken as around 10 MPa, although a significantly greater range is sometimes quoted.

The filter paper and the soil specimen are placed in a large container that is sealed with plastic electrical tape. The container is then stored in an insulated, temperature-controlled box during equilibration. The filter paper is allowed to equilibrate for typically around 7 days. The suction is obtained from a calibration curve using the measured equilibrium water content of the filter paper.

The accuracy of suction measurements using the filter paper method depends on the quality of filter paper, the sensitivity of the weighing balance, the accuracy of calibration, temperature and equilibration time. Several researchers (Fawcett and Collis-George, 1967; Hamblin, 1981; Chandler and Gutierrez, 1986; Swarbrick, 1995; Navaneethan *et al.*, 2005)have evaluated the use of filter papers to measure suction. The most commonly employed filter papers are Whatman No. 42 and Schleicher & Schull No. 589-WH (Bulut and Leong (2008) noted that in the USA this latter paper is now identified as grade 989-WH). The Whatman No. 42 filter paper appears to give more consistent results than the Schleicher & Schull No. 589-WH filter paper. Calibration of the filter paper to be used can be carried out by either equilibrating it on a pressure plate brought to a known suction or by enclosing it in a sealed container with a salt solution of known vapour pressure. Further information on calibration of filter papers is provided by Houston *et al.* (1994), Bulut *et al.* (2001), Leong *et al.* (2002) and Bulut and Wray (2005) amongst others. The now withdrawn ASTM (D5298–94) (1997) also provides calibration curves. The calibration curves for total suction and matric suction are different, as shown in Figure 2.15, and there are some differences in filter paper calibration curves produced by different investigators that according to Bulut and Leong (2008) are attributable to factors such as the suction source used for calibration, the thermodynamic definitions of suction components and equilibrium time.

The calibration curve for filter paper is always bi-linear with the change in sensitivity for matric suction occurring at a water content of about 45% for Whatman No. 42 paper and about 54% for Schleicher & Schuell No. 589-WH paper. The lower part of the curve represents the high range of filter paper water contents where the water is believed to be held by the influence of capillary forces. The upper part of the calibration curve represents

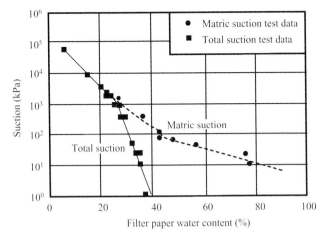

Figure 2.15 Total and matric suction calibration curves for Whatman No. 42 filter paper (modified from Leong *et al.*, 2002). Reproduced by permission of ASTM

lower water contents in the filter paper where the water is held in absorbed water films within the filter paper.

The time taken to reach equilibration is important. Swarbrick (1995) suggested that equilibrium times vary from 2 days to several weeks depending on the soil type. For in-contact filter paper measurements, the time required to reach equilibrium is generally less than for the non-contact filter paper method. Houston *et al.* (1994) and Ridley (1995) suggested that true equilibrium may never be reached for non-contact filter paper measurements at low suction values.

The filter paper technique is generally considered suitable only for laboratory measurements. However, Crilley *et al.* (1991) introduced an approach to measuring in situ suctions using a non-contact filter paper technique. In this approach, the filter paper is wrapped around a vertical aluminium mandrel that is placed down a borehole and supported by PVC tubing. The filter paper protected in a sealed tube is extracted from the borehole and taken away for measurement.

2.2.11 Osmotic suction measurement techniques

Osmotic suction can be indirectly determined by measuring the electrical conductivity of the pore water. The greater the dissolved salts, the greater the osmotic suction and the electrical conductivity. The datum for osmotic suction is that of pure water. The osmotic suction may thus be calibrated against electrical conductivity. Pore water can be extracted using a pore fluid squeezer (Fredlund and Rahardjo, 1993; Manheim, 1996; Leong *et al.*, 2007), although there is evidence that the results are influenced by the magnitude of the extraction pressure.

Alternatively, the osmotic suction is often determined by subtracting independent measurements of matric suction from total suction in accordance with Equation 1.14 (e.g. Tang *et al.*, 2002). However, Richards and Ogata (1961) reported that measured total suctions were not the same as the sum of the matric and osmotic suctions, though Krahn and Fredlund (1972) reported good agreement between osmotic suction measurements, using the squeezing technique, and the difference between the total suction using a psychrometer and the matric suction using a null-type axis translation apparatus. Leong *et al.* (2007) suggested that although the results of Krahn and Fredlund showed good agreement for Regina clay, they exhibited poor agreement for a Glacial Till. Leong *et al.* (2007) carried out suction measurement tests on three different soils. They used the chilled-mirror dew point approach to measure total suctions, a null-type axis translation technique to measure matric suction and measurements of electrical conductivity of extracted pore water to give the osmotic suction. They reported that at low osmotic suctions the sum of the matric and osmotic suctions was less than the measured total suction, while at high osmotic suctions the sum could exceed the total suction measurement. They concluded that caution should be exercised when inferring either osmotic or matric suction from total suction measurements.

2.3 Control of suction in laboratory tests

Published literature outlines numerous methods and refinements to methods of controlling the suction in laboratory tests. The most important principles of these approaches to tests in apparatus such as the triaxial cell, shear box and oedometer are briefly described.

2.3.1 Axis translation technique

Hilf (1956) proposed the axis translation technique to control suction in unsaturated soils. Several researchers have successfully used this approach to control the matric suction in laboratory tests on unsaturated soils (Matyas and Radhakrishna, 1968; Fredlund and Morgenstern, 1977; Ho and Fredlund, 1982; Escario and Saez, 1986; Wheeler and Sivakumar, 1995) and notably in the triaxial cell tests reported later in the book. Figure 2.16 illustrates the triaxial cell set-up adopted by Wheeler and Sivakumar (1993).

The basic principle of the axis translation technique is to elevate the pore air pressure in unsaturated soils so that positive values of pore water pressure are measured. The difference between the pore air pressure and the pore water pressure is the soil matric suction. A saturated high air entry filter is used to prevent air from getting into the pore water measuring system. Normally, a ceramic disc of sintered kaolin is used as a high air entry filter. The air entry value of the filter must be higher than the matric suction to be measured to prevent pore water entering the measuring system. The air entry value is inversely proportional to the pore size of the filter since it controls the radii of curvature of the air–water menisci at the filter boundary. The air entry value can be increased by reducing the pore size of the ceramic filter during manufacture. Ceramic filters with an air entry value up to 1500 kPa are available commercially.

The saturated pores of the ceramic filter provide a connection between the soil pore water and the pore water measuring system. By increasing the pore air pressure and the total stress, the water pressure can be maintained at the desired positive value throughout the system. The main advantage of this method is that there is no chemical used in the process of controlling suction, and therefore no risk of change in chemistry of the pore fluid. While elevating pore water pressure from negative to positive values prevents cavitation within the soil pores, the in situ stress conditions are not replicated in the laboratory tests. Another disadvantage of this technique is the risk of air diffusion through the high air entry filter into the pore water pressure system. This leads to the formation of air bubbles,

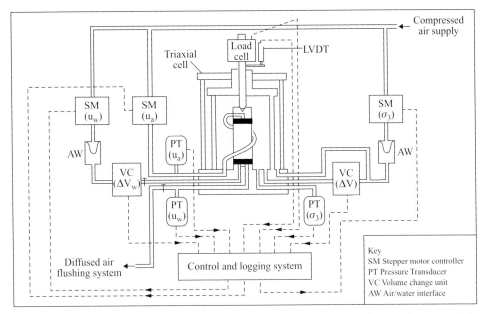

Figure 2.16   Triaxial cell set-up using the axis translation technique (after Sivakumar, 1993).

thus influencing the measurements made. A flushing system is needed to overcome this problem (Bishop and Donald, 1961; Fredlund, 1975; Sivakumar, 1993).

2.3.2 Osmotic technique

Historically, suction control using the osmosis phenomenon originated in the field of soil science (Painter, 1966; Zur, 1966; Waldron and Manbeian, 1970). Kassiff and Ben Shalom (1971) were the first researchers to introduce this approach to geotechnical engineering. Subsequently, several researchers have successfully used this technique to control suction in oedometers (Delage *et al.*, 1992; Dineen and Burland, 1995; Delage, 2002), the shear box (Boso *et al.*, 2005) and the triaxial apparatus (Delage *et al.*, 1987; Cui and Delage, 1996; Ng *et al.*, 2007). The triaxial cell set-up adopted by Cui and Delage (1996) is shown in Figure 2.17. Blatz *et al.* (2008) have discussed the merits and limitations of the osmotic technique for controlling matric suction, and the vapour equilibrium method for controlling total suction.

The osmotic suction method achieves control of the matric suction by allowing the pore water to equilibrate with a salt solution of known osmotic potential, separated from the soil specimen by a semi-permeable membrane (Zur, 1966). The semi-permeable membrane is permeable to water molecules but impermeable to the salt molecules. Different values of suction are applied using different concentration of the salt solution. Normally, PEG is the salt solution used due to its large molecular size. The salt solution generates an osmotic pressure gradient across the membrane that has the effects of drawing water from the soil until the suction and osmotic pressure are in equilibrium. The main advantage of this method is that cavitation, the formation of air bubbles within the soil pores, is not inhibited, as the pore water pressure within the soil is maintained at its negative

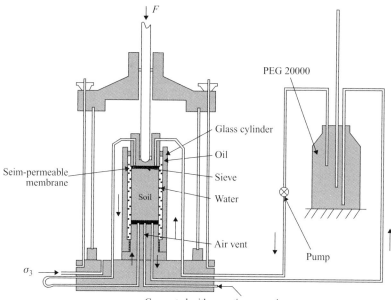

Figure 2.17 Triaxial cell system using the osmotic technique (after Cui and Delage, 1996). Reproduced by permission of the authors and the Institution of Civil Engineers

value. However, the fragility of the semi-permeable membrane can affect the suction measurements in long duration tests (Cunningham *et al.*, 2003). Another disadvantage is the possible migration of soil salts, dissolved in the soil water, entering the salt solution and thus affecting the soil water chemistry and the soil properties. Delage *et al.* (2008) discussed the advantages and disadvantages of the osmotic technique and calibration requirements. The maximum applied suction is determined by the PEG solution, and according to Delage *et al.* (1998), it is up to about 10 MPa. However, Tarantino and Mongiovi (2000) concluded that the maximum applied pressure was limited by chemical breakdown of the semi-impermeable membrane.

2.3.3 Vapour equilibrium technique

The vapour equilibrium technique for controlling total suction was again originally developed by soil scientists. Traditional methods of controlling relative humidity include the isopiestic (or same pressure) method and the two-pressure method. The isopiestic approach relies on attaining vapour pressure equilibrium for salt or acid solutions in a closed thermodynamic environment. The two-pressure approach relies on manipulation of relative humidity by either varying pressure or by mixing vapour-saturated gas with dry gas. According to Delage *et al.* (1998), the vapour equilibrium approach was first applied to control total suction in geotechnical testing by Esterban and Saez (1988). The control of humidity by achieving vapour pressure equilibrium has been employed in geotechnical research by Oteo-Mazo *et al.* (1995), Delage *et al.* (1998), Al-Mukhtar *et al.* (1999), Cunningham *et al.* (2003) and Blatz and Graham (2003). In this approach, a soil specimen is placed in a thermodynamically sealed system controlled by the headspace of a desiccator where an aqueous solution results in a controlled partial vapour pressure generated by the known concentration salt solution. The soil specimen undergoes water exchange with the vapour until the suction in the specimen is in equilibrium with the partial vapour pressure. Normally, solutions of sulphuric acid, sodium and potassium chloride salt, or glycerine and water are used to generate the partial vapour pressure. The control of humidity by the two-pressure approach on the other hand appears to have received less attention though. Kunhel and Van Der Gaast (1993) reported on total suction measurements obtained by mixing vapour-saturated gas with dry gas via a feedback system. Nishimura and Fredlund (2003) used an air regulator and air bubble system in controlling the relative humidity. In this latter approach, the independent assessment of applied suction was achieved by a relative humidity sensor embedded in the circulatory system removing the need for a sensor in the soil specimen.

The vapour equilibrium technique has been used to study the combined effect of high suction and stress level on the mechanical properties of soils, notably the more plastic, swelling soils. Al-Mukhtar et al. (1999) reported on oedometer results on smectite under suctions up to 298 MPa and applied vertical stress up to 10 MPa. Blatz and Graham (2000) described a triaxial cell set-up using the vapour equilibrium approach and report limited tests on a 50:50 sand-to-bentonite mixture, where total suctions were up 7 MPa and imposed isotropic stresses up to 3 MPa. A schematic of the set-up for triaxial cell testing is presented in Figure 2.18. Thermocouple psychrometers were used to measure suctions during testing.

The vapour equilibrium technique has all the advantages ascribed to the osmotic technique with an additional advantage that this approach can be used to apply very high suction values ranging from approximately 4 to 600 MPa. However, the limitation of this method is that equilibration of suction within the soil is very slow and has been reported as often taking 1–2 months. However, testing times can be significantly reduced using an

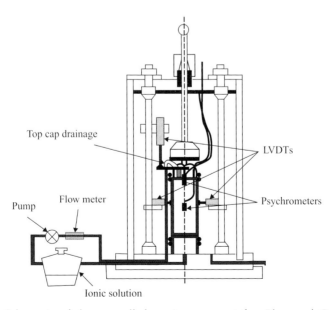

Figure 2.18 Schematic of the controlled suction system (after Blatz and Graham, 2000). Reproduced by permission of the authors and the Institution of Civil Engineers

air circulation technique employing an air pump (Blatz and Graham, 2000; Cunningham *et al.*, 2003; Lloret *et al.*, 2003; Dueck, 2004; Oldecop and Alonso, 2004; Alonso *et al.*, 2005).

2.4 Conclusions

The following are the two main conclusions that we can draw from this chapter:

- There are a large number of techniques for measuring and controlling suction in unsaturated soils. Each technique has different advantages and disadvantages as well as limitations in application, range of measurable suctions (and negative pore water pressures) and sensitivity.
- Care must be exercised in selection of the technique to be adopted for a particular problem, paying attention to whether matric, total or osmotic suction is being measured.

In later chapters, most test results reported are for tests in the triaxial cell for specimens tested through various stress paths. The axis translation technique has generally been adopted with a limited number of tests reported where suctions have been recorded using a thermocouple psychrometer.

Chapter 3
Laboratory Techniques

3.1 Introduction

The validation of a theoretical framework, or assessment of a hypothesis, requires high-quality experimental data collected from a carefully executed testing programme. In geotechnical engineering, such data may be obtained from full-scale field testing or laboratory-based research. Within the context of unsaturated soils, methodologies for obtaining definitive data from full-scale field testing are still the subject of development, and such data are inevitably affected by many factors, the influences of which may not be readily isolated to reveal the underlying principles governing the soil behaviour. While field measurements are important, and the main aim of research is to provide a basis for prediction of field deformations and strength, laboratory-based research has the distinct advantage of allowing careful control of material properties and other influencing factors.

Considerable progress has been made in laboratory-based testing techniques for unsaturated soils. Such testing requires meticulous adherence to testing protocols, including specimen recovery and preparation, the testing equipment, the testing programme, and data handling and analysis. As elsewhere in the book, the term *specimen* is used to describe soils carefully prepared for testing, while the term *sample* describes volumes of material that are to be prepared as specimens. The term *sampling* identifies the process of cutting and trimming samples recovered from the field or prepared as larger compacted or consolidated samples in the laboratory.

This chapter provides information and guidance on laboratory testing techniques in the investigation of the stress–strain behaviour, volume changes and shear strength of unsaturated soils. It outlines the testing techniques adopted in providing the experimental data used in later chapters to investigate and validate the theoretical concepts developed from considerations of the thermodynamic principles governing the behaviour of unsaturated soils. We will be principally interested in triaxial cell stress path testing but much of the discussion is equally applicable to other testing techniques. First, however, the importance of material selection, material and specimen preparation, specimen size and sampling is discussed.

3.2 Material selection and specimen preparation

Laboratory-based research is fundamental to developing a quantifiable understanding of the response of unsaturated soils to changing conditions. While the testing programme must reasonably replicate the conditions to be appraised, material selection and specimen preparation are equally important and forms a prerequisite to obtaining meaningful, reproducible results.

3.2.1 Material selection

Selection of the soil to be tested depends on the purpose of the research project. The researcher may have a range of options for selecting a suitable soil type in a situation where the research is to establish general principles of soil behaviour. However, for problem-based research of site condition, the researcher is likely to have no freedom in the selection of the material. For example, recent years have witnessed increased research into the performance of sand–bentonite barriers exposed to an aggressive ground environment (Stewart *et al.*, 2001; Tang and Graham, 2002; Blatz and Graham, 2003).

The 'testing time' is influenced by the soil to be tested and is an important consideration in any research programme, as it influences decisions on the size of the specimen and the ambitions of the research within the available time scale. Normal testing programmes involve testing between equilibrium (or steady-state) conditions. Non-establishment of such conditions leads to determinations of phase volumes and distributions at odds with those where the specimen is allowed to equilibrate. However, when a soil is in an unsaturated state, the time required to reach an equilibrium state can be significantly longer than that of the soil in a saturated state. Thus, protracted consolidation times can be experienced under elevated suctions. The collection of data on kaolin by the authors has shown that the coefficient of consolidation of unsaturated kaolin can be around 10 times lower than that under saturated conditions. Nevertheless, using kaolin has advantages in research programmes designed to establish general behaviour characteristics. It has a relatively high rate of consolidation, compared to more plastic lower-permeability clays, and consistent, uniform bulk samples are readily available from commercial sources. However, the researcher should be aware that kaolin is a processed material from natural resources and there can be differences between the supplies obtained over a period of time. Simple checks such as index testing and particle size analysis will usually confirm the consistent characteristics of the raw material.

3.2.2 Specimen size

The most common laboratory test procedures employed in investigating the behaviour characteristics and strength of unsaturated soils are one-dimensional consolidation testing, shear box testing and triaxial cell testing. Whichever test method is proposed, the size of the specimen to be tested will dictate the duration of the test. The appropriate specimen size is influenced by the soil's grading and the need to test a representative volume of material. In uniform clay, such as kaolin, relatively small specimens can be used. In triaxial cell testing, typical specimen diameters are around 38–100 mm with lengths of around 76–200 mm. If the material to be tested includes larger particles, such as in a glacial till where a proportion of sand and gravel size particles may exist, consideration must be given to the use of a larger specimen size.

The accurate and representative measurement of strains and volume changes is crucial to interpretation of soil behaviour. Methods of measurement differ for the different tests and the monitoring methodology also influences the appropriate specimen size. Measurement of strains and volume changes, particularly those employed in triaxial cell testing, will be discussed later in this chapter. However, each method involves a degree of unavoidable error and consideration needs to be given to the percentage error compared to the overall volume of the soil in deciding on the appropriate specimen size; larger specimens lead to less percentage error.

3.2.3 Preparation of material

Laboratory preparation of saturated specimens often involves mixing of dry powdered soil with water. A sample wetted to high water content is usually taken as having no preferred particle structure though subsequent compaction or consolidation can lead to anisotropic properties. However, preparation of unsaturated specimens often involves mixing the soil with water at less than the plastic limit. This inevitably leads to the formation of *aggregates* (or peds) of particles in fine-grained soils. The aggregates are generally saturated, particularly at water contents close to optimum water content, and are stiff and of markedly variable size. Unless careful control measures are adopted, there can be significant differences in the pore size distributions of specimens prepared under essentially similar conditions, and this influences the subsequent performance under testing. In order to avoid problems, the authors have found that careful sieving of the aggregated material allows a relatively uniform aggregate and pore size distribution to be achieved. Figure 3.1 shows a sample of kaolin that has passed through a 1.12-mm sieve. It has been found appropriate to seal sieved samples of kaolin in a plastic bag and to store them in a cool place for at least 3 days, to ensure moisture equilibrium before continuing specimen preparation. If a large amount of material is required, the processing time can be long and may invite unwanted evaporation of water. If more than 1 kg of dry material needs to be prepared, it is recommended that the material is processed in batches and at the end of the process the material is thoroughly mixed before storing.

3.2.4 Sampling method and specimen preparation

The sampling technique plays an important role in the preparation of unsaturated soil specimens (Alonso *et al.*, 1992; Maâtouk *et al.*, 1995; Sivakumar and Wheeler, 2000). In the field, sampling of soils, whether from compacted fill or natural material, is fraught with difficulties. Small-diameter driven tube samples from in situ material generally exhibit significant disturbance. Best practice suggests that, where possible, block samples are obtained, allowing specimens to be carefully cut from the blocks under laboratory conditions. Reduction in imposed in situ stresses as a result of sampling must be taken into account when deciding on the testing regime and assessing specimen behaviour.

In the laboratory, where specimens are prepared from disturbed soils, the method of specimen preparation must reflect the aims of the research and the likely field conditions. If it is proposed to assess the performance of compacted fills, attempt must be made to replicate the field compaction. This is not easy and the researcher is often forced to assume standard Proctor compaction or other compaction or compression procedure reasonably reproduces the degree of compaction and soil structure in the field. It is also often the

Figure 3.1   Sieved kaolin aggregates.

case that the size of the sample produced is not compatible with the size of the specimen needed for one-dimensional consolidation testing, shear box testing or triaxial testing, and sub-sampling is necessary leading to possible specimen disturbance. But this is not always the case. Sivakumar (1993) used a compaction mould of 50 mm in diameter and 150 mm in height (Figure 3.2) and found that a weight of 0.45 kg with a head diameter of 22 mm,

Figure 3.2   Compaction mould (50 mm diameter and 150 mm height).

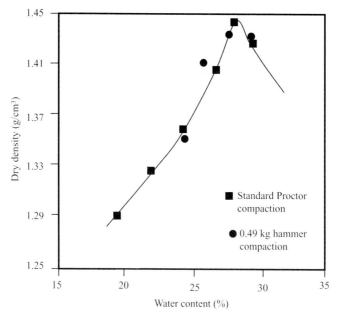

Figure 3.3 Comparison of dynamic compaction characteristics (after Sivakumar, 1993).

falling 300 mm onto soil in the mould, could be used to produce specimens of kaolin that were reasonably compatible with standard Proctor compaction of larger samples (Figure 3.3). In this procedure the specimens were dynamically compacted in nine layers with nine blows of each layer. Unsaturated, one-dimensional compaction specimens prepared in this way were used in the triaxial cell tests reported by Wheeler and Sivakumar (1995, 2000) and Sivakumar and Wheeler (2000).

Caution should be exercised in preparing specimens using dynamic compaction as experience indicates that these can exhibit unacceptable variations in voids ratio and degree of saturation, both within and between specimens. Research has shown that more consistent, uniform specimens are obtained by static compression, which can be used to achieve overall end conditions of density compatible with those achieved under dynamic compaction. Sivakumar (1993) reported excellent repeatability of soil properties for kaolin statically compressed to a pre-selected compression pressure, in similar size small-diameter moulds to those used in preparing the dynamically compacted specimens described above. Axial pressure on samples of wetted kaolin was increased at a constant rate with the samples compressed in nine layers (Figure 3.4) as for dynamic compaction. The compression was terminated when the target vertical pressure was achieved. A disadvantage of this methodology of specimen preparation is the potential interface between the layers. It is possible to improve this by scarifying each layer before a further layer is added. Sivakumar (2005) reported that for kaolin, 1200 kPa of vertical pressure closely replicated standard Proctor compaction – though this pressure may be different for different soil types. It should also be noted that the optimum water content based on standard Proctor, or other modified dynamic compaction, may not be identical with that of static compaction since each method involves a different deformation mechanism. Dynamic compaction involves severe shear deformation of the aggregated structure of unsaturated soils, whereas shear deformation is less evident under static compaction.

Figure 3.4 Frame used for static compression.

Specimens prepared using either one-dimensional dynamic compaction or static compression are subjected to anisotropic deformations that affect the subsequent stress–strain behaviour (Sivakumar *et al.*, 2010a, 2010b). The influence of anisotropy can be overcome by isotropic compression of samples in a standard triaxial cell. Isotropically prepared saturated specimens of 100 mm in diameter were produced by Sivakumar *et al.* (2002). The technique was subsequently modified by Sivakumar (2005) to produce isotropically prepared unsaturated specimens of kaolin (Figure 3.5). In this approach a rubber membrane is placed around the pedestal in a triaxial cell and sealed to the base using two O-rings. A membrane stretcher is then placed around the membrane and the top of the membrane folded at the top of the stretcher (Figure 3.6). A 100-mm diameter dry porous disc is then placed above the pedestal and sieved, aggregated material poured slowly into the membrane. After filling the membrane with material, a top cap with a dry porous stone is placed on the soil sample and the membrane unfolded. At this stage the membrane stretcher is removed and the top cap sealed using two O-rings. The triaxial cell can then be assembled and pressurised. Any excess air pressure in the specimen, developed as a result of the application of the external pressure, should be allowed to dissipate from the top and bottom of the specimen. Practice using kaolin suggests that the system should be allowed to stand for at least 3 days.

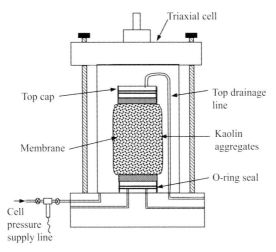

Figure 3.5 Triaxial cell set-up used for isotropic compression of samples (after Sivakumar, 2005).

A specimen of lesser dimensions can be obtained by carefully pushing a sampler into the isotropically compressed sample (Figure 3.7). The sampler employed by Sivakumar (2005) had a cutting shoe with the required diameter of the specimen, but with the inner diameter of the sampler beyond the cutting shoe of a slightly larger diameter, reducing friction and other disturbing influences on the specimen. The advancement of the sampler can be eased by trimming the excess material from the specimen tube using a thin wire saw. It was found that a rate of penetration of 5 mm/min ensured minimal specimen disturbance.

Figure 3.6 Placement of aggregates in membrane.

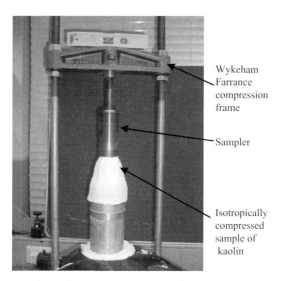

Figure 3.7   Progression of sampler into isotropic sample.

3.3 Experimental techniques for volume change and strength measurements

Experimental techniques in testing unsaturated soils are significantly more complex than in saturated soils. In the following, an introduction is provided to the one-dimensional consolidation test, the shear box test (or direct shear test), the simple shear apparatus, triaxial cell testing and the true triaxial cell. While other testing techniques, such as the Bromhead ring shear apparatus (Merchán *et al.*, 2008), the hollow cylinder triaxial test (Komornik *et al.*, 1980), the modified 'floating ring' oedometer test with lateral pressure measurement (Habib *et al.*, 1992; Habib, 1995; Habib *et al.*, 1995) and the K_0 compression apparatus (Rahardjo and Fredlund, 1995) – all modified to suction measurement or control – may find increasing usage in investigating unsaturated soils, we will restrict our discussion to the test apparatus with the stress states illustrated schematically in Figure 3.8. Initially only an outline of triaxial testing techniques is presented in Section 3.3.4, but a more detailed discussion on the triaxial cell and stress path testing is presented in Section 3.5. It is the results from triaxial cell tests that are reported and analysed in subsequent chapters. Most of the techniques and measurement requirements described for triaxial testing are equally applicable, or provide guidance, in carrying out the other consolidation and shearing tests.

In the oedometer, tests are carried out to determine the one-dimensional consolidation or swelling characteristics when the specimen is subjected to variations in vertical stress or suction. Consolidation or swelling tests under isotropic or variable axial and radial loading can be carried out in the triaxial cell and true triaxial cell. In the direct shear box, simple shear apparatus, triaxial cell and true triaxial cell, consolidation of soil specimens may be undertaken prior to determination of the shearing strength characteristics. The shearing stages in these tests are undertaken by independently maintaining or recording the pore air and pore water pressures under drained or undrained test conditions. Tests carried out on unsaturated soils without measurement of pore air and pore water pressures provide little useful data.

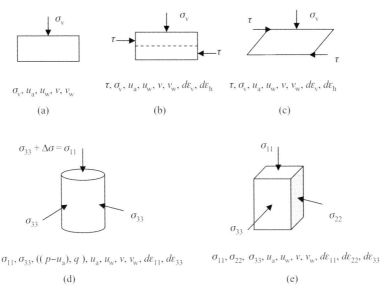

Figure 3.8 Stress states in (a) one-dimensional consolidation test, (b) direct shear box test, (c) simple shear test, (d) triaxial cell test and (e) true triaxial test.

3.3.1 One-dimensional consolidation apparatus

One-dimensional, or uni-axial, consolidation (Figure 3.8(a)) tests have been used to investigate the volume change characteristics of unsaturated soils for many years (Matyas and Radhakrishna, 1968; Escario and Saez, 1973; Fredlund and Morgenstern, 1976; Alonso et al., 1990). The compression, shrinkage and swelling characteristics of various unsaturated soils under variable vertical loading and suction conditions have been reported. In the system used by Matyas and Radhakrishna (1968) the suction was controlled using the axis translation technique. The oedometer has, however, been constantly modified to perform tests on unsaturated soils. Romero et al. (1995) described the development of a suction- and temperature-controlled oedometer to investigate the swelling behaviour of Boom clay at different temperatures. Kassiff and Ben Shalom (1971) adapted the osmotic technique described in Section 2.3.2 to control matric suction in the oedometer cell. Subsequently, Delage et al. (1992), Dineen and Burland (1995), Monroy et al. (2008a,b) and Cuisinier and Deneele (2008) utilised osmotic oedometer cells in the investigation of soil volume change behaviour. A schematic of the oedometer set-up adopted by Monroy et al. (2008a,b) is presented in Figure 3.9. Cuisinier and Masrouri (2004) discussed the use and issues relating to the vapour equilibrium technique, outlined in Section 2.3.3, in oedometer testing.

In one-dimensional consolidation testing, the vertical pressure, pore air and pore water pressures, volume change (determined from vertical deformation measurements of the specimen) and water volume change (from water uptake or drainage from the specimen) should be recorded. Such measurements allow the net vertical stress (difference between the total vertical stress and the pore air pressure) to be determined, along with the suction, specific volume and specific water volume throughout the compression or swelling processes under examination.

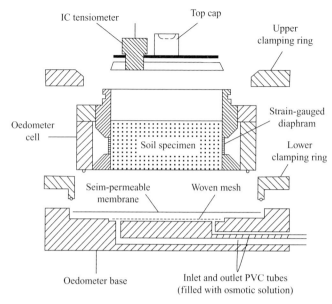

Figure 3.9   Osmotic oedometer cell (after Monroy *et al.*, 2008). Reproduced by permission of Taylor and Francis Group

3.3.2 Direct shear box apparatus

Donald (1956) modified the traditional direct shear box and reported a series of shear tests on unsaturated fine sands and coarse silts. The pore air was maintained at atmospheric pressure by leaving the top of the shear box exposed to the atmosphere, while the pore water pressure was controlled. Escario (1980) also modified the traditional shear box to allow shear testing of unsaturated soils. The major change was the provision of a system to control and record both pore air and pore water pressures, allowing the suction to be controlled using the axis translation technique. In shear box tests (Figure 3.8(b)), horizontal shear stress is recorded along with the vertical applied pressure. As for oedometer consolidation tests, the specimen volume change is measured by recording the change in vertical height of the specimen and the water volume change measured from the water uptake or drainage from the specimen. The measurements allow the shear strength of unsaturated soils to be investigated in relation to net vertical stress, specific volume and specific water volume throughout the shearing process.

Gan *et al.* (1988) have described a modified direct shear apparatus where the shear box was set within an air pressure chamber allowing control of the air pressure in the soil specimen. Control of the pore water pressure was through a high air entry disc at the base of the specimen. The set-up allowed elevated air pressures and the axis translation technique to be employed in testing a glacial till. The apparatus was subsequently used by Gan and Fredlund (1996), Oloo and Fredlund (1996) and Vanapalli *et al.* (1996) to investigate the shear behaviour of various unsaturated soils. Han *et al.* (1995) and DeCampos and Carrillo (1995) describe a similar test set-up as illustrated in Figure 3.10.

Tarantino and Tombolato (2005) have presented the results from direct shear tests on unsaturated statically compacted specimens of speswhite kaolin. The shear box had the facility to monitor suction using Trento tensiometers. A schematic of the shear box described in detail by Caruso and Tarantino (2004) is presented in Figure 3.11.

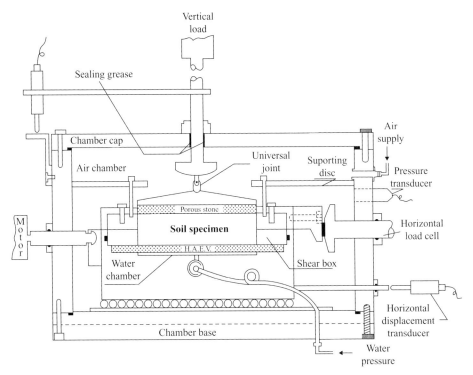

Figure 3.10 Modified direct shear apparatus (after De Campos and Carrillo, 1995). Reproduced by permission of Taylor and Francis Group

3.3.3 Simple shear apparatus

Tombolato *et al.* (2008) described the simple shear apparatus (SSA) designed to investigate the shear strength of unsaturated soils by imposing a constant volume and constant water content mode of deformation. A schematic of the SSA is shown as Figure 3.8(c). The

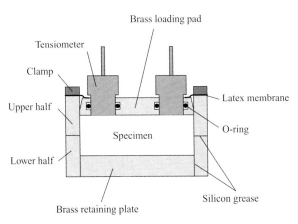

Figure 3.11 Schematic layout of the monitored-suction shear box (after Caruso and Tarantino, 2004). Reproduced by permission of the authors and the Institution of Civil Engineers

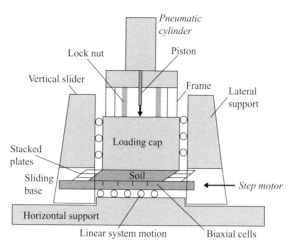

Figure 3.12 Schematic layout of the simple shear apparatus (after Tombolato *et al.*, 2008). Reproduced by permission of Taylor and Francis Group

apparatus used by Tombolota *et al.* (2008) is illustrated in Figure 3.12. Suction was measured at the top surface of the soil specimen by five pairs of Trento high-capacity tensiometers (Tarantino and Mongiovi, 2002). The vertical and shear forces were measured using five pairs of biaxial load cells at the bottom surface of the specimen.

Boso *et al.* (2005) described an osmotic shear box (Figure 3.13) where a polyethylene glycol (PEG) solution below the specimen and two tensiometers installed in the loading pad allowed the independent control and measurement of suction.

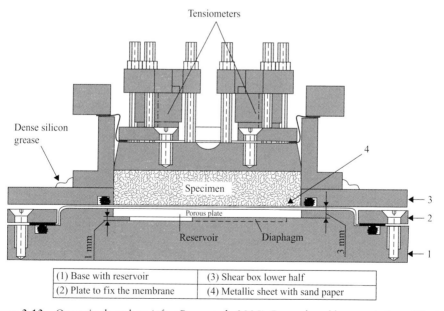

| (1) Base with reservoir | (3) Shear box lower half |
| (2) Plate to fix the membrane | (4) Metallic sheet with sand paper |

Figure 3.13 Osmotic shear box (after Boso *et al.*, 2005). Reproduced by permission of Taylor and Francis Group

3.3.4 Triaxial cell and stress path apparatus

In the axi-symmetrical tests carried out in the triaxial cell (Figure 3.8(d)), the cell pressure and deviator stress (the difference between the total axial stress and the cell pressure) are recorded along with the pore air and pore water pressures, the axial compression and volume change of the specimen, as well as any water volume change. The soil specimen is contained in a rubber sheath and measurements allow the stresses imposed on the specimen and the suction within the specimen to be determined, as well as the axial strain, axi-symmetrical volume change and the air and water volume changes throughout a test. Strength results are often reported in terms of the mean net stress $(p - u_a)$ and the deviator stress q. This approach has the advantage of not confining the specimen to one-dimensional compression as in the oedometer test, or dictating the direction of the shear plane as in the shear box test. However, it has the drawback that it is not truly 'triaxial'. The cell pressure is applied radially around the whole specimen, and radial deformation of the specimen under test is not uniform, with specimens generally exhibiting a degree of 'barrelling'. Nevertheless, the triaxial cell, as with the oedometer and shear box tests, has been shown to produce reliable, reproducible data that can be used to investigate unsaturated soil behaviour.

In normal triaxial cell tests, constant rates of strain are adopted. The strain rates are selected depending on material type and size of specimen to ensure that the specimen under test is close to equilibrium at all stages of the test. Conventionally, these are usually carried out as fully drained or undrained tests. However, more detailed investigation of soil behaviour requires different stress paths to be followed such as constant suction, constant net stress and constant volume tests. These stress path tests are carried out in modified triaxial test equipment capable of modelling any axi-symmetrical stress path with changes often in small increments between equilibrium conditions.

3.3.5 True triaxial apparatus

Hoyos *et al.* (2008) described a true triaxial apparatus for testing cubic specimens of unsaturated soil (Figure 3.8(e)). This was a development of work reported by Hoyos (1998) and Hoyos and Macari (2001). The apparatus allowed testing of soils under constant suction conditions, using the axis translation technique, over a wide range of stress paths. The soil specimen under test was seated on a high air entry disc and encased in flexible (latex) membranes on the remaining five sides. The apparatus can thus be described as a mixed-boundary device. Matsuoka *et al.* (2002) also reported a true triaxial apparatus but with three pairs of rigid loading plates in three orthogonal directions.

3.4 Essential measurements

There are a number of important measurements and recordings that should be made in tests on unsaturated soils if a meaningful interpretation of the data is to be obtained. These are discussed below.

3.4.1 Initial conditions and imposed stresses and pressures

The initial conditions of a specimen are dictated by the test programme and the material to be tested. These include accurate measurement of the specimen dimensions, water content, dry density and determination of the air voids content. The measurements are readily made using conventional laboratory techniques. The imposed stresses are determined from the applied loads and, in the case of the triaxial cell, also by the cell pressure. In the oedometer and shear box tests, radial confining stresses are unknown unless special measuring devices are employed and the specimens are tested under conditions of no lateral strain.

3.4.2 Suction, pore water pressure and pore air pressure

The accurate measurement and interpretation of soil suction is vital to understanding the behaviour of unsaturated soils. Various techniques are available, as discussed in Chapter 2, and further reading is given by Hilf (1956), Matyas and Radhakrishna (1968), Fredlund and Morgenstern (1977), Ho and Fredlund (1982), Escario and Saez (1986), Wheeler and Sivakumar (1995), Cui and Delage (1996), Cunningham *et al.* (2003) and Blatz and Graham (2003). The use of the axis translation technique, the osmotic technique and the vapour equilibrium techniques for control of suctions are further discussed in Section 3.5.1.

3.4.3 Volume changes

This includes measurements of the total volume changes, water volume changes and air volume changes. The lack of lateral straining in the oedometer and shear box due to rigid confinement of the specimens simplifies the total volume change measurements, which are determined by recording changes in the vertical height of the specimen under test. The overall specimen volume change dV is given by:

$$dV = dh \times A_p \qquad [3.1]$$

where dh is the vertical compression (or expansion) and A_p is the cross-sectional area of the specimen.

However, simple vertical, axial measurements to determine volume change are not appropriate in triaxial testing where lateral straining occurs. This and the dependence of volume change of unsaturated soils on the change in volumes of both the air and water phases require special procedures to be employed.

Volume change of water in a soil specimen, whether in the oedometer, shear box or triaxial cell, can be readily determined by recording the water uptake or the water being expelled from a specimen, using common laboratory techniques, as water is usually deemed incompressible. However, direct measurement of air volume change presents particular difficulties principally because of the compressibility of air. The approach adopted is to determine the air volume change as the difference between the total volume change dV and the water volume change dV_w. This is expressed mathematically as:

$$dV = dV_a + dV_w \qquad [3.2]$$

where dV_a is the air volume change.

3.5 Further details of triaxial and stress path testing techniques

Triaxial testing encompasses both the conventional approaches of fully drained and undrained strain-controlled tests, and more detailed stress path testing techniques where the deviator stress can be applied using either a stress-controlled or a strain-controlled approach. Recent developments now allow both stress-controlled and strain-controlled tests to be performed using a standard compression frame, where the controlling program interacts with the digitalised speed system built into the compression frame[1]. Using a normal stress path apparatus, or a digitalised compression frame approach, a soil specimen can be subjected to any stress path in $q : (p - u_a) : s, q : (p - u_a) : v$ or other stress–volume space (Sivakumar, 2005). Figure 3.14 presents an example of a stress path in $q : (p - u_a) : s$ space.

3.5.1 Suction control

Most reported academic research has utilised the axis translation technique to control soil suction (Hilf, 1956). In this method, the pore air pressure is elevated to a target value and the pore water pressure is kept above zero to avoid the influences of cavitation in the measuring system. On change of suction, water moves into the soil or drains out of the soil depending on the change from the initial conditions. Figure 3.15 shows the results of wetting and drying tests where the axis translation technique was employed to control suction. The figure shows the effect of water flow into a kaolin specimen (wetting or swelling path) under a mean net stress of 50 kPa, where suction was reduced from 1000 kPa (generated by the compaction process) to 50 kPa, and the drainage of water from similar specimens (drying or shrinkage paths), where the suctions were subsequently increased to values of 100, 200, 300 and 450 kPa (Sivakumar *et al.*, 2006b).

If the axis translation technique is used to control the suction in a specimen, success is dependent on the quality of the high air entry filter separating the specimen from the pore water monitoring system as well as on the saturation of the filter with water. The

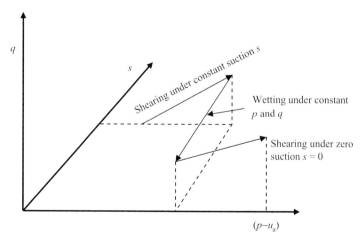

Figure 3.14   Stress path in $q : (p - u_a) : s$ stress space.

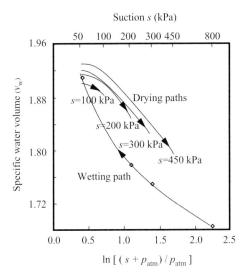

Figure 3.15 Specific water volume against suction characteristic during wetting and drying (after Sivakumar *et al.*, 2006b).

term 'high air entry' means that the filter can withstand a differential pressure (i.e. the difference between air pressure and water pressure) up to the stated value. Proprietary high air entry filters can be purchased from Soilmoisture Equipment Corporation[2]. The filters are available at various air entry values up to 15 bar (1.5 MPa). The higher the air entry value, the lower the permeability of the filter. The filters are ceramic and extremely brittle. Although the filters are stamped with the air entry value, it is desirable to check their acceptable range before embarking on detailed soil testing.

Where the high air entry filter is to be glued inside a stainless steel ring, as in triaxial cell testing (Figure 3.16), the use of slow-healing, rigid epoxy resin, such as Araldite (in the UK), is recommended. Any excess glue on the faces of the filter and the ring can subsequently be machined off, but every attempt must be made to prevent excess reduction in the thickness of the stone. It is also advised not to use any form of coolant during the machining process as this can block the pores of the filter.

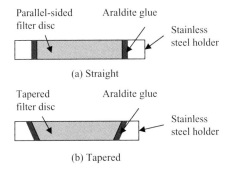

Figure 3.16 High air entry filter disc holder arrangements (after Brown, 2009).

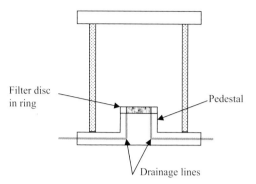

Figure 3.17   Chamber for saturating a filter disc (after Brown, 2009).

Figure 3.17 shows a chamber that can be used in saturating a filter. The chamber and the filter ring containing the filter should be secured to the pedestal in the chamber with a facility to flush water under the filter to remove any trapped air. The drainage lines should be flushed prior to the attachment of the filter ring to avoid build-up of water pressure under the filter as it is being fastened. Complete saturation of the filter is essential before its use in a test. Sivakumar (1993) recommended a procedure for this purpose. The saturation chamber should be filled with de-aired water and pressurised to a value higher than the air entry value of the filter. Pressurisation can be achieved using a hydraulic pressure control system such as those commercially available from GDS Instruments[3] or VJ Technology Limited. An air–water interface should not be used in pressurising the chamber as the water in the system is fully saturated with dissolved air and the saturation of the filter cannot be ensured. After pressurising, the system should be left under pressure for at least 2 days during which time no drainage from the chamber should be allowed. At the end of the second day, one of the drainage lines should be opened and the set-up left for a further day. At the end of the third day, the pressure should be brought to zero and the saturation chamber dismantled and reassembled with freshly de-aired water. The procedure should then be repeated. Experience indicates that it is sufficient to leave the filter in the chamber during this repeat process for a day, rather than the 2 days of the initial part of the saturation process. The pressure in the saturation chamber should be brought to zero gradually over a few hours before removing the filter for subsequent use. The filter attached on the pedestal should be carefully removed but it is important to leave one of the drainage lines open to avoid water pressure under the filter ring dropping below zero.

Saturation chambers are typically constructed of aluminium, stainless steel or brass. If aluminium is used, the chamber should be anodised to avoid corrosion of the metal surfaces as the formation of aluminium oxide on such chambers is common if the water is not properly de-ionised. The authors have experienced considerable difficulty saturating filters in aluminium chambers that have become pitted with corrosion spots (Brown, 2009). These problems are not encountered when the saturation chamber is constructed of stainless steel or brass.

The surface of the filters can become fissured and fragile after saturation (Figure 3.18) and be no longer suitable for use. Care is needed in handling the filters on removal from the saturation chamber and on subsequent use in testing.

There must be rigorous adherence to established procedures in setting up a triaxial cell for testing. It is particularly important to ensure continuity of the water phase from the drainage line to the filter and the specimen, or erroneous water pressure and volume change

Figure 3.18   Fissured and fractured high air entry filter disc (after Brown, 2009).

measurements will ensue. Before testing begins, the drainage lines should be flushed with de-aired water and the saturated filter mounted carefully on the pedestal. The thickness of the O-ring used to seal the specimen membrane to the pedestal supporting the filter is very important. Too thin an O-ring can lead to leakage of air from the specimen into the drainage lines (Figure 3.19(a)). On the other hand, too thick an O-ring can lead to the situation where a thin cavity exists behind the filter, which may invite significant setting up and measurement difficulties (Figure 3.19(b)). For example, during the setting up procedure, the drainage lines should be left closed. When the specimen is placed on the filter, it attempts to draw water from the filter and the pressure under the filter ring can fall below zero. If the water pressure is sufficiently low, it may lead to cavitation, and subsequently the formation of a thin air sheet under the filter. If this occurs, then the required continuity of the water phase will be disrupted. It is possible to avoid contact of the specimen with the filter during the initial setting up procedure by placing a thin wire loop (Figure 3.19(c)) on the upper surface of the filter. The assumption is made that the wire will be pushed into the specimen when the cell pressure and deviator stress are applied. However, if the wire does not properly push into the specimen then an air cavity will form between the filter and the specimen. If the separating wire approach is used, the apparent strains as a result of the wire pushing into the specimen should be taken into account in determining the actual straining of the specimen under test (Brown, 2009).

A further problem can occur if excess glue is used to secure the metal ring to the stone filter, and the glue is not set rigidly, resulting in it extruding under pressure (Figure 3.19(d)). This can lead to inadequate contact between the specimen, filter and pedestal, again leading to erroneous measurements.

Suction can also be measured or controlled using psychrometers, as discussed in Chapter 2. Thom *et al.* (2008) employed a thermocouple psychrometer to measure suction in unsaturated kaolin specimens subjected to repeated loading. The suctions measured in these tests were in the range of 300–1000 kPa. In such an approach it is essential that the psychrometer tip is carefully embedded in the specimen to ensure good contact of material and measuring instrument. Figure 3.20 shows a typical top cap arrangement for the

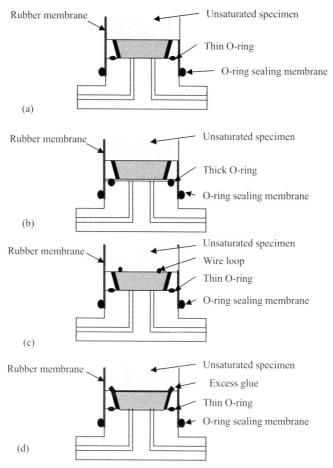

Figure 3.19   (a) The problem caused by a small O-ring, (b) the problem caused by a big O-ring, (c) the fuse wire technique and (d) the problem caused by excess glue (after Brown, 2009).

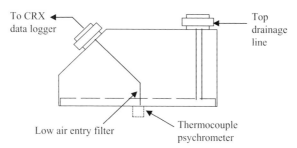

Figure 3.20   Psychrometer top cap (after Thom *et al.*, 2008). Reproduced by permission of ASTM

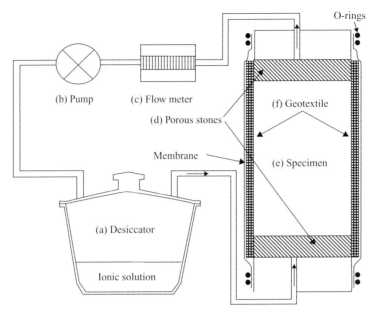

Figure 3.21 The vapour equilibrium method for triaxial testing (after Blatz and Graham, 2000). Reproduced by permission of the authors and the Institution of Civil Engineers

triaxial cell. The success of this approach depends on the rigorous control of temperature in the laboratory.

As discussed in Sections 2.3.2 and 2.3.3, several researchers have reported the use of the osmotic technique for matric suction control in triaxial testing, while others have reported testing using the vapour equilibrium approach to control total suction. Blatz and Graham (2000, 2003) employed the vapour equilibrium method to control suction in triaxial testing on compacted sand–bentonite. The suction in the test specimen was altered by circulating vapour above a salt solution and the specimen. The specimen was wrapped with a special porous geosynthetic filter fabric so that the vapour flowed from one end of the specimen to the other (Figure 3.21). Using this method the specimen was subjected to very high suction values of several megapascals.

3.5.2 Measurement of specimen strains and volume change

The measurement of volume changes in unsaturated specimens in the triaxial cell is a challenging issue and various approaches have been employed over the years to improve measurement accuracy. Volume change of water in a soil specimen can be readily determined from water content changes. However, measurement of air volume change presents particular difficulties as air is highly compressible and can also slowly diffuse into the surrounding cell fluid through the rubber membrane surrounding the specimen. The change in volume of air in a specimen under test can be evaluated in accordance with Equation 3.2 from the difference between the total volume change and the change in water content. While overall volume change can be evaluated from change in fluid volume in the cell, corrections need to be applied for possible sources of error. Improvements to triaxial

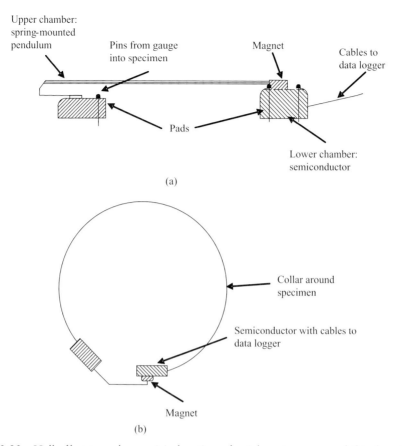

Figure 3.22 Hall effect transducers: (a) elevation of axial strain gauge and (b) plan view of radial strain gauge (after Thom *et al.*, 2008). Reproduced by permission of ASTM

testing apparatus, which minimise the problem of inaccuracy in volume measurement, are discussed below but first we will outline the approach of determining strains and volume change measurement from strain gauges mounted internal to the triaxial cell.

Under triaxial stress conditions, internal strain gauges mounted on a specimen can be employed in order to evaluate specimen volume change (Maâtouk *et al.*, 1995; Zakaria *et al.*, 1995; Blatz and Graham, 2003; Thom *et al.*, 2008; Brown, 2009). Figure 3.22(a) and (b) shows a Hall effect transducer used to measure volume change (Thom *et al.*, 2008). Rigorous and delicate procedures need to be employed in attaching strain gauges to directly measure specimen strains. The locations on the membrane where the pads are to be attached should be lightly roughened with 'zero'-grade sand paper before applying glue. Brown (2009) employed radial strain gauges (linear voltage differential transducers) and inclinometers (travel length 5 mm) to determine lateral and axial strains (Figure 3.23). Internal strain gauges record local strains and not overall strains, and though valuable in determining variations in specimen deformation, the measured strains are not as reliable as large strain deformation measurements in giving volume change performance. In addition, experience has shown that the glue may not adequately secure a gauge in long-term tests, which are often required in the testing of unsaturated soils.

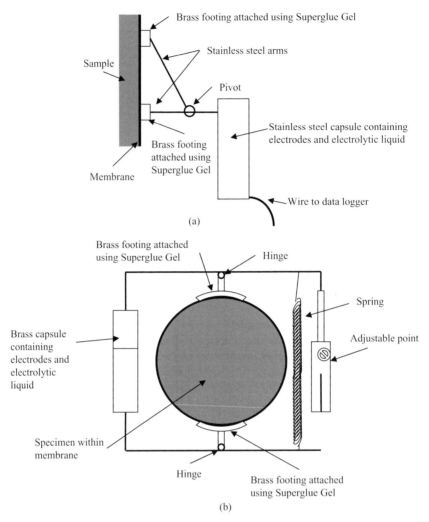

Figure 3.23   (a) Axial and (b) radial inclinometers (after Brown, 2009).

Various indirect approaches have been developed over the years for the measurement of specimen volume change. A particular problem to be overcome is the influence on measurements of the change of volume of the cell itself under changing pressures. Bishop and Donald (1961) developed a modified triaxial cell to monitor the volume change of specimens in assessing the effective stress concepts for unsaturated soils (Figure 3.24). The modified triaxial cell included an additional inner cylindrical cell sealed to the cell base. Mercury was used as the cell fluid. The overall volume change of the unsaturated specimen was monitored using the movement of a stainless steel ball floating on the mercury. However, the use of mercury requires special safety precautions because of its toxicity. Cui and Delage (1996) used a similar technique to measure the total volume change of an unsaturated soil specimen, but coloured water was used as the cell fluid

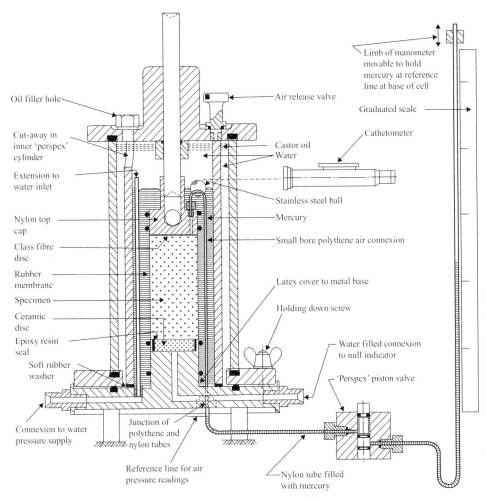

Figure 3.24 Modified triaxial cell (after Bishop and Donald, 1961).

rather than mercury and water levels were measured using high-precision cathetometer based readings. In order to avoid the absorption of air by the water and to reduce water evaporation, a thin layer of silicon oil was placed above the water. Rampino *et al.* (1999) developed a double-cell triaxial set-up for measuring volume changes of unsaturated soils using the methodology proposed by Okochi and Tatsuoka (1984) and Tatsuoka (1988). A number of researchers including Ng *et al.* (2002), Vassallo *et al.* (2007) and Papa *et al.* (2008) have successfully utilised the double-cell approach. The triaxial and oedometer set-ups used in suction-controlled tests by Aversa and Nicotera (2002) are illustrated in Figure 3.25. Rojas *et al.* (2008) suggested further modifications to the triaxial set-up to reduce testing times. Rather than utilising level measurements to determine specimen volume change, Ng *et al.* (2002) described the measurement of specimen volume change by recording the difference in pressure between the water inside the reference tube and the open-ended bottle-shaped inner cell using a high-accuracy differential pressure transducer.

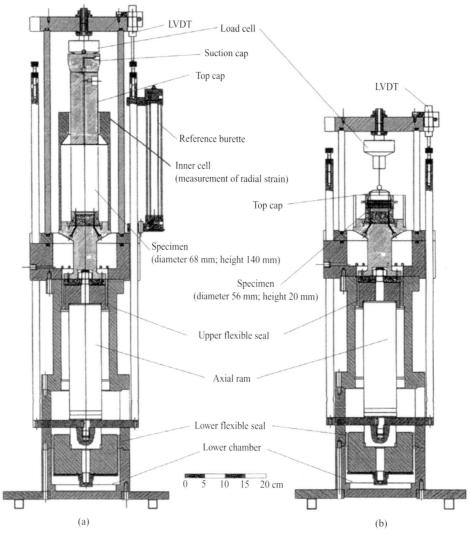

Figure 3.25 (a) Triaxial and (b) oedometer apparatus for suction-controlled tests on unsaturated soils (after Aversa and Nicotera, 2002). Reproduced by permission of ASTM

Aluminium and steel were used in place of acrylic to construct the inner cell in order to minimise the effects of creep and hysteresis and to overcome the problem of the absorption of water by acrylic. Ng *et al.* (2002) also employed paraffin – floating on top of the water, as recommended by Tatsuoka (1988) – to overcome problems of evaporation of water with time. However, Sivakumar (1993) pointed out that allowing paraffin to float on the top surface of the water can cause problems when the direction of the flow is changed. In addition, the outer and inner cells (above the reference water levels) pressurised with air can be dangerous when operating the system under high pressures.

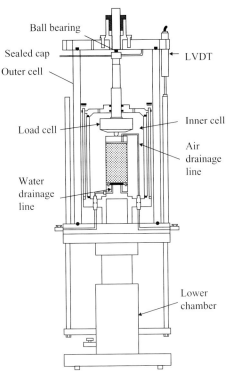

Figure 3.26 Schematic diagram of the twin-cell stress path apparatus (after Sivakumar, 2005, and Sivakumar *et al.*, 2006a).

3.5.3 Twin-cell stress path apparatus

Wheeler (1986) developed a modified triaxial apparatus, referred to as the double-walled triaxial cell, to monitor specimen volume change. The volume change of the unsaturated specimen was measured by monitoring the amount of water draining into or out of the inner cell by pressurising the inner and the outer cells to the same pressure. Although this method produced valuable information (Wheeler and Sivakumar, 1995), the approach suffered from a number of disadvantages, particularly absorption of water by the inner acrylic cell wall with time. In addition, setting up the specimen in the triaxial cell required it to be performed under water. The effect of temperature variations on the volume change measurement was also of concern. The problems of absorption of water by the acrylic of the cell and the effects of temperature variations in the testing environment were successfully overcome in the new twin-cell stress path system for testing unsaturated soils developed by Sivakumar (2005) and further described by Sivakumar *et al.* (2006a).

Figure 3.26 shows a cross-sectional view of the twin-cell stress path apparatus composed of: (a) an outer cell comprising a standard stress path cell supplied by VJ Technology Limited; and (b) an inner cell comprising a small triaxial cell capable of testing specimens with a diameter of 50 mm. The inner cell wall is made of high-quality glass ground to have parallel ends. The use of glass overcomes the problem of absorption of water by acrylic. A cylindrical cavity at the base of the inner cell allows it to be directly located on the pedestal

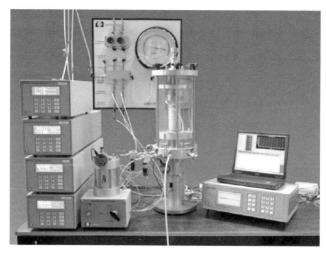

Figure 3.27   Twin-cell stress path system (after Sivakumar, 2005, and Sivakumar *et al.*, 2006a).

of the standard stress path cell. Water and air drainage lines are fitted to the bottom of the inner cell and passed through the outer cell in order to connect them to the relevant pressure systems. When the system is assembled, the outer cell fully encloses the inner cell, which on equalisation of pressures of the inner and outer cells prevents expansion of the inner cell on raising the cell pressure. The load cell is located inside the inner cell. A tight O-ring seal is used to prevent any leak of inner cell fluid into or out of the outer cell through the loading ram bushing. The deviator stress on the specimen can be applied by controlling the pressure in the lower chamber. The lower chamber is that of the usual Bishop and Wesley stress path cell. The assembled system is shown in Figure 3.27[4].

The twin-cell stress path apparatus must be assembled with care. The entrapment of air bubbles in the inner cell must be avoided and can be overcome by the assembly of the cell under water, though this approach is an intricate undertaking. It has been found that the use of an O-ring can lead to air entrapment around the upper ring of the cell if the cell is filled in the conventional manner. The risk of air entrapment and the need to assemble the inner cell under water is minimised if the seal is provided by a ring of 'square' cross section.

With the specimen mounted on the pedestal and sealed within a membrane, the air drainage line should be connected and the other ends of the air drainage line, water drainage lines and the inner cell line should be closed. A small vacuum, applied by reducing the air pressure, stabilises the specimen during the setting up procedure.

The flow of water in or out of the inner cell and the flow of water in or out of the specimen can be measured using traditional volume change units. Hydraulic pressure controllers (see Section 3.5.1) can be used to apply relevant pressures and control the volume changes. However, it should be noted that the in and out lines of the inner cell should be pressurised equally and therefore an additional volume change unit should be located on the pressure line to the inner cell. It should also be noted that if an Imperial College type volume change unit is used, there must be a minimum pressure of 20 kPa for the bellows to perform satisfactorily. The zero setting should therefore be established when the pressure of 20 kPa is applied to the in and out lines of the inner cell.

Even though the inner and outer cells are pressurised to the same value, the inner cell experiences volume change due to a number of effects. Factors contributing to this

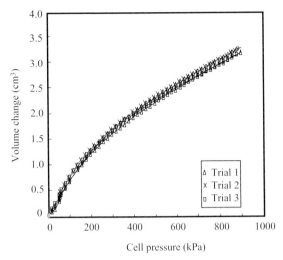

Figure 3.28 Apparent volume change of inner cell with increasing cell pressure (after Sivaku-mar, 2005, and Sivakumar *et al.*, 2006a).

volume change are the flexibility of the fittings and connections, the expansion of the volume change device, the compressibility of water, the deformation of the inner cell and the deformation of the load cell. While these effects are relatively small, calibration is necessary. Account should also be taken of the volume of water replaced by movement of the loading ram into the inner cell during shearing. This latter correction can be achieved by multiplying the cross-sectional area of the loading ram by its movement.

Figure 3.28 shows the volume change of the inner cell plotted against the cell pressure. In this case the cell pressure was ramped at a rate of 15 kPa/min to a target pressure of 900 kPa. The calibration is shown as repeated three times and in each case the cell was reassembled. The calibration shows that the volume change of the volume system is small and repeatable even at high cell pressures.

Figure 3.29 presents the results of a calibration test to examine the hysteresis of the system as a result of reversal of the cell pressure. The apparent volume change of the inner cell is shown for increase of the cell pressure from 0 to 900 kPa, subsequently reduced to 200 kPa, followed by reloading to 900 kPa. The reproducibility of the volume change on loading reversal is good. However, a reduction of cell pressure to low values might trigger problems with dissolved air coming out of solution, giving rise to inaccuracies in volume change measurements even though initially de-aired water is employed in the cell. This problem potentially arises when a specimen is present in the cell and the axis translation technique is used to control suction. Diffusion of air from the specimen to the inner cell fluid can take place through the rubber membrane. This dissolved air subsequently comes out of solution when the cell pressure is reduced significantly and needs to be addressed in the test protocol.

Since the measurement of specimen volume change is made by detecting the flow of water in or out of the inner cell, it is important to minimise temperature variations in the laboratory as this affects volume change measurements. The effect of unavoidable temperature variation can be minimised by calibrating the cell volume change against changes in temperature measured using a temperature probe (LM35 precision temperature

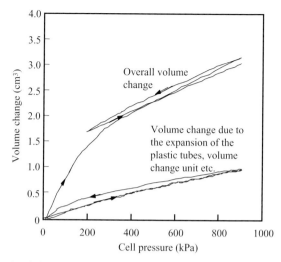

Figure 3.29 Reproducibility of the inner cell volume change during loading and unloading (after Sivakumar, 2005, and Sivakumar *et al.*, 2006a).

sensor) in the outer cell. Calibration for the effect of temperature variation on the inner cell volume change can be examined by observing the flow of water into or out of the inner cell when the cell pressure is maintained constant. Figure 3.30 shows the volume change of the inner cell plotted against time with the cell pressure maintained at 900 kPa for approximately 10 days. Also included in this figure is the recorded change in

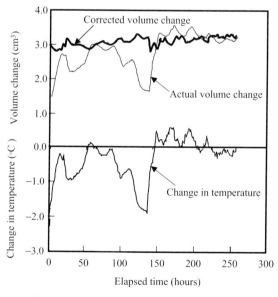

Figure 3.30 Variation of measured and corrected volume change with elapsed time (after Sivakumar, 2005, and Sivakumar *et al.*, 2006a).

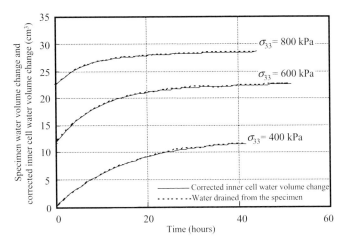

Figure 3.31   Specimen volume change and inner cell water volume change against time (after Sivakumar, 2005, and Sivakumar *et al.*, 2006a).

temperature. During the first 6 days of the above assessment, the temperature in the laboratory was allowed to vary. Up to 2°C variation was recorded. During the last 4 days of the assessment, the temperature control system was set to its default setting of 20°C and a maximum fluctuation of ±0.5°C was recorded. There is a close correlation of variation in the inner cell volume with temperature allowing calibration correction to be established.

It is possible to check the accuracy of the specimen volume change measurements made using the twin-cell stress path system. This can be done by performing a test on a saturated specimen. The volume change of the inner cell (after calibration for apparent volume change) should be equal to the volume of water leaving or entering the saturated specimen. Figure 3.31 shows the consolidation data for a specimen initially isotropically consolidated to 200 kPa of effective confining pressure and then subsequently subjected to increased consolidation pressures of 400 and 600 kPa. These pressures corresponded to cell pressures of 400, 600 and 800 kPa. The specimen was allowed to consolidate fully at each pressure. The volume change of the specimen was assessed from the amount of water that drained out of the specimen and the water that flowed into the inner cell to compensate for the reduction in the specimen volume. In the latter approach, relevant calibration was applied to eliminate the apparent volume change of the volume system caused by the increase in cell pressure. The volume change curves of Figure 3.31 are essentially coincident. The agreement between the two volume change measurement approaches is excellent.

3.6 Conclusions

The following outlines the main conclusions of this chapter:

- The importance and significance of material selection, methods of specimen preparation and the need to carefully address the aims of a testing programme are discussed and are prerequisite to obtaining meaningful, reproducible results.

- Laboratory testing of unsaturated soils generally involves protracted time periods and without careful control and calibration protocol errors can arise leading to invalid test data.
- Laboratory-based techniques for investigating the volume change, strains and stresses in unsaturated soils are highly advanced and allow repeatable and reliable test data to be obtained. In particular, the twin-cell triaxial cell stress path apparatus allows the greatest flexibility of soil testing with the minimum of error.

Notes

1. A commercially available controlling program for a standard compression frame is available from VJ Technology Limited, 11 The Metro Business Centre, Toutley Road, Wokingham, Berkshire RG41 1QW, UK.
2. Soilmoisture Equipment Corporation, 801 S. Kellogg Avenue, Goleta, CA 93117, USA.
3. GDS Instruments, Unit 32, Murrell Green Business Park, London Road, Hook, Hampshire RG27 9GR, UK.
4. There are two control systems available in the market: *Triax*, developed by D. Toll, University of Durham, and *WinClip*, developed by VJ Technology Limited. Both these control programs operate on Windows and offer a wide range of potential stress paths.

Chapter 4
Background to the Stresses, Strains, Strength, Volume Change and Modelling of Unsaturated Soil

4.1 Introduction

Quantifying the controlling stresses is essential if the strength of unsaturated soils is to be adequately assessed for practical problems such as bearing capacity, earth pressure on retaining walls and slope stability. The prediction of volume change, differential straining and yielding also requires the stress regime under changing conditions to be determined. As discussed in Chapter 1, an essential component of the stresses in unsaturated soils is the matric suction. The concept of matric suction (or negative pore water pressure) in unsaturated soils was originally developed in soil physics in relation to plant growth and moisture deficiency. The role of suction as a major factor influencing the mechanical behaviour of unsaturated soils was subsequently investigated at the Road Research Laboratory, UK (Croney and Coleman, 1948, 1954, 1960). Developments in the mechanics of unsaturated soils since this time can be divided into four overlapping thematic stages:

- In the first stage, researchers tried to find a single 'effective' stress to explain mechanical behaviour, as had previously been established for saturated soils. However, comparison with experimental data has shown that unsaturated soil behaviour cannot be adequately modelled using a single 'effective' stress.
- In the second stage, researchers developed constitutive frameworks to explain the shear strength and volume change characteristics of unsaturated soils in terms of two independent stress state variables. The independent stress state variables employed have generally been the net stress $(\sigma - u_a)$ and the matric suction $(u_a - u_w)$. This approach has resulted in some success.
- In the third stage, researchers have attempted to analyse unsaturated soil behaviour in terms of constitutive relationships linking volume change, shear strength and shear deformation in a single elasto-plastic model. While considerable progress has been made through this approach, and researchers are now able to explain many features of soil behaviour, there are also anomalies that are not readily explained.
- The fourth stage covers increasing research into the introduction of hydraulic hysteresis and anisotropic yielding into an elasto-plastic framework.

This chapter presents the concepts behind current understanding and interpretation of unsaturated soil behaviour and generally, though not strictly, follows the four stages described above noting obvious overlaps and conflicting ideas where necessary.

4.2 Stresses in soils

Knowledge of the stresses controlling the deformation and shear behaviour of soils is essential to the development of a predictive framework. Classical soil mechanics generally assumes that for most practical purposes the behaviour of a water-saturated soil is governed by a single stress state variable given by Terzaghi's (1936) effective stress, which can be written in tensor form as:

$$\sigma'_{ij} = \sigma_{ij} - u_w \delta_{ij} \qquad [4.1]$$

where σ'_{ij} is Terzaghi's effective stress tensor, σ_{ij} the total stress tensor and δ_{ij} the Kronecker delta ($\delta_{ij} = 1$ if $i = j$ and $\delta_{ij} = 0$ if $i \neq j$).

This form of presentation allows the shear stress $\tau = \sigma_{ij}$, when $i \neq j$, to be included in a single equation with the effective stress $\sigma'_{ij} = \sigma_{ij} - u_w \delta_{ij}$, where $i = j$. The effective stress equation is independent of any volumetric terms for the solid and water phases as they are shown from the equilibrium analysis in Chapter 1 to cancel out. The stress state variables (effective stress) given by the difference between the total stress and the pressure of the pore fluid (water), together with the shear stresses, are sufficient to define the stress regime in a saturated soil.

For a perfectly dry soil with zero suction, where air fills the pore spaces, the following equation applies in place of Equation 4.1:

$$\overline{\sigma}_{ij} = \sigma_{ij} - u_a \delta_{ij} \qquad [4.2]$$

where $\overline{\sigma}_{ij}$ is the net stress tensor and u_a the pore air pressure.

This is compatible with the soil behaving as a dry, granular material. The net stress is the effective stress for such a material. As for saturated soils, no volumetric terms appear in Equation 4.2. This is also an effective stress equation and the stress state variables, given by the difference between the total applied stress and the pressure of the pore fluid (air), together with the shear stresses, are sufficient for the stress regime in a perfectly dry soil to be defined.

Engineers have faced challenging problems where the assumptions of saturated or perfectly dry conditions are unreasonable. In order to appraise shear strength, stress–strain behaviour, shrinkage and swelling, yielding and collapse in unsaturated soils there has been concerted research into the controlling stress regime. Bishop (1959) proposed a single stress variable equation for unsaturated soils. This formulation, given by Equation 4.3, includes the parameter χ in taking account of the difference between pore air pressure u_a and pore water pressure u_w on the mechanical behaviour:

$$\sigma'_{Bij} = \overline{\sigma}_{ij} + \chi s \delta_{ij} \qquad [4.3]$$

where σ'_{Bij} is Bishop's stress tensor, χ is generally assumed to be a function of degree of saturation and is zero for a dry soil and unity for a saturated soil, and $s = (u_a - u_w)$ is the matric suction.

Equation 4.3 couples the effect of the so-called net stress $\overline{\sigma}_{ij}$ and the matric suction s in an equation for the controlling stress. However, the equation has proved to be unreliable in predictive analysis principally due to the uncertainty over the parameter χ, though much research is still centred on its usage. Other forms of effective stress equations for unsaturated soils have been proposed, such as those by Croney et al. (1958), Aitchison (1961) and Jennings (1961), though Bishop's equation reduces to the same form as these equations when the pore air pressure in Equation 4.3 is taken as the datum of atmospheric pressure. Bishop et al. (1960) concluded that there was no unique relationship between the degree of saturation and χ, while Jennings and Burland (1962) suggested that Bishop's equation did not provide an adequate relationship between effective stress and volume change. Bishop and Blight (1963) and Brackley (1971) amongst others have also cast doubt on the use of a single stress equation such as Bishop's in governing volume changes in unsaturated soils. Importantly, Blight (1965) pointed out that the value of χ depends on the type of process to which a soil is subjected, and is a non-linear function that depends on stress level and strain. The conclusion from such observations prompted many researchers to accept that the search for a single effective stress for unsaturated soils, similar to that for saturated soils, was unlikely to be successful.

Nevertheless, Richards (1966) and Aitchison (1965, 1973) proposed extensions to the effective stress equation to incorporate solute or osmotic suction, which is important where there is a chemical imbalance and the thermodynamic chemical potential is likely to lead to the exchange of mass. The equation of Aitchison (1965, 1973), which differs slightly from that of Richards (1966), can be written as:

$$\sigma'_{Aij} = \sigma_{ij} + (\chi_m s + \chi_s \phi_s)\sigma_{ij} \qquad [4.4]$$

where σ'_{Aij} is Aitchison's stress tensor, χ_m the matric suction parameter, χ_s the osmotic suction parameter and ϕ_s the osmotic suction.

The parameters χ_m and χ_s are complex and, as with the parameter χ in Bishop's equation, are dependent on the test conditions and stress path though normally their values are within the range of 0–1.

As a consequence of the lack of success with finding a single stress state variable for unsaturated soils, Burland (1964, 1965) suggested that the behaviour should be related independently to the net stress and matric suction. This formed a change of emphasis in research activities with researchers using the uncoupled stress state variables to investigate strength and volume change characteristics. In accordance with the findings of Fredlund and Morgenstern (1977)[1] from theoretical considerations, any two of the three stress state variables $\bar{\sigma} = (\sigma - u_a)$, $\sigma' = (\sigma - u_w)$ and $s = (u_a - u_w)$ can be used to describe the behaviour of an unsaturated soil. The experimental evidence from Tarantino et al. (2000) supports this contention. The independent stress state variables have been used to develop constitutive equations from experimental evidence and in this way tentative steps have been made towards formulating a general constitutive framework for unsaturated soils.

From examination of the power input into unsaturated soils, Houlsby (1997) developed the concept of work conjugate stress and strain-increment variables. A primary aim of the paper was to determine the conjugate variables that would allow constitutive modelling to be more productive. However, the analysis ignored the surface tension effect of the contractile skin and did not account for the fluid pressure acting through the volume of the solids (as discussed in Chapter 1). An equation emerged from the analysis which had a form similar to Equation 4.2 of Bishop (1959) but with χ replaced by the degree of saturation S_r. Equation 4.5 presents this equation in tensor form:

$$\sigma''_{ij} = \sigma_{ij} - [S_r u_w + (1 - S_r)u_a]\,\delta_{ij} \qquad [4.5]$$

where Houlsby (1997) described σ''_{ij} as the average soil skeleton stress tensor.

Jommi (2000) examined the use of Equation 4.5 under triaxial stress conditions but without significant advancement in interpretation of soil behaviour. One of the problems with Equation 4.5 is that the volumes of all three phases are not represented by the term S_r. The solid particles are not represented, yet it is these that give a soil its strength.

While there would be advantages if the stress conditions in an unsaturated soil could be formulated in terms of a single stress state variable, this avenue of research has proved unfruitful. The use of two independent stress state variables has resulted in advances in our understanding of unsaturated soil behaviour, but it is inconsistent with thermodynamic principles not to include conjugate volumetric and strain-increment terms in such analysis. In Chapters 6 and 7 a formulation of the dual stress regime controlling unsaturated soils is developed. The controlling stresses are coupled in an equation linking the stresses with conjugate volumetric variables. In Chapter 9 the analysis is extended to the work input into unsaturated soils, which provides the conjugate stress and strain-increment variables in triaxial cell tests.

4.3 Strains in soils

On the basis of three orthogonal extensible (and compressible) fibres embedded in a soil specimen, Schofield and Wroth (1968) identified total deformation as made up of body displacement, body rotation and body distortion. The directions of embedment 1, 2 and 3 are not necessarily principal strain or strain-increment directions. The body distortion is defined as including compressive strains and shear strains. The analysis isolates the influences of body displacement and body rotation to identify generalised strain variables capable of describing body distortion. These variables comprise the compressive strains $\varepsilon_{11}, \varepsilon_{22}, \varepsilon_{33}$ and the shear strains[2] $\varepsilon_{23} = \varepsilon_{32}, \varepsilon_{31} = \varepsilon_{13}, \varepsilon_{12} = \varepsilon_{21}$ as in Equation 4.6:

$$\varepsilon_{ij} = \begin{bmatrix} \varepsilon_{11} & \varepsilon_{12} & \varepsilon_{13} \\ \varepsilon_{21} & \varepsilon_{22} & \varepsilon_{23} \\ \varepsilon_{31} & \varepsilon_{32} & \varepsilon_{33} \end{bmatrix} \qquad [4.6]$$

Figure 4.1 shows the simple case of two-dimensional body distortion for the extensible fibres of initial lengths x and y in the 1 and 2 directions respectively. The compressive strains ε_{11} in the 1 direction and ε_{22} in the 2 direction are given by $-dx/x$ and $-dy/y$. Similarly, in the orthogonal 3 direction for a fibre of initial length z, the compressive strain ε_{33} is given by $-dz/z$. Here we have defined compressive strains as positive consistent with the work input analysis of Chapter 9.

In Figure 4.1, the shear strain $\varepsilon_{12} = \varepsilon_{21}$ is given by the angle ω between the distorted fibres and the original positions. The other shear strains are defined in a similar manner.

Isolation of the influences of body displacement and body rotation allows the strain variables to be used to define the overall strain measurements of specimens in the triaxial cell and other laboratory tests, where the compressive and shear strains are sufficient to describe overall specimen straining. However, in Chapter 9 when discussing the straining of the aggregates and between the aggregates in an unsaturated soil, it is necessary to take account of not only body distortion but also other strain components that give rise to energy dispersion under the application of loading.

For the simple case of body distortion and no body displacement or rotation, if the extensible fibres in the soil specimen are orientated to the principal strain directions, there are only compressive strains and Equation 4.6 simplifies to:

$$\varepsilon_{ij} = \begin{bmatrix} \varepsilon_{11} & 0 & 0 \\ 0 & \varepsilon_{22} & 0 \\ 0 & 0 & \varepsilon_{33} \end{bmatrix} \qquad [4.7]$$

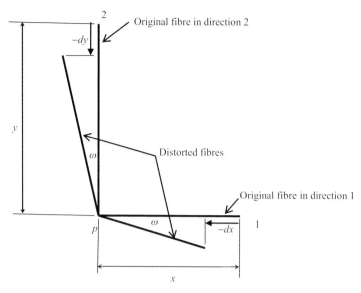

Figure 4.1 Definitions of strains.

The volumetric strain ε_v is given by the sum of the compressive strains, thus:

$$\varepsilon_v = \varepsilon_{11} + \varepsilon_{22} + \varepsilon_{33} \qquad [4.8]$$

We will be particularly concerned with triaxial cell testing where the axial and radial directions are principal stress directions, and axi-symmetrical conditions exist, giving $\sigma_{22} = \sigma_{33}$, as in Figure 4.2.

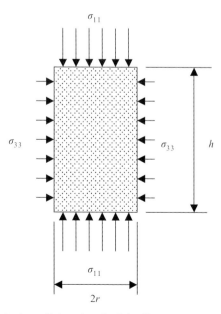

Figure 4.2 Axi-symmetrical conditions in triaxial cell.

For incremental changes dh in the length of the specimen h and dr in the radius of the specimen r:

$$d\varepsilon_{11} = -\frac{dh}{h} \quad \text{and} \quad d\varepsilon_{33} = -\frac{dr}{r} \qquad [4.9]$$

where $d\varepsilon_{11}$ is the axial strain-increment and $d\varepsilon_{33}$ the radial strain-increment.[3]

The volumetric strain-increment is accordingly given by:

$$d\varepsilon_v = d\varepsilon_{11} + 2d\varepsilon_{33} = -\frac{dv}{v} \qquad [4.10]$$

where $dv = 1 + de = dV/V_s$ is the increment of change in specific volume, $v = 1 + e = V/V_s$ is the original specific volume, V is the total volume of the soil specimen, V_s is the volume of the solid phase, e is the voids ratio and de is the incremental change in voids ratio.

The fixed mass of soil particles provides the reference datum for volume changes as a result of changes in the volumes of water and air in a specimen[4].

The deformation state variables must be consistent with the continuity requirements of continuum mechanics. This requires that the variables when integrated should give the total deformations and applies to a single-phase or multi-phase material. In an unsaturated soil, the total volume change is made up of the summed changes in volume of the phases. Accordingly:

$$dv = dv_w + dv_a \qquad [4.11]$$

where dv_w is the increment of change in the specific water volume and dv_a is the increment of change in the specific air volume.

A further important strain parameter employed in the analysis of triaxial cell test data is the deviator strain-increment $d\varepsilon_q$ given by:

$$d\varepsilon_q = 2\left(d\varepsilon_{11} - d\varepsilon_{33}\right)/3 \qquad [4.12]$$

In addition to the requirement that the variables are consistent with the continuity requirements of continuum mechanics, it is also necessary that when the stresses and strain-increments are multiplied together they are conjugate and give the incremental work input per unit volume performed by the imposed stresses. It is not intuitively apparent why there is the 2/3 term in Equation 4.12, but it is a consequence of the summed energy in the definitions of stresses and strains. The ideas behind conjugate stresses and strain-increments under triaxial stress conditions, where the mean stress and deviator stress are often employed in place of the axial and radial stresses, are developed in later chapters. The thermodynamic concepts and analysis of undrained and drained loading in the triaxial cell are dealt with in Chapter 6, and the conjugate stresses and strain variables (and strain-increment variables) for saturated and unsaturated soils are determined in Chapter 9. The continuity and work input relations for the straining of the phases in a multi-phase material are shown to be more complex than in the foregoing analysis.

4.4 Constitutive modelling

The use of independent stress state variables in constitutive modelling is an empirical but practical approach to a complex problem that has found a good degree of success. Constitutive modelling is used in other science disciplines, such as climatology, where

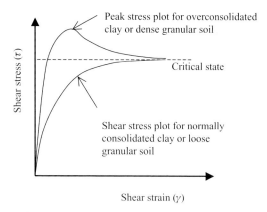

Figure 4.3 Behaviour of specimens in drained shear tests.

complex behaviour makes a rigorous scientific analysis impossible. However, in such an approach, unless account is taken of all the relevant variables, parameters derived from experimental observations incorporate the influence of undefined variables which render the analysis to be highly dependent on test conditions. The result is that extrapolation of conclusions from a particular test to other test conditions leads to anomalies. This is the case with Bishop's stress (Equation 4.3), particularly when tested against volume change behaviour in unsaturated soils. In examination of currently available constitutive models it is by no means clear that all the relevant variables, notably the volumetric and strain terms, are adequately and appropriately represented.

4.4.1 Shear strength

Figure 4.3 shows generalised plots of shear stress τ against shear strain for drained tests on clay and sand. Overconsolidated clays and dense sands exhibit a peak strength followed by a reduction to the critical state strength. Normally consolidated clays and loose sands achieve maximum strength at the critical state. The critical state thus represents a stress condition that a soil specimen subjected to shearing must experience irrespective of its initial state. It is generally assumed that shear strength and volumetric conditions remain relatively unchanged subsequent to achieving the critical state. However, continued shearing to very large strains can result in reduction of soil strength and further volumetric changes as a soil approaches its residual state. Saturated plastic clays can exhibit a significant reduction in shear strength beyond the critical state to the residual state. A low residual strength is characterised by alignment of particles (Lupini *et al.*, 1981). However, soils that comprise rotund particles do not exhibit alignment and may not experience significant change beyond the critical state. In such soils, turbulent behaviour and randomly oriented particles are evidenced on shearing. Unsaturated fine-grained soils comprise aggregations of particles that can be expected to exhibit a degree of rigidity and resilience dependent on the level of suction. Their behaviour may thus be perceived as complying more closely with that of a rotund particle structure. We will not be concerned with the residual state but with the behaviour up to the critical state. First we will deal with the general strength of soils and concentrate on the critical state framework for unsaturated soils in Section 4.5.

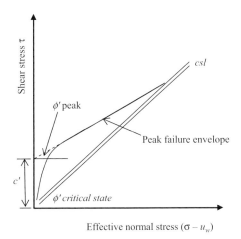

Figure 4.4   Peak and critical state strength envelopes for saturated soil.

The strength of a saturated soil as given by a Mohr–Coulomb failure envelope can be expressed by the equation below, which incorporates Terzaghi's single stress state variable at failure (Equation 4.1):

$$\tau = c' + \sigma' \tan \phi' \qquad [4.13]$$

where τ is the shear stress at failure, c' is the intercept of the failure envelope with the shear stress axis, $\sigma' = (\sigma - u_w)$ is the effective normal stress on the failure plane at failure and ϕ' is the effective angle of friction associated with changes in σ'.

The failure envelopes for peak and critical states given by Equation 4.13 are illustrated in Figure 4.4.

As discussed in Section 4.2, any two of the three stress state variables can be used to describe the stress regime in unsaturated soils. An equation for the shear strength must reflect the duality of the stress regime implicit in this statement. Fredlund and Morgenstern (1977) proposed that the two uncoupled, independent stress state variables $\bar{\sigma}$ and s were the most appropriate. On this basis, Fredlund et al. (1978) proposed an extension to the Mohr–Coulomb failure envelope for saturated soils (given by Equation 4.14) to incorporate the suction:

$$\tau = c' + \bar{\sigma} \tan \phi' + s \tan \phi^b \qquad [4.14]$$

where ϕ^b is the friction angle associated with changes in s alone.

The failure envelope for an unsaturated soil can be plotted in a three-dimensional manner, as illustrated in Figure 4.5. The three-dimensional plot consists of shear stress τ and two stress state variables: net normal stress $\bar{\sigma}$ and matric suction s. The shear strength envelope is planar if ϕ^b is constant or curved if ϕ^b varies with matric suction. As the soil becomes saturated, the matric suction reduces to zero and the pore water pressure approaches the pore air pressure. As a result, the three-dimensional failure envelope reduces to the two-dimensional Mohr–Coulomb failure envelope for a saturated soil.

Escario (1980) examined the shear strength of unsaturated soils by conducting drained direct shear and triaxial tests on unsaturated Madrid grey clay. The tests were performed under controlled suction by adopting the axis translation technique. Test specimens were statically compacted and brought to a pre-selected suction value prior to shearing. The

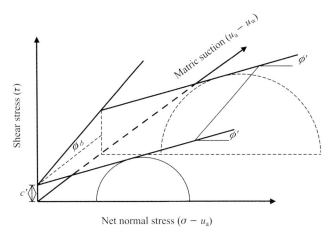

Figure 4.5   Failure envelope for unsaturated soils (Fredlund and Morgenstern, 1977).

direct shear and triaxial test results obtained by Escario (1980) produced a series of curved failure envelopes that were in general in agreement with Equation 4.14 and confirmed an increase in shear strength with increase of suction. Subsequently, Ho and Fredlund (1982) performed a series of multi-stage triaxial tests on undisturbed specimens of two residual soils from Hong Kong under consolidated drained test condition. The suction was again controlled using the axis translation technique. When the triaxial test results were analysed using Equation 4.14, a planar failure envelope resulted. Ho and Fredlund (1982) concluded that shear strength increased linearly with respect to matric suction. A similar trend was observed by Rahardjo *et al.* (1994) for residual clay from Singapore. However, there is also a significant amount of data which contradict the foregoing. Gan *et al.* (1988) conducted multi-stage direct shear tests on an unsaturated glacial till. Suction was again controlled using the axis translation technique during shearing. The results showed non-linearity of the failure envelope in the shear stress versus the matric suction plane. Fredlund *et al.* (1995) came to a similar conclusion as shown in Figure 4.6. Non-linearity in the shear strength versus matric suction relationship was also observed by Escario and Saez (1986). Oloo and Fredlund (1996) explained the non-linearity of the shear strength envelope with respect to matric suction, and the decrease in ϕ^b as suction increased, as being the result of the diminishing contribution of matric suction to the shear strength as the water content of the soil approached the residual water content.

Escario and Saez (1986) proposed the following relationship between ϕ^b and ϕ':

$$\tan \phi^b = \chi \tan \phi' \qquad [4.15]$$

where χ is the parameter from Bishop's stress (Equation 4.3).

Thus, while it is true that in cohesive soils there is a general increase in strength with increase in suction, the value of ϕ^b is not necessarily constant and may decrease. However, there is some evidence that in cohesionless soils, in particular, there may be an initial increase in shear strength followed by a decrease as suction is progressively increased (Escario and Juca, 1989; Fredlund *et al.*, 1995). Leroueil (1997) outlined a number of limitations in the use of Equation 4.14, as others have done (Escario and Saez, 1986; Fredlund *et al.*, 1987), and suggested that while applicable for a variety of unsaturated soils, most of the time, this was often within a limited range of matric suction and net

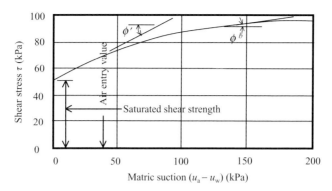

Figure 4.6 Failure envelope shear stress τ against matric suction (after Fredlund *et al.*, 1995). Reproduced by permission of Taylor and Francis Group

normal stress. Of note, as in Figure 4.6, tests on a range of materials suggest a marked reduction in ϕ^b on change from a saturated or near-saturated soil to an unsaturated soil. There is an inherent assumption in the equation of Fredlund *et al.* (1978) of a smooth transition between unsaturated and saturated behaviour. It is shown in Chapter 8 that this assumption is often unjustified.

Vanapalli *et al.* (1996) developed an empirical, analytical model to predict the shear strength in terms of soil suction, with validation of the approach based on statically compacted glacial till tested in a modified direct shear apparatus. The approach incorporated the soil water characteristic curve and the saturated shear strength parameters and resulted in Equation 4.16, which may be compared with the expression (Equation 4.14) proposed by Fredlund *et al.* (1978):

$$\tau = \left[c' + \overline{\sigma} \tan \phi' \right] + s \left[\Theta^\kappa \tan \phi' \right] \quad [4.16]$$

where $\overline{\sigma} = (\sigma - u_a)$ is the net normal stress on the shear plane at failure, Θ is the normalised volumetric water content $= \theta_w / \theta_s$, κ is a fitting parameter, θ_w is the volumetric water content at current matric suction and θ_s is the volumetric water content when $S_r = 100\,\%$.

Vanapalli *et al.* (1996) also proposed an equation for the shear strength which removed the need for the fitting parameter:

$$\tau = \left[c' + \overline{\sigma} \tan \phi' \right] + s \left[\left(\frac{\theta_w - \theta_r}{\theta_s - \theta_r} \right) \tan \phi' \right] \quad [4.17]$$

where θ_r is the residual volumetric water content.

The authors suggested that soils such as highly plastic clays were resistant to desaturation and could exhibit essentially linear shear strength behaviour over a relatively large range of soil suctions.

4.4.2 Volume change characteristics

Researchers have examined the possibility of describing the deformation behaviour of unsaturated soils using Bishop's (1959) single-valued effective stress equation, but a unique relationship between volume change and Bishop's stress has been shown not to exist (Bishop and Blight, 1963; Burland, 1965; Blight, 1965; Aitchison, 1967; Morgenstern, 1979). As discussed in Section 4.2, Burland (1965) proposed that the volume change should be independently related to $\bar{\sigma}$ and s, and a number of researchers subsequently developed frameworks that treated the volume change and shear strength behaviour of unsaturated soils in terms of the two independent stress state variables (Matyas and Radhakrishna, 1968; Barden *et al.*, 1969; Aitchison and Woodburn, 1969; Aitchison and Martin, 1973; Fredlund and Morgenstern, 1977). This section reviews the most important developments in establishing a constitutive framework for the volume change characteristics of unsaturated soils.

Matyas and Radhakrishna (1968) appear to be amongst the first researchers to have introduced the concept of 'state variables'. The state variables of $\bar{\sigma}$, s and e (or S_r) were used to plot the volume change data as three-dimensional plots, thus defining 'state paths' and 'state surfaces'. In their study, Matyas and Radhakrishna (1968) determined the volume change characteristics in isotropic and K_0 consolidation tests on statically compacted specimens of unsaturated kaolin-flint. The plots emerged as having a consistent warped surface as illustrated in Figure 4.7. The warped surface illustrated the swelling that occurred on wetting at a low value of net stress and the collapse that occurred on wetting at a high value of net stress. The collapse as a result of suction reduction indicated a meta-stable soil structure. The surfaces were found not to be wholly unique suggesting hysteresis on wetting and drying.

Subsequently, the significance of the uncoupled net stress and suction was examined in a number of research projects on the volume change behaviour of unsaturated soils. In particular, Fredlund and Morgenstern (1976) proposed mathematical expressions relating the voids ratio and the water content changes of an unsaturated soil to the independent stress state variables $\bar{\sigma}$ and s. The proposed constitutive relations were supported by a series of tests on undisturbed Regina clay and compacted kaolin under K_0 and isotropic loading conditions. During the tests the total pressure, pore air pressure and the pore water pressure were controlled independently. The proposed mathematical expressions for the state surfaces for voids ratio e and water content w were:

$$e = e_0 - C_t \log \bar{\sigma} - C_m \log s \qquad [4.18]$$

$$w = w_0 - D_t \log \bar{\sigma} - D_m \log s \qquad [4.19]$$

where C_t is the compression index with respect to $\bar{\sigma}$, C_m is the compression index with respect to s, D_t is the water content index with respect to $\bar{\sigma}$, D_m is the water content index with respect to s, e_0 is initial voids ratio and w_0 is the initial gravimetric water content.

Equation 4.18 defines a planar surface in $e : \log \bar{\sigma} : \log s$ space and Equation 4.19 a planar surface in $w : \log \bar{\sigma} : \log s$ space, as illustrated in Figure 4.8(a) and (b) respectively. Similar forms of equation can be used for unloading surfaces. The analysis assumed that the plots were linear over a relatively large stress range. The formulation does not explicitly include both wetting-induced swelling and wetting-induced collapse and Equations 4.18 and 4.19 can be used only over a limited range of suction values because they do not satisfy saturated conditions when the suction approaches zero. However, Fredlund (1979) suggested that these effects could be incorporated if the compression and water content indices were stress-state-dependent. Ho *et al.* (1992), Fredlund and Rahardjo (1993) and Ho and Fredlund (1995) further described their measurement and usage.

Amongst other relevant approaches, Lloret and Alonso (1985) proposed expressions for the state surfaces for voids ratio e and degree of saturation S_r. The equations, though not reproduced here, defined warped surfaces so that wetting-induced swelling and wetting-induced collapse were both included.

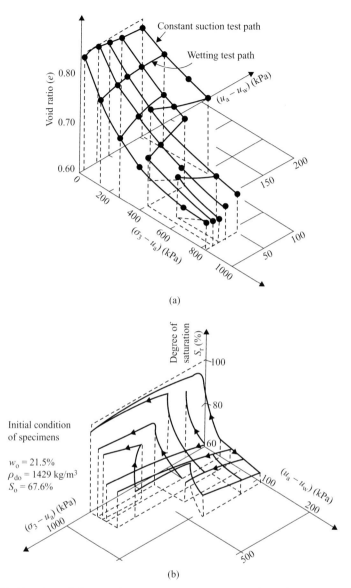

Figure 4.7   Void ratio and degree of saturation constitutive surfaces for a mixture of flint and kaolin under isotropic loading conditions: (a) void ratio constitutive surface; and (b) degree of saturation constitutive surface (after Matyas and Radhakrishna, 1968). Reproduced by permission of the authors and the Institution of Civil Engineers

4.5 Critical state framework for saturated soils

An alternative approach to defining the strength and deformation behaviour of a saturated soil is provided by critical state soil mechanics. This forms the basis of the Original Cam clay model described by Schofield and Wroth (1968) and the Modified Cam clay model described by Roscoe and Burland (1968), as well as other modifications to extend the ideas

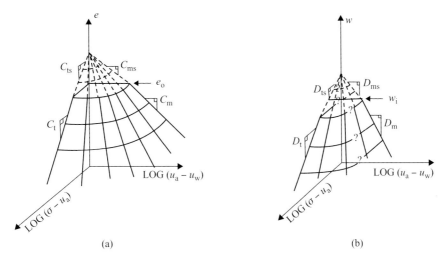

Figure 4.8 Schematic soil structure and water content constitutive surfaces: (a) void ratio constitutive surfaces for monotonic volume change – solid lines indicate compression, dashed lines indicate swelling or rebound; (b) water content constitutive surface for monotonic volume change – solid lines indicate water content decrease, dashed lines indicate water content increase (after Ho and Fredlund, 1995). Reproduced by permission of Taylor and Francis Group

to a more general class of soil. In this respect it is important to recognise that the Cam clay models were developed from test data on isotropically consolidated specimens and assumed that elastic behaviour was isotropic (Graham *et al.*, 1989; Graham and Houlsby, 1983). Those aspects of critical state soil mechanics and the Cam clay models for saturated soils that are of particular relevance to our current understanding of unsaturated soils will be briefly described and later in the chapter will be extended to encompass unsaturated conditions.

There are a number of elements to the models of saturated soil behaviour based on critical state soil mechanics:

- An isotropic normal compression line (*iso-ncl*);
- A one-dimensional normal compression line (*1d-ncl*);
- A critical state line (*csl*);
- A state boundary surface which links isotropic normal compression states, one-dimensional compression states and critical states;
- Yield surfaces which lie on the state boundary surface and which act as plastic potential surfaces;
- The 'hardening' or 'softening' of a soil as it yields and moves over the state boundary surface;
- Elastic straining within the yield surface and a combination of elastic and plastic straining when the yield surface is reached.

The basic theoretical constructs within the Cam clay models are essentially similar, the main differences being the form of the yield surface.

The critical state of a soil is the condition when it exhibits relatively constant stress and volume conditions under shearing. Wood (1990) expressed this mathematically as follows:

$$\frac{\partial p'}{\partial \varepsilon_q} = \frac{\partial q}{\partial \varepsilon_q} = \frac{\partial v}{\partial \varepsilon_q} = 0 \qquad [4.20]$$

where $p' = (p - u_w)$ is Terzaghi's mean effective stress, $p = (\sigma_{11} + 2\sigma_{33})/3$ is the mean total stress and $q = (\sigma_{11} - \sigma_{33})$ is the deviator stress.

The first term in Equation 4.20 defines the rate of change of Terzaghi's mean effective stress with shear strain, the second term defines the rate of change of deviator stress with shear strain and the third term defines the rate of change of specific volume with shear strain. All should equate to zero. A further condition that should be satisfied is $\partial u_w / \partial \varepsilon_q = 0$, i.e. the rate of change of pore water pressure with shear strain should be zero.

Figure 4.9 illustrates a unique relationship between the stress and volumetric variables in $p' : q : v$ space based on the *csl*, *iso-ncl* and *1d-ncl*. A premise of the critical state concept is that a soil will, on shearing, end up on a unique critical state line irrespective of its initial condition. Often the same intrinsic critical state concepts are applied to sands as to clays though this appears not always to be justifiable and there are recognisable differences in behaviour characteristics. There are also soils that do not appear to conform to the critical state principles. Ferreira and Bica (2006) reported on the effects of structure on a residual soil from Botucatu sandstone. While a unique *ncl* and *csl* could be determined for the natural residual soil, remoulded compacted specimens did not exhibit a unique *ncl* and *csl*. The authors described the soils as exhibiting a transitional behaviour between clays and sands and concluded that the *csl* was dependent on initial voids ratio at odds with the concept of a unique *csl*. There was also no evidence that the structure of the natural soil was sufficiently altered either by compression or shearing to bring the state towards that of the equivalent remoulded soil. Nevertheless, the critical state concepts appear justifiable

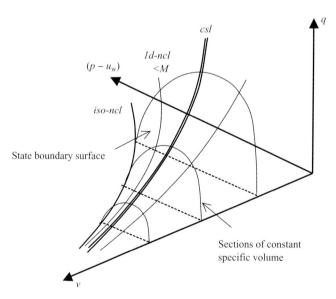

Figure 4.9   State boundary surface linking the *iso-ncl*, *1d-ncl* and *csl* for a soil.

for a wide class of saturated materials and the following discusses first the strength and then the deformation characteristics within the critical state framework.

4.5.1 Shear strength

Experimental evidence suggests that the strength envelope at critical state when plotted in the $\tau : \sigma'$ plane is a straight line passing through the origin and is the Mohr–Coulomb failure criterion with the cohesion $c' = 0$. From Equation 4.13, the critical state strength can thus be represented by:

$$\tau = \sigma' \tan \phi' \qquad [4.21]$$

The equivalent relationship between deviator stress q and mean effective stress p' at the critical state for a saturated soil under triaxial stress conditions is given by Equation 4.22 and represented in Figure 4.10. The *csl* of Figure 4.10 is the projection of the *csl* of Figure 4.9 on the $p' : q$ plane.

$$q = Mp' \qquad [4.22]$$

where M is the stress ratio parameter.

Equations 4.21 and 4.22 describe the same condition and the parameters are related by the following equation:

$$M = \frac{6 \sin \phi'}{3 - \sin \phi'} \qquad [4.23]$$

4.5.2 Volume change characteristics

The idealised compression characteristics for saturated soils are represented in Figure 4.11 as plots in $v : \ln p'$ space. The isotropic normal compression line (*iso-ncl*) and the critical state line (*csl*) are taken as parallel lines in $v : \ln p'$ space though they are curved in three-dimensional $p' : q : v$ space in Figure 4.9. Inside the *iso-ncl*, one-dimensional normal consolidation lines (*1d-ncl*) parallel to the foregoing lines are also shown in Figure 4.11.

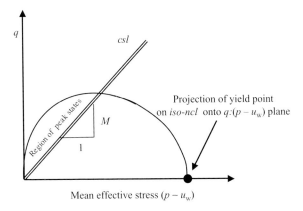

Figure 4.10   Critical state strength under triaxial stress conditions.

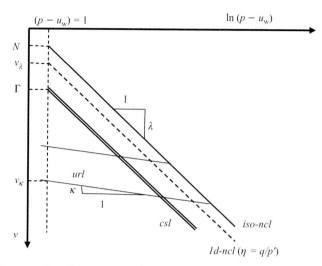

Figure 4.11   Compression characteristics for a saturated soil.

For the *iso-ncl* the specific volume is related to the mean effective stress by the following equation:

$$v = N - \lambda \ln p' \qquad [4.24]$$

For a *1d-ncl* the following straight-line relationship is defined:

$$v = v_\lambda - \lambda \ln p' \qquad [4.25]$$

The unloading and reloading pressure–volume characteristic lines are represented by *url* in Figure 4.11 and defined by the following relationship:

$$v = v_\kappa - \kappa \ln p' \qquad [4.26]$$

If the soil is subjected to shearing, the ultimate state of the soil regardless of initial condition is taken as being on the *csl* represented by:

$$v = \Gamma - \lambda \ln p' \qquad [4.27]$$

where N, v_λ, v_k and Γ are the intercepts of the *iso-ncl*, the *1d-ncl*, the *url* and the *csl* respectively at $p' = 1.0$ kPa. λ is the slope of the *iso-ncl*, the *1d-ncl* and the *csl*. κ is the slope of the *url*.

The *1d-ncl* can be defined by the value of the mobilised stress ratio η given by the following expression:

$$\eta = \frac{q}{p'} \qquad [4.28]$$

For $m = M$ the *1d-ncl* lies on the *csl* and for $m = 0$ the *1d-ncl* corresponds to the *iso-ncl*. The stress ratio η has a value less than M for a normally or lightly overconsolidated specimen, but a value greater than M for a heavily overconsolidated specimen. The stress ratio η is useful in identifying qualitatively the volumetric behaviour during shearing. For $\eta < M$ on the so-called 'wet side' of the *csl* (between the *csl* and the *iso-ncl*), compressive plastic hardening occurs on yielding under shearing. For $\eta > M$ on the 'dry side' of the *csl* (opposite side of *csl*), plastic softening occurs on yielding under shearing.

For a heavily overconsolidated soil specimen the peak strength is given by (Wood, 1990; Atkinson, 1993):

$$\frac{q}{p'} = M - \frac{d\varepsilon_v}{d\varepsilon_q} \qquad [4.29]$$

The expression shows that the peak deviator stress ratio q/p' is the sum of the critical state stress ratio M and the rate of dilation $-d\varepsilon_v/d\varepsilon_q$ (the negative sign is required because volumetric strain $d\varepsilon_v$ is negative for dilation). This is in accordance with the argument of Schofield (2005, 2006) that the peak strength of both granular and cohesive deposits is the sum of the critical state strength and interlocking (or the rate of dilation).

On the *iso-ncl*, a soil specimen is normally consolidated and an increase in isotropic mean stress p' leads to both elastic and plastic volumetric strainings. However, a decrease in isotropic stress p' for the specimen entails the specimen following a *url* and the specimen becoming overconsolidated. On the *url*, the specimen is assumed to experience only elastic volumetric strain. To draw further conclusions it is necessary to recognise the three-dimensional relationship of the $p' : q : v$ plot of Figure 4.9. The state boundary surface is shown as linking the *iso-ncl*, the *1d-ncl* and the *csl*. This surface marks the boundary to soil conditions although cemented soils can exhibit unstable states beyond the state boundary surface. However, other than for cemented soils, a soil specimen under test that reaches a condition represented by a point on the state boundary surface will experience a combination of elastic and plastic strainings and the soil can be considered to be yielding. If the stress state of the soil is brought inside the boundary surface the strains are usually assumed to be purely elastic. It is important to recognise that this is an idealisation of soil behaviour, and inelastic behaviour inside the state boundary surface can occur, particularly if time-dependent (viscous) behaviour is present.

The existence of a unique state boundary surface for saturated clay soils has been demonstrated by data reported by Atkinson and Bransby (1978) and Wood (1990). That part of the state boundary surface on the 'wet side' of the critical state is often referred to as the *Roscoe surf*ace and that part on the 'dry side' is often called the *Hvorslev surface*. The Roscoe and Hvorslev surfaces are shown in Figure 4.12. A tensile fracture cut-off surface is also shown. The Hvorslev and the tensile fracture surfaces are generally represented as straight lines on constant volume sections.

The principles of critical state soil mechanics can be extended to the volume change behaviour of unsaturated soils though the critical state characteristics of unsaturated soils are still the subject of much research. While there is a growing body of experimental evidence on the behaviour of soils in an unsaturated state, there is as yet insufficient definitive information and agreed principles to draw all-encompassing conclusions.

4.5.3 Yielding

Figure 4.13 shows an isotropic swelling and recompression line or unload–reload line *url*. On this line a soil experiences only elastic behaviour. If the volumetric and shear deformations are uncoupled, then volumetric changes are related only to the mean effective stress p' and are unrelated to changes in deviator stress q. Similarly, shear strains are related only to q with no influence from p'. This relies on an assumption that the soil is isotropic. The uncoupling of the influences of p' and q is discussed further in Chapter 6. Nevertheless,

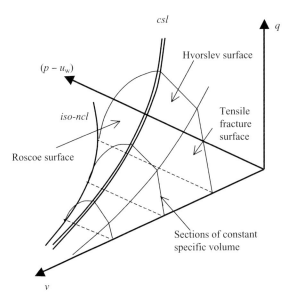

Figure 4.12   State boundary surface limiting states.

this means that on the isotropic *url*, only volumetric straining occurs and no shear strain-ing. It also means that inside the state boundary surface, vertically above the *url*, a plane can be drawn on which only elastic straining occurs. This is referred to as an *elastic wall*. Since the soil will yield when the state reaches the state boundary surface, the intersection of the elastic wall with the state boundary surface represents a *yield surface*. The shape of the projection of the yield surface onto the $q : p'$ plane differs from the projection of

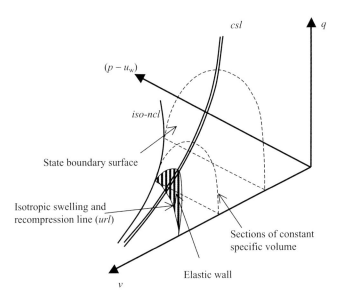

Figure 4.13   The elastic wall.

the sections of constant specific volume v, also shown in Figure 4.13. It is the assumption of the shape of the projection of the yield surface that often distinguishes approaches to yielding in critical state soil mechanics and other forms of elasto-plastic modelling.

Yielding is usually considered as controlled by the mean effective stress and deviator stress and forms a locus in the $p':q$ plane. The Original Cam clay model (Schofield and Wroth, 1968) assumed the yield surface locus for isotropic specimens to be a series of logarithmic spirals symmetrical about the p'-axis in the $p':q$ plane as illustrated in Figure 4.14. The equation for the logarithmic yield surfaces can be written as:

$$\ln\left(\frac{p'}{p'_{cs}}\right) = 1 - \frac{\eta}{M} \qquad [4.30]$$

where p'_{cs} is the value of the mean effective stress at the critical state.

The Modified Cam clay model (Roscoe and Burland, 1968) assumed the yield surfaces in the $p':q$ plane to be ellipses, as illustrated in Figure 4.15. The equation for the yield surfaces for isotropic specimens can be written as Equation 4.31 and are again symmetrical about the p'-axis.

$$\frac{p'}{p'_{cs}} = \frac{2M^2}{M^2 + \eta^2} \qquad [4.31]$$

The concept of yielding in saturated soils has been explored by a number of researchers including Mitchell (1970), Crooks and Graham (1976), Tavenas and Leroueil (1977) and Graham *et al.* (1983). This constitutes relatively large-strain yielding at a macro-mechanical level and it is important to realise that yielding and the onset of plastic deformations are likely to start at much smaller strains. Nevertheless, the concept of large-strain yielding provides idealised behaviour characteristics that have allowed advancement of our understanding of soil behaviour.

The yield surfaces shown in Figures 4.14 and 4.15 are for isotropic soil specimens. In real soils the assumption of isotropic characteristics is unlikely to be realised. Consequently, there have been attempts to examine the yielding of one-dimensionally compressed, K_0,

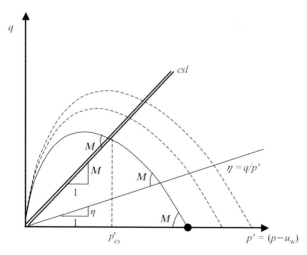

Figure 4.14   Logarithmic yield curves in $p' - q$ plane for Original Cam clay model.

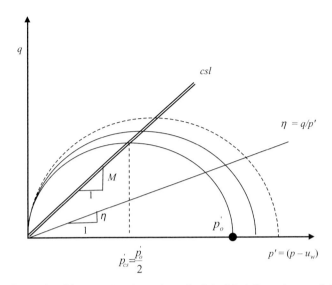

Figure 4.15   Elliptical yield curves in $p' : q$ plane for Modified Cam clay model.

specimens. For this more general class of anisotropic materials there is evidence to suggest that the yield curves are rotated around the K_0 line in the $p' : q$ plane, as depicted in Figure 4.16 (Crooks and Graham, 1976; Graham and Houlsby, 1983; Graham *et al.*, 1983, 1988; Sivakumar *et al.*, 2001).

Graham *et al.* (1989) suggested that asymmetry of the yield loci about the p'-axis may be a function of the plotted variables and the authors presented limited data to suggest that normalising the axis may result in a more symmetrical yield surface around the normalised p'-axis. Nevertheless, the importance of the yield locus is that it separates relatively stiff, pseudo-elastic pre-yield behaviour from the large-strain post-yield behaviour. The shape and magnitude of a yield locus depend on the composition, anisotropy and stress history of the soil (Graham *et al.*, 1988; Sivakumar *et al.*, 2002). However, it is important to realise that while yielding is classically associated with a sharp change from purely

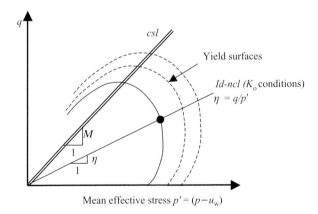

Figure 4.16   Idealised yield curves in $p' : q$ plane for one-dimensionally consolidated saturated soil specimens.

elastic to plastic behaviour, in soils yielding does not take place abruptly but over a range of pressures, leading to a transition from elastic behaviour to elasto-plastic behaviour (Smith *et al.*, 1992).

Yielding in unsaturated soils is incorporated in the ideas behind the Barcelona Basic Model and the amendments and additions to the concepts as discussed in the following.

4.5.4 The plastic potential and normality

A common way of describing the flow rule for plastic straining is to define a plastic potential envelope that is orthogonal to the vectors of plastic straining. Often it is assumed that the plastic potential is synonymous with the yield curve, which results in an associated flow rule and normality of the plastic strain-increments to the local stresses on the yield curve. This is illustrated in Figure 4.17. However, normality of plastic yielding to the yield curve is an assumption that is not always justifiable as demonstrated by Cui and Delage (1996).

Hardening (increase in the size of the yield curve) or softening (decrease in the size of the yield curve) occurs at the yield surface. Depending on the form of the yield curve or plastic potential, a hardening (or softening) law or flow rule giving the relationship between the change in yield stress and the plastic strain-increment can be written.

4.6 The constitutive Barcelona Basic Model for unsaturated soils

Schofield (1935) appears to be the first to have introduced the concept of suction in unsaturated soils to research in soil mechanics. Brady (1988) and Murray and Geddes (1995) drew on previously published work on unsaturated soils, in particular the work reported by Croney and Coleman (1954, 1960), to extend the concepts of critical state soil mechanics from saturated to unsaturated soils. The Barcelona Basic Model (BBM) proposed by Alonso *et al.* (1990) utilises a critical state framework, extended to include

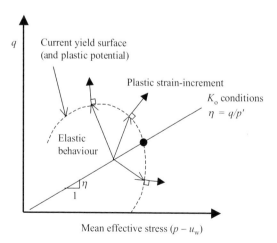

Figure 4.17   Normality of plastic straining on yield surface when it acts as the plastic potential.

suction as a separate stress state variable. The BBM follows earlier work by Alonso *et al.* (1987) who proposed an elasto-plastic model based on experimental data from suction-controlled triaxial tests on compacted clay. Significantly more research on the behaviour of soils under unsaturated conditions has now been undertaken and a number of constitutive models have been proposed. However, the BBM forms the basis for most elasto-plastic constitutive modelling for unsaturated soils (Gens, 2010). The model was defined in terms of the mean net stress $\bar{p}$ and suction s, along with the deviator stress q for triaxial test conditions. The model was intended for the cases of slightly or moderately expansive soils but is able to account in a simplistic way for most of the mechanical characteristics of unsaturated soils.

4.6.1 Shear strength

It is assumed that under shearing a soil will ultimately reach a critical state. It is further assumed that there is a linear increase of critical state shear strength q with both $\bar{p}$ and s. A smooth transition between unsaturated and saturated conditions is also an inherent assumption and thus all the critical state lines are considered parallel to the constant slope M line for a saturated soil as depicted in Figure 4.18. The critical state line for an unsaturated soil can be written as:

$$q = M\bar{p} - Mks \qquad [4.32]$$

where $-ks$ is the intersect on the $\bar{p}$-axis at which $\bar{p} = -\bar{p}_s = -ks$ in Figure 4.18 and k is a negative constant as in Figure 4.20.

4.6.2 Volume change characteristics

In accordance with Figure 4.19, the volume change behaviour on an *iso-ncl* for an isotropically prepared unsaturated soil under constant suction conditions can be expressed

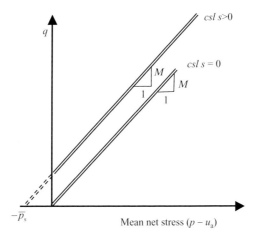

Figure 4.18   BBM critical state strength.

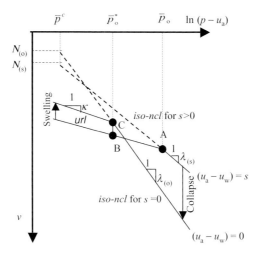

Figure 4.19 Isotropic stress–volume characteristics (*iso-ncl* and *url*) for unsaturated soils (modified from Alonso *et al.*, 1990). Reproduced by permission of the authors and the Institution of Civil Engineers

mathematically as:

$$v = N_{(s)} - \lambda_{(s)} \ln \left[\frac{\bar{p}}{\bar{p}^c} \right] \qquad [4.33]$$

where $\bar{p}^c$ is the net stress at a reference stress state for which $v = N_{(s)}$. $\lambda_{(s)}$ is the stiffness parameter for changes in $\bar{p}$ for virgin states of the soil (both $N_{(s)}$ and $\lambda_{(s)}$ are functions of suction).

The following incremental expression for compression under constant s can be written:

$$dv = -\lambda_{(s)} \frac{d\bar{p}}{\bar{p}} \qquad [4.34]$$

The unloading and reloading pressure–volume characteristic lines are represented by *url* in Figure 4.19 and it is assumed that the slope κ of the lines is independent of s. The lines at inclinations λ and κ for $s = 0$ and $p' = \bar{p}$ correspond to the *iso-ncl* and the *url* respectively in Figure 4.11, and if $\bar{p}^c = 1$, $N_{(0)}$ corresponds to N. Equation 4.33 for an unsaturated soil thus reduces to Equation 4.24 for a saturated soil. This is not to say that there is a physically smooth transition from unsaturated behaviour to saturated behaviour or vice versa, but only that the equations are compatible.

From experimental data, Alonso *et al.* (1990) proposed an empirical equation for the slope $\lambda_{(s)}$ of the *iso-ncl* for unsaturated soils under constant suction:

$$\lambda_{(s)} = \lambda_{(0)} \left[(1 - r_c) \exp(-\beta s) + r_c \right] \qquad [4.35]$$

where r_c is a constant related to the maximum stiffness of the soil (for infinite suction), β is a parameter that controls the rate of increase of soil stiffness with suction, and $\lambda_{(0)}$ is the stiffness parameter for the *iso-ncl* for $s = 0$.

For unloading and reloading at constant s, elastic swelling and compression are given by:

$$dv = -\kappa \frac{d\bar{p}}{\bar{p}} \qquad [4.36]$$

Assuming a linear dependence between v and $\ln(s + p_{atm})$ both in the elasto-plastic and elastic range for suction shrinkage and swelling at constant $\bar{p}$, the following expressions emerge for virgin compression and swelling/shrinkage respectively:

$$dv = -\lambda_{(s)} \frac{ds}{(s + p_{atm})} \qquad [4.37]$$

$$dv = -\kappa_{(s)} \frac{ds}{(s + p_{atm})} \qquad [4.38]$$

where $\lambda_{(s)}$ is the stiffness parameter for changes in suction for virgin states and $\kappa_{(s)}$ is the elastic stiffness parameter for changes in suction.

4.6.3 Yielding

The BBM assumes elastic behaviour if the state of the soil remains inside a yield surface defined in $\bar{p} : q : s$ space, with plastic strains commencing once the yield surface is reached. Thus, for stress paths remaining inside the yield surface, an increase (or decrease) of mean net stress $\bar{p}$ produces elastic compression (or expansion, as along path A–B in Figure 4.19), and a decrease (or increase) in suction s as along path B–C produces elastic swelling (or shrinkage).

To depict yielding under triaxial stress conditions in the BBM, the following yield loci are defined, which are likely to be simplifications of the interlinked yield conditions for real soils:

- A load-collapse yield locus (LC);
- A suction increase (SI) yield locus.

These are depicted in Figures 4.20–4.22. For isotropic stress states, the intersection of the yield surface with the $q = 0$ plane defines the LC yield locus shown in Figure 4.20 and given by the following expression:

$$\left(\frac{\bar{p}_0}{\bar{p}^c}\right) = \left(\frac{\bar{p}_0^*}{\bar{p}^c}\right)^{[\lambda_{(0)} - \kappa]/[\lambda_{(s)} - \kappa]} \qquad [4.39]$$

where $\bar{p}_0$ is the pre-consolidation stress and $\bar{p}_0^*$ is the pre-consolidation stress for saturated conditions.

The SI yield curve is also shown in Figure 4.20 and is taken as a straight line at constant s in the $q = 0$ plane.

The deviator stress q is incorporated in the model to account for triaxial stress states, and the LC yield curve for the case of constant suction is described by an ellipse as in the Modified Cam clay model for saturated soils. A family of elliptical yield loci may be drawn in the constant suction plan, and two ellipses ($s = 0$ and $s > 0$) are shown in Figure 4.21. These are assumed to have a constant aspect ratio. The equation of the ellipses is

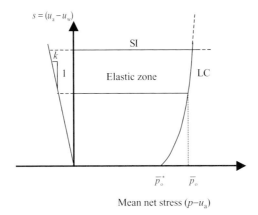

Figure 4.20 LC and SI yield loci in $\overline{p}:s$ plane for $q = 0$ (modified from Alonso *et al.*, 1990). Reproduced by permission of the authors and the Institution of Civil Engineers

given by:

$$q^2 - M^2\,(\overline{p} + \overline{p}_s)\,(\overline{p}_0 - \overline{p}) = 0 \qquad [4.40]$$

It should be noted that the description of yielding in $\overline{p}:q:s$ space in the BBM does not include any volumetric terms. In unsaturated soils the relative volumes of the phases are considered an important consideration in examining and attempting to predict behaviour patterns such as yielding. This will be discussed further in Section 8.4.

4.6.4 The plastic potential and normality

Normality of the plastic strain-increment to the current yield surface is an assumption frequently made but not always justifiable. Alonso *et al.* (1990) suggested a non-associated

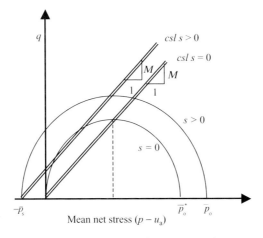

Figure 4.21 Elliptical LC yield curves and critical state strength in $q:\overline{p}$ plane (modified from Alonso *et al.*, 1990). Reproduced by permission of the authors and the Institution of Civil Engineers

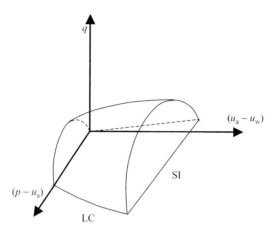

Figure 4.22   Three-dimensional view of yield surfaces in $\bar{p} : q : s$ space (modified from Alonso *et al.*, 1990). Reproduced by permission of the authors and the Institution of Civil Engineers

flow rule for elliptical LC yield loci in s constant planes in developing the BBM. For yielding on the SI locus in the constant $\bar{p}$ plane, normality was assumed. Cui and Delage (1996) presented results for an Aeolian silt which demonstrated a number of important features. As discussed further in Section 4.7.5, the LC yield locus was rotated about the K_0 line in the $\bar{p} : q$ plane. It was also noted that when suction hardening extended the LC yield curve, the direction of the plastic strain-increment was approximately radial, suggesting an isotropic suction hardening phenomenon. When hardening occurred due to the action of net stress, non-associated plastic strain-increments were recorded. A hyperbolic plastic potential was found satisfactory in modelling the non-associated plastic strains.

4.7 Extended constitutive and elasto-plastic critical state frameworks for unsaturated soils

There are a large number of assumptions in the BBM that continue to be challenged by researchers. This has led to different though comparable constitutive approaches in an effort to develop a more comprehensive model able to deal with inconsistencies in the BBM. However, the framework of the BBM for an idealised initially isotropic soil is considered by many to provide a good starting point even though experimental research has highlighted the need to modify and develop the model if more general features of unsaturated soil behaviour are to be adequately addressed. Shortcomings of the model have been highlighted by a number of researchers. In particular, as pointed out by Lloret *et al.* (2008) amongst others, the BBM does not intrinsically include a term (or terms) for the volumes of the phases and cannot fully describe the important features associated with hydraulic hysteresis associated with wetting and drying paths. Badonis and Kavvadas (2008) pointed out that the BBM does not address the range of suctions over which shrinkage occurs. The BBM also assumes constant κ for the *url* independent of the suction and that for increasing suction there is a linear increase in shear strength and a linear decrease of the slope of the normal compression line λ_s. Badonis and Kavvadas (2008) suggested that a more general class of models should allow for the non-continuous increase of shear strength with increasing suction, while Georgiadis *et al.* (2008) suggested a non-linear increase in shear strength with increasing suction. Additionally, while the monotonic decrease of the compression line was supported by Cui and Delage (1996), it was challenged by Wheeler and Sivakumar (1995). The issue remains unresolved.

Based on ongoing experimental research, a number of modified unsaturated soil models have been suggested in order to accommodate a more general class of material behaviour. Some of those to be found in the literature are described in the following along with discussions of their significance in the light of ongoing research. Also discussed are aspects of hydraulic hysteresis and yielding.

4.7.1 Constitutive model of Toll (1990) and Toll and Ong (2003)

The papers describe a constitutive framework for unsaturated soils in terms of the stress state variables $\bar{p}$, q and s and the volumetric variables v and S_r. Validation of the framework was provided by results of triaxial tests on compacted Kiunyu gravel (Toll, 1990) and residual Jurong soil (Toll and Ong, 2003). The triaxial tests were carried out in a standard displacement-controlled triaxial cell with suction controlled by the axis translation technique. Most of the tests were conducted under constant water mass.

The authors proposed the following equation for the strength at the critical state under triaxial stress conditions:

$$q = M_a \bar{p} + M_b s \qquad [4.41]$$

where M_a and M_b are the total and suction stress ratio parameters respectively.

Equation 4.41 can be compared with Equation 4.32 of the BBM. Both equations incorporate the three independent stress variables $\bar{p}$, q and s identified by Matyas and Radhakrishna (1968) as necessary to define the stress state in unsaturated soils. However, the interpretation of the stress ratios differs. Values of M_a and M_b in Equation 4.41 were back-calculated from the experimental results using an iterative process based on multiple linear regression assuming that each value was a function of S_r and consequently of the matric suction s. Figure 4.23 presents the interpreted values for Kiunyu gravel. It can be seen that the value of M_a increases and the value of M_b decreases from the saturated stress ratio M when S_r reduces and s increases.

The values of M_a and M_b are related to ϕ and ϕ^b of Equation 4.14 by the following expressions:

$$M_a = \frac{6 \sin \phi'}{3 - \sin \phi'} \qquad [4.42]$$

$$M_b = \frac{6 \cos \phi' \tan \phi^b}{3 - \sin \phi'} \qquad [4.43]$$

The stress ratio parameters M_a and M_b describe the components of shear strength associated with the uncoupled stress state variables $\bar{p}$ and s respectively. As M_a and M_b vary with S_r, Equations 4.42 and 4.43 indicate that ϕ' as well as ϕ^b vary with the degree of saturation. While Equation 4.41 recognises the dual stress regime present in unsaturated soils, it has no theoretical basis and the choice of values for M_a and M_b, in matching the equation to experimental data, is somewhat arbitrary. Toll *et al.* (2008) examined three alternative ways of determining M_a and M_b for constant water mass triaxial tests

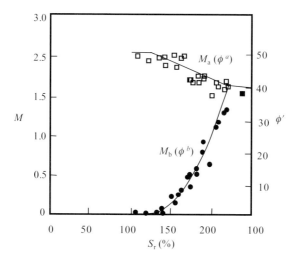

Figure 4.23 Variation of critical state stress ratios M_a and M_b with degree of saturation S_r for Kiunyu gravel (after Toll, 1990). Reproduced by permission of the authors and the Institution of Civil Engineers

on artificially bonded sand using the axis translation technique without reaching a firm conclusion as to the most appropriate way.

Wheeler (1991), Alonso *et al.* (1992) and Wheeler and Sivakumar (1995) proposed strength equations of a form somewhat similar to Equation 4.41. In particular, Wheeler (1991) proposed an alternative form of critical state relationship for unsaturated soils based on the experimental results of Toll (1990). He suggested that the critical state equation for the deviator stress q can be represented by the following equation:

$$q = M\bar{p} + f(s) \tag{4.44}$$

where M is the critical state stress ratio for a saturated soil and is constant for a given material, and the function f can be found by plotting $q - M\bar{p}$ against s. The function f is thus matched to the experimental data and not linked to fundamental physics.

As with the proposal of Fredlund *et al.* (1978) encapsulated in Equation 4.14, a smooth transition between unsaturated and saturated conditions is assumed in determining the values of the stress ratio parameters of Equation 4.41. This relies on the progressive breakdown of the aggregated structure associated with unsaturated conditions to the more dispersed condition associated with saturated soils. Chapter 9 shows that Equation 4.41 can be derived on a theoretical basis without imposing any assumptions on the values of M_a and M_b. This alternative approach allows the significance of the stress ratios to be clearly demonstrated and validated using experimental data.

Toll (1990) included in his constitutive model an empirical equation relating the dual stress regime to the specific volume at the critical state. The equation can be written as:

$$v = \Gamma_{ab} - \lambda_a \ln \bar{p} - \lambda_b \ln s \tag{4.45}$$

where λ_a is the compression coefficient as a result of the influence of $\bar{p}$ and λ_b is the compression coefficient as a result of the influence of s.

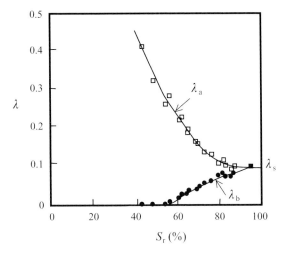

Figure 4.24 Variation of critical state compressibilities λ_a and λ_b with degree of saturation S_r for Kiunyu gravel (after Toll, 1990). Reproduced by permission of the authors and the Institution of Civil Engineers

Toll also proposed the following relationship for Γ_{ab}:

$$\Gamma_{ab} = 1 + \frac{\Gamma - 1}{S_r} \qquad [4.46]$$

where Γ is the intercept of the csl at $p' = 1.0$ kPa for a saturated soil and Γ_{ab} is the intercept of the csl at $p' = 1.0$ kPa for an unsaturated soil.

The values of λ_a and λ_b were back-calculated using an iterative process as for the stress ratios in Equation 4.41. The values for Kiunyu gravel are shown in Figure 4.24. The value of λ_a increases from the saturated value λ with reducing S_r, and the value of λ_b decreases from the saturated value λ with reducing S_r. From analyses of experimental data on the critical state of unsaturated Kiunyu gravel and Jurong soil, Toll and Ong (2003) concluded that the following five variables were necessary to describe the critical state: $\bar{p}$, q, s, v and S_r. They also concluded that five parameters were required: M_a, M_b, Γ_{ab}, λ_a and λ_b.

Due to the large number of independent variables and parameters, Wheeler (1991) suggested that the proposed framework could not be used for any form of prediction. Based on the results of Toll (1990), Wheeler (1991) proposed a critical state volumetric relationship as a component of a constitutive framework to complement the strength relationship given by Equation 4.44. The relationship between the stress state variables $\bar{p}$ and s and the specific water volume v_w was given by:

$$v_w = \Gamma - \lambda \ln \bar{p} + fs \qquad [4.47]$$

where Γ and λ are the intercept and slope of the csl for the saturated soil.

The form of the last term in Equation 4.47 can be examined by plotting $v_w - \Gamma + \lambda \ln \bar{p}$ against s.

4.7.2 Constitutive elasto-plastic model of Wheeler and Sivakumar (1995)

The authors proposed an elasto-plastic critical state framework based on a series of suction-controlled triaxial tests on specimens of compacted speswhite kaolin. The framework was defined in terms of five state variables: mean net stress $\overline{p}$, deviator stress q, suction s and specific volume v, although the significance of the specific water volume v_w was also examined. The framework consists of an isotropic normal compression hyperline, a critical state hyperline and a state boundary hypersurface. The term hyperline is used to describe a locus of states within four-dimensional space $(\overline{p} : q : s : v)$ and the term hypersurface describes a four-dimensional surface joining the normal compression and critical state hyperlines. If the state of the soil lies inside the state boundary hypersurface, the soil behaviour is assumed to be elastic, with movement over the state boundary hypersurface corresponding to plastic expansion of the yield surface in stress space. The framework enhances the approach advocated in the BBM by inclusion of the specific volume v as an intrinsic component.

Wheeler and Sivakumar (1995) defined the isotropic normal compression hyperline in the plane $q = 0$ by the following equation:

$$v = N_{(s)} - \lambda_{(s)} \ln (\overline{p}/p_{atm}) \qquad [4.48]$$

where $\lambda_{(s)}$ and $N_{(s)}$ are functions of suction and p_{atm} is the atmospheric pressure taken as a reference pressure.

Equation 4.48 can be compared with Equation 4.33 of the BBM. Both equations represent the *iso-ncl* and they are the same when the reference states are the same, i.e. when $p_{atm} = \overline{p}^c$. The BBM assumes a monotonic decrease of $\lambda_{(s)}$ with increasing suction but the results of Wheeler and Sivakumar (1995) showed relatively little variation in $\lambda_{(s)}$ with s for suctions between 100 and 300 kPa, but a significant drop when s was reduced to zero. They argued that differences in experimentally determined values of $\lambda_{(s)}$ from the behaviour assumed in the BBM were due to the soil fabric produced by the initially one-dimensionally compacted specimens. Futai and Almeida (2005) reported decreasing values of $\lambda_{(s)}$ with decreasing suction for a natural, intact gneiss residual soil, but different stratigraphic horizons produced different relationships.

The influence of soil fabric was investigated by Sivakumar and Wheeler (2000) and Wheeler and Sivakumar (2000). They investigated the influence of compaction pressure, compaction water content and method of compaction (static or dynamic) for speswhite kaolin. The following conclusions were reached for specimens isotropically compressed following initial one-dimensional compression or compaction:

- Compaction pressure influenced, to a limited degree, the position of the normal compression line for different values of suction suggesting an influence from the initial compaction-induced soil fabric.
- A more radical effect on subsequent soil behaviour resulted from a change in compaction water content. The initial state of the soil, the value of suction and the positions of the normal compression line for different suctions were strongly influenced. This suggested that it may be necessary to treat samples compacted at different water contents as fundamentally different materials.
- A change from static compression to dynamic compaction (with no change in compaction water content or compaction-induced dry density) had no apparent effect on subsequent soil behaviour. This finding was tempered with a note that this may not be applicable for all clays and at all water contents.

These findings must be recognised in interpretation of the results of tests reported in the literature as they are generally on specimens compacted or compressed one-dimensionally into a mould (Sivakumar, 1993; Cui and Delage, 1996; Sharma, 1998; Wang *et al.*, 2002; Wheeler and Sivakumar, 2000; Tan, 2004). The process of one-dimensional compression inevitably introduces a degree of stress-induced anisotropy in specimens and results in a complex soil fabric that affects subsequent behaviour and normal compression data. However, Sivakumar *et al.* (2010b) reported evidence to support the principle of a unique *iso-ncl* at a given suction for unsaturated kaolin specimens initially prepared under truely isotropic conditions using the methodology outlined in Section 3.2. The results for IS(A) and IS(B) in Figure 4.25 are for specimens prepared at different initial specific volumes and appear to suggest it reasonable to assume a linear *iso-ncl* in the $\bar{p} : v$ plane for specimens with true isotropic stress history.

Wheeler and Sivakumar (1995) defined the critical state hyperlines by the following equations:

$$q = M_{(s)}\,\bar{p} + \mu_{(s)} \tag{4.49}$$

$$v = \Gamma_{(s)} - \Psi_{(s)} \ln\left(\frac{\bar{p}}{p_{\text{atm}}}\right) \tag{4.50}$$

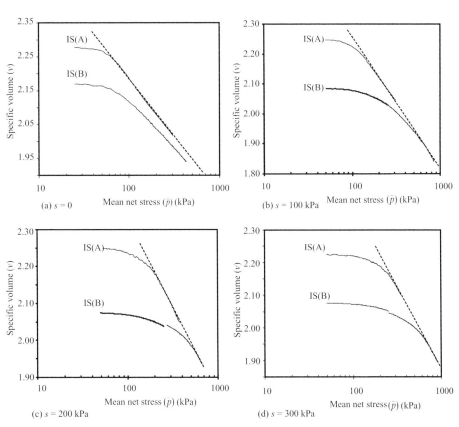

Figure 4.25   Pressure–volume relationship for specimens of kaolin with isotropic stress history. From Sivakumar, *et al.*, in press. Geotechnique [doi: 10.1680/geot.8.P.007]. Reproduced with permission.

$$v_w = A_{(s)} - B_{(s)} \ln\left(\frac{\bar{p}}{p_{atm}}\right)$$ [4.51]

where the parameters $M_{(s)}$, $\mu_{(s)}$, $\Gamma_{(s)}$, $\Psi_{(s)}$, $A_{(s)}$ and $B_{(s)}$ are dependent on suction. The authors suggested that the variation of $M_{(s)}$ with suction is relatively small and can be assumed as constant slope M as for a saturated soil. However, the value of $\mu_{(s)}$ varies with suction in a non- linear fashion.

Wheeler and Sivakumar (2000) examined the effects of initial conditions resulting from different compaction efforts and water contents on the position of the *csl* and concluded that:

- A change in compaction water content produced a marked effect not only on the positions of the normal compression line for different values of suction, but also on the locations of the *csl* in the $\bar{p} : v$ plane and the $\bar{p} : v_w$ plane.
- The *csl* in the $\bar{p} : q$ plane was not affected by the compaction water content.
- Difference in soil fabric caused by a change of compaction water content, such as a change from an aggregated fabric to a more dispersed fabric, continued to influence the volumetric soil behaviour even on shearing to the critical state.

While Wheeler and Sivakumar (2000) concluded that M and $M_{(s)}$ were largely unchanged by the initial conditions and $M_{(s)} = M$ for suctions up to 300 kPa, Sivakumar *et al.* (2010a) reported a small increase in $M_{(s)}$ as the suction increased, with a more significant increase in $M_{(s)}$ as the suction increased beyond a critical value. Futai and Almeida (2005) on the other hand reported increasing values of $M_{(s)}$ with increasing suction from saturated to air-dried conditions for natural, intact gneiss residual soil. Values of $\mu_{(s)}$ also increased with suction but then dropped for very high suctions (air-dried conditions). A further challenging aspect of the findings of Sivakumar *et al.* (2010a) was that the *csl* and *iso-ncl* were not parallel. Adding to the complexity was that the *csl* for $s = 100$ kPa in the $\bar{p} : v$ plane was below the *csl* for other values of suction, including zero as discussed further in Section 8.2.1. In addition, the shearing of the specimens resulted in the uptake of water (i.e. increase in v_w) and there were question marks on whether the specimens reached a true critical state condition.

Maâtouk *et al.* (1995) examined the critical state of unsaturated soils by conducting a series of controlled suction triaxial tests on unsaturated silty material (rock powder) and showed that the shear strength was independent of suction when $\bar{p}$ was larger than about 200 kPa. The *csl* at different values of suction converged towards a point corresponding to a mean net stress of 450 kPa (Figure 4.26). A similar observation was made when the voids ratio obtained in the shear tests at critical state were plotted in $\bar{p} : e$ space. The compression lines, at different values of suction, converged towards the saturated line at a $\bar{p}$ value of about 700 kPa (Figure 4.27) and supported the model proposed by Wheeler and Sivakumar (1995).

4.7.3 Constitutive model of Wang et al. (2002)

The authors examined the critical state of unsaturated soils by conducting suction-controlled triaxial tests on unsaturated Botkin silt. In order to simplify the soil structure and stress history, the authors prepared the soil specimens by consolidating soil slurry in a cylindrical mould. Three critical state equations for unsaturated soils were proposed:

$$q = M\bar{p} + q_{os}$$ [4.52]

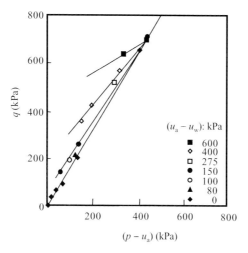

Figure 4.26 q against $\overline{p} = (p - u_a)$ critical state strength envelopes at various suctions (after Maâtouk *et al.*, 1995). Reproduced by permission of the authors and the Institution of Civil Engineers

$$v = v_{os} - \lambda \ln \overline{p} \tag{4.53}$$

$$v_w = v_{wos} - \lambda_w \ln \overline{p} \tag{4.54}$$

where M, λ and λ_w are the slopes of the *csl* in the $q : \overline{p}$, $v : \overline{p}$ and $v_w : \overline{p}$ planes respectively; q_{os} is the intercept of the *csl* with the q-axis; v_{os} is the specific volume of the soil at critical state when $\overline{p} = 1$ kPa and v_{wos} is the specific water volume of the soil at critical state when $\overline{p} = 1$ kPa.

Wang *et al.* (2002) concluded that the parameters M and λ in Equations 4.52 and 4.53 respectively were independent of suction and corresponded to the values for a saturated

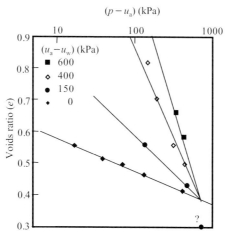

Figure 4.27 Void ratio e against $\overline{p} = (p - u_a)$ at critical state at various suctions (after Maâtouk *et al.*, 1995). Reproduced by permission of the authors and the Institution of Civil Engineers

soil, while the values of q_{os}, v_{os} and v_{wos} were functions of suction. The adoption of a constant M independent of s agrees with the results obtained by Alonso *et al.* (1990) and Wheeler and Sivakumar (1995). However, the value of λ was independent of suction, which differs from the results obtained by Alonso *et al.* (1990), Wheeler and Sivakumar (1995) and Maâtouk *et al.* (1995). The authors concluded that the value of λ was independent of suction because of the simplified soil fabric and stress history.

4.7.4 Hydraulic hysteresis

Non-reversibility of volume change behaviour during wetting and drying cycles within the LC yield locus of Figures 4.20–4.22 is a recognised phenomenon that is not modelled in the BBM. The irreversible behaviour is illustrated in Figure 3.15. The BBM assumes that soil behaviour inside the LC yield locus is elastic with any volume change recoverable if the stress state is reversed. This is not the case in practice. In accordance with the discussion in Section 1.7, there are three mechanisms which dictate the overall swelling behaviour of unsaturated soils with the addition of water: (a) swelling of individual aggregates due to water uptake, (b) aggregate slippage at the inter-aggregate contacts and (c) distortion of aggregates into the inter-aggregate pore spaces (Alonso *et al.*, 1995; Sivakumar *et al.*, 2006b). The relative influence of these factors determines the overall response of clays during wetting. Not all these effects are reversible and subsequent drying results in different magnitudes of micro-mechanical and inter-aggregate changes from those during wetting, resulting in different overall effects. The difference between wetting and drying cycles is known as hydraulic hysteresis.

Wheeler *et al.* (2003) discussed hydraulic hysteresis and collapse phenomena in a coupled elasto-plastic constitutive framework and used a mechanical model to outline the reasons why a soil can have different degrees of saturation at a given value of suction depending on whether the soil is wetting or drying. It was argued that the suction in the *bulk water* within water-filled voids influences both the normal and tangential (shear) forces at particle contacts, while the suction within *meniscus water* at inter-particle or inter-aggregate contacts influences only the normal forces. Hysteresis and collapse were described as influenced by the components of contact forces generated by suction via two elasto-plastic physical phenomena:

- The first was the mechanical straining of the soil. This comprised elastic deformation of soil particles and aggregates under the controlling stress regime, along with plastic straining as a consequence of slippage at inter-particle and inter-aggregate contacts. Slippage was described as controlled by the tangential and normal components of forces at points of contact and inhibited by the presence of meniscus water which provides a stabilising effect through an additional component of normal force at contacts without adding to the tangential force. Slippage at particle contacts can be expected to produce both plastic volumetric and plastic shear strains and not to follow a reversible path during wetting and drying cycles.
- The second elasto-plastic phenomenon considered by Wheeler *et al.* (2003) to influence hysteresis and collapse was the hydraulic process of water inflow and outflow to the void spaces. As discussed in Section 1.4.3, surface tension forces resist the filling of air-filled void spaces during wetting and resist the emptying of water-filled void spaces during drying. This contributes to the complex hysteretic effect in a soil where there are interconnected void spaces of variable dimensions.

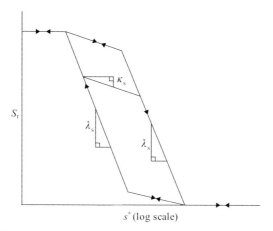

Figure 4.28 Model for water retention behaviour (after Wheeler *et al.*, 2003). Reproduced by permission of the authors and the Institution of Civil Engineers

Within this framework, the authors examined the transition between saturated and unsaturated conditions, irreversible compression during the drying stages of wetting and drying cycles, and the influence of a wetting and drying cycle on subsequent behaviour during isotropic loading. The model for the hysteretic wetting and drying behaviour of an unsaturated soil is shown in Figure 4.28. In the figure the modified suction component s^* comes from considerations of the work input analysis of Houlsby (1997) and is given by the following equation:

$$s^* = ns \tag{4.55}$$

where n is porosity.

The significance of hysteresis and collapse and the link with the thermodynamic potential are discussed in Section 7.10.

4.7.5 Yielding

There is a large volume of experimental evidence to support the concept of a load collapse yield locus LC, and to a lesser extent suction increase and suction decrease yield loci SI and SD respectively. However, the forms of the loci and their interaction are subject to much research. Support for the concept of an LC yield locus has been provided by Wheeler and Sivakumar (1995), Sivakumar and Ng (1998), Tang and Graham (2002) and Blatz and Graham (2003). Zakaria *et al.* (1995) carried out a series of suction-controlled triaxial tests on compacted kaolin using the axis translation technique that suggested that the LC yield curve was qualitatively consistent with the model of Alonso *et al.* (1990). Figures 4.20–4.22 illustrate the yield surfaces within the BBM. The size of the constant-suction cross sections of the yield surfaces increases with increasing suction. The yield loci are reasonably represented by ellipses that are symmetrical about the $\bar{p}$-axis. The symmetry was attributed to the isotropic stress history of the specimens. Sivakumar and Wheeler (2000) provided experimental evidence and discussed the expansion of the LC yield locus for speswhite kaolin arising from increasing compactive effort. Wheeler *et al.* (2003) used the modified suction component s^* of Equation 4.48 and the mean Bishop's stress with

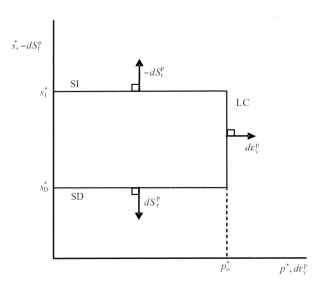

Figure 4.29 LC, SI and SD yield curves for isotropic stress states (after Wheeler *et al.*, 2003). Reproduced by permission of the authors and the Institution of Civil Engineers

$\chi = S_r$ to examine yielding and developed simple shapes for the LC and SI yield curves as well as an SD yield curve, along with associated flow rules for all three yield curves for isotropic stress states. The curves are illustrated in Figure 4.29 for isotropic stress states and in Figure 4.30 for anisotropic stress states. Importantly, Sivakumar *et al.* (2010a) have questioned the shape of the LC locus proposed by Alonso *et al.* (1990) and have provided

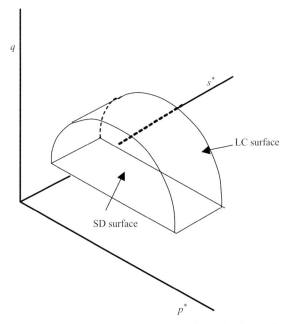

Figure 4.30 Yield surfaces for anisotropic stress states (after Wheeler *et al.*, 2003). Reproduced by permission of the authors and the Institution of Civil Engineers

experimental data on kaolin to indicate that the LC yield locus for truly isotropically prepared specimens is a straight line in the $\bar{p} : s$ plane. The curved nature of the LC locus reported by Alonso *et al.* (1990) was attributed to the initial anisotropic, one-dimensional, preparation of the specimens.

As for saturated anisotropic soils discussed in Section 4.5.3, there is evidence of rotation of the LC yield surface in the $\bar{p} : q$ plane for unsaturated anisotropic soils. This is supported by the evidence from tests on compacted silt carried out by Cui and Delage (1996), which showed that after one-dimensional compaction, constant suction yield curves were inclined in the $\bar{p} : q$ plane so that the major axis of each curve coincided approximately with the K_0 line. The significance of the rotation of the yield locus for anisotropically prepared soil specimens was discussed by Wheeler and Sivakumar (2000). For an unsaturated soil specimen initially one-dimensionally compressed, unloaded and then isotropically compressed to point A as in Figure 4.31, the yield curve is given by Y1. On increasing the isotropic loading, yield will occur at point B in accordance with the rotated yield surface. Figure 4.31(b) shows that the specific volume at point B lies considerably below that of the true *iso-ncl* at the same suction as well as that of the corresponding *1d-ncl*. On loading from point B to point C, the yield curve expands and rotates to that of Y2 as the influence of the isotropic loading starts to overshadow the anisotropic stress history. In the $v - \bar{p}$ plane of Figure 4.31(b), point C lies below the *iso-ncl* but subsequent compression under increasing $\bar{p}$ means that the compression curve gradually converges on the *iso-ncl* as the yield curve gradually rotates in a clockwise manner. The rotation of the yield surface is important in the interpretation of experimental data on soils.

Further complexity is apparent in the form of the SI yield locus for which limited experimental evidence exists. The original proposal by Alonso *et al.* (1990) for an isotropic soil under isotropic loading conditions was of a straight-line yield locus parallel to the $\bar{p}$-axis. This conflicts with the principles of thermodynamics at the point where the LC and SI yield loci meet, and the hypothesis that the loci are continuous. On the basis that increasing net pressure and increasing suctions both cause structural changes and produce plastic hardening, Delage and Graham (1995) argued that the LC and SI yield loci should be coupled. Figure 4.32 shows the possible linking of the LC and SI yield surfaces as a smooth curve line in $\bar{p} : s$ space. Tang and Graham (2002) and Blatz and Graham (2003) examined the possible coupling by conducting triaxial tests on a sand–bentonite mixture but the results were inconclusive. While Sivakumar *et al.* (2002) and Sivakumar *et al.* (2010a) have provided some evidence in support of the continuity of the LC and SI yield loci, further experimental evidence is required.

Interestingly, Gens and Alonso (1992), Alonso (1998), Alonso *et al.* (1999) and Sánchez *et al.* (2010) have proposed refinements to the BBM that allow a degree of coupling between the micro- and macro-structure in unsaturated soils. This provides a simple mechanism for visualising mechanical behaviour up to and including yielding. Micro- and macro-mechanical fabric levels have been linked in a simple manner with the complex behaviour characteristics of suction-controlled oedometer tests on compacted bentonite by Lloret *et al.* (2003). The aggregate structure of London clay subject to wetting and loading in conventional and osmotic shear box tests has also been discussed by Monroy *et al.* (2010) in terms of a double-structure elasto-plastic model. The interplay of different levels of structure and straining is an important feature in the assessment of the strain mechanisms in unsaturated kaolin examined in Chapter 9. There is evidence to suggest that the onset of yielding based on overall specimen deformation measurements is likely to mask irreversible, plastic straining at lower stress levels at a micro-mechanical level.

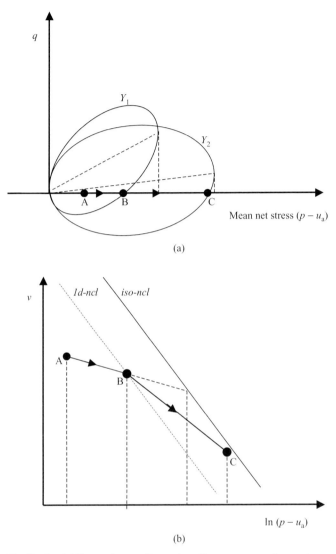

Figure 4.31 Idealised yielding of one-dimensionally compressed specimen subsequently isotropically compressed: (a) rotation of yield locus and (b) compression characteristics (modified from Wheeler and Sivakumar, 2000).

4.8 Concluding remarks

The chapter provides a review of historic and current concepts surrounding our understanding of unsaturated soil behaviour. The significance of stresses and strains is initially discussed along with the widely utilised concept of constitutive modelling, which generally employs the stress state variables of mean net stress $\overline{p}$ and suction s along with the deviator stress q to describe the controlling stress regime. The background to critical state concepts in saturated soils is extended to the critical state concepts within the BBM for unsaturated

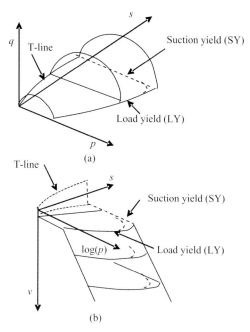

Figure 4.32   Conceptual elastic-plastic framework for unsaturated soils: (a) three-dimensional stress state space and (b) three-dimensional volume state space (modified after Delage and Graham, 1995). Note: the authors use SY to signify the SI yield locus and LY to signify the LC yield locus. Reproduced by permission of Taylor and Francis Group

soils. This includes the concepts behind the strength, volumetric change characteristics and yielding, along with a brief introduction to the concepts of a plastic potential and an associated and non-associated plastic strain-increment. Advances in current thinking including the introduction of hydraulic hysteresis and rotated yield loci are discussed.

The story is far from complete and our understanding of unsaturated soils is not yet on a firm theoretical footing. Advancements in constitutive models have undoubtedly paved the way to a better understanding of unsaturated soils, but intriguing questions in relation to anomalies in observed behaviour still need answering. Notably, further detailed experimental evidence is required on the slopes and continuity of the *iso-ncl*, *1d-ncl* and *csl* and how they are affected by suction for different materials. The influence of initial state on the strength and behaviour characteristics also needs further experimental investigation as does the significance of macro- and micro-levels of straining and yielding. The validity and improvement in modelling of cyclic wetting and drying must also be addressed. What becomes obvious is that while there is a degree of agreement amongst researchers in terms of research tactics, the models are incomplete and there is no firm consensus, often on even the basic concepts, as the inherent variability of materials and their multifaceted behaviour are not readily confined to simple behaviour patterns.

Constitutive frameworks for unsaturated soils must adhere to the general principles of thermodynamics. The development of critical-state-based models, which forms a major line of current research, relies on elasto-plasticity principles that are intrinsically linked to thermodynamics. In Chapters 5 and 6, thermodynamic concepts relating to soil behaviour, and in particular soil behaviour in the triaxial cell, are introduced before the significance of the controlling stresses and conjugate volumetric variables are described in detail in Chapters 7 and 8. These chapters present an alternative view of the strength and deformation characteristics of unsaturated soils, and a model incorporating both stresses and volumetric terms is developed. The analysis is extended in Chapter 9 to examine the macro- and micro-mechanical stress–strain behaviour based on conjugate stress and strain-increments for soil specimens under triaxial stress conditions, and provides insight into the controlling stresses and the meaning of strains.

Notes

1. See Clifton *et al.* (1999) for a compilation of papers by Fredlund which expands this topic as well as providing background reading to a number of topics on unsaturated soils covered in the book.
2. The shear strains are half the magnitude of the definition of shear strain often taught to engineers and of opposite sign: $\varepsilon_{ij} = -\gamma_{ij}/2$, where $i \neq j$.
3. In the triaxial cell tests, the choice of variables entails the assumption of the coincidence of the principal axes of stress, strain and strain-increment.
4. We have used the prefix 'd' rather than 'δ' to indicate small displacements and strains, as in later chapters, when considering thermodynamic concepts, it is necessary to distinguish between changes that are path-dependent and not exact differentials and given the prefix 'δ' and changes that are considered independent of path and exact differentials (only dependent on initial and final conditions) and given the prefix 'd'.

Chapter 5
Thermodynamics of Soil Systems

5.1 Introduction

In general, soils are three-phase materials comprising a mix of solid particles, liquid and gas. The liquid is water and the gas usually deemed to be air. However, the chemistry of not only the soil particles but also the water and the percentages of the different gases making up the air may vary. Nevertheless, the behaviour of soils must be compliant with the general laws of *thermodynamics*. This is true irrespective of the degree of saturation, chemistry of the solid, liquid and gas phases as well as the shape and properties of the soil particles. The description and development of the general principles of thermodynamics vary widely from discipline to discipline, and texts in soil mechanics can be expected to differ significantly from, say, those in chemistry. Nevertheless, soils must adhere to the basic thermodynamic rules describing the specifics of heat transfer, work and the movement of substances in any process undergone by the soil, though the influence of past stress history and the dispersal of energy due to work input needs particularly careful attention, as does the definition of the system's boundaries. In essence we will primarily deal with principles that can be more correctly described under the term 'equilibrium thermostatics', but we will use the all encompassing term thermodynamics since the former terminology is in less general usage.

The variability of soils means they exhibit behavioural trends that present difficulties in formulating rigorous analysis. Thermodynamics, with its global principles to which material behaviour must adhere, provides a basis for understanding and interpreting generalised soil characteristics. In this chapter, the principles of thermodynamics are outlined, particularly as they relate to a multi-phase material, and in subsequent chapters, these principles are used to examine the mechanical behaviour of unsaturated soils. Chapter 6 examines the fundamental considerations and assumptions in laboratory tests on soils in the triaxial cell. Chapter 7 develops an equation describing the controlling stress regime in an unsaturated soil. Chapter 8 develops a theoretical approach to the interpretation of soil strength and volume change. Finally, Chapter 9 uses work input analysis to establish the conjugate stresses and strains under triaxial stress conditions. The principles of thermodynamics are thus used to provide a framework for the interpretation of experimental soil behaviour based on a sound theoretical basis.

5.2 Outline of thermodynamic principles and systems

Thermodynamics treats soil as either a mass with no distinction as to its composition or as an assemblage of solid particles, liquid and gas. In either case, the system is acted on by physical fields, such as gravity, and bounded by a surface of arbitrary shape. Thermodynamics is concerned with energy. Although all matter is 'frozen energy', in the sense that it was created due to expansion and associated cooling of the universe following the 'big bang', the general principles of thermodynamics are taken as generally relating to the macroscopic or large-scale properties of matter and are more difficult to relate to the small-scale or microscopic structure; this is the province of quantum mechanics. From the principles of thermodynamics it is possible to derive general relations between variables such as the imposed pressures (and stresses) and strains and to examine expansion and compression characteristics as well as the influence of heat exchange. The principles of thermodynamics also tell us the properties of a system that need to be determined experimentally. In soils, these include coefficients such as the critical state parameters M, λ and κ. The principles are used in later chapters to define strength and compression parameters for unsaturated soils that need to be determined experimentally.

   A thermodynamic process occurs when there is a change in the value of one or more of the thermodynamic variables used to describe the system. Thermodynamic texts generally deal with *reversible processes*. However, because natural materials often exhibit behaviour that is not strictly reversible, it is usual to consider changes as occurring as a series of infinitesimal steps (Houlsby, 1997; Li, 2007a, 2007b), each of which can be reversed so that the system variables have the same values as they had before the infinitesimal change took place. The assumption of reversible increments of change means the changes can be described by exact differentials that are dependent only on the end conditions and not on the path followed. Thus, integration is not required. A reversible, infinitesimal process must be *quasi-static* and is a process in which the system is in essence at or very close to thermodynamic *equilibrium* or *meta-stable equilibrium*, the meanings and significances of which are discussed in Section 5.3 and explored further in terms of the thermodynamic potentials in Section 5.8. Soils generally exhibit hysteresis when subject to change and the assumption of process reversibility to describe their behaviour is not easily reconciled. However, the assumption of reversibility is deemed reasonable in soil tests with small incremental changes between equilibrium conditions. This allows examination of the significance of equilibrium and hysteresis as well as allowing the most appropriate variables to be defined. The significance of hysteresis and stress history will be discussed in greater detail in Section 7.10.

   An accurate thermodynamic description of the *system* under investigation is important in order to reach meaningful conclusions. In triaxial cell testing, for instance, the system may be described by a soil specimen and the triaxial cell in which the specimen is contained (as analysed in Chapter 6); or it may be described by the soil specimen alone. Both systems are taken as separated from their surroundings but with which they may interact in a specific manner. In order to investigate how such systems interact with their surroundings, it is necessary to consider the systems as surrounded by a thermodynamic *wall*, the properties of which influence the interactions that can take place. Common thermodynamic walls are outlined below, though slightly different definitions are given in different references depending on the discipline being studied:

- A *rigid wall* prevents a system from changing its volume or shape so that no mechanical work is undertaken on or by the system.

- An *adiabatic wall* prevents a system from gaining or losing heat by interaction with the surroundings and is thus thermally isolated.
- An *isolating wall* prevents any interaction between the system and the surroundings. These walls are rigid, perfectly insulated and impermeable to matter.
- An *open wall* permits the free transfer of both matter and thermal energy.

In any thermodynamic analysis, it is important to describe carefully the system and restrictions placed on behaviour by the thermodynamic wall employed so that the system represents as closely as possible the conditions to be investigated. Walls are important because they simplify complex situations and allow analysis without knowing everything about a system. In essence, the walls restrict as many degrees of freedom as possible and allow a study of the effect of changing the few degrees of freedom left unchecked. On the downside, they may not wholly replicate actual conditions.

5.3 Introduction to equilibrium and meta-stable equilibrium

When an arbitrary system is *isolated* and there is balance between applied and resisting forces, all movements should eventually cease. The system is then said to be in a state of mechanical equilibrium. This does not mean that pressures are the same everywhere, for example in a vertical column of liquid, the pressure increases with decreasing elevation due to earth's gravitational field. However, there is a balance of forces. For soil specimens in the triaxial cell, the pedestal on which a specimen sits counters the gravitational effect of the specimen's weight, and the specimen is considered sufficiently small that the change of gravitational field from top to bottom of the specimen can be ignored at the stress levels normally imposed on the specimen. However, in thermodynamics it is necessary to consider more than purely mechanical equilibrium. If there are variations in temperature, this leads to changes. When temperatures equilibrate and heat exchange ceases, thermal equilibrium is established. There is also a need to consider chemical imbalance where substances can react. In general, in a soil, this might include change of chemical composition as well as phase changes such as phase water to water vapour. When sufficient time has elapsed and all possible chemical reactions have taken place, the system can be said to be in chemical equilibrium. In general, other changes to a system can occur, such as those dictated by electrical and magnetic potentials, but these are outside the scope of this book.

We will restrict ourselves to considering thermal, mechanical and chemical equilibrium, and when each of these is in balance, a system is said to be in thermodynamic equilibrium. It is important to note that in a real soil, non-equilibrium of any of the potentials is likely to have a knock-on effect on the others. However, if sufficient time elapses, an equilibrium (or meta-stable equilibrium[1]) is achievable. When examining and discussing equilibrium, this can be taken to include meta-stable equilibrium unless we distinguish between stable equilibrium and meta-stable equilibrium where this is important, as in soils where a meta-stable equilibrium may be established, such as in quick clays, but the soil structure could suffer a collapse to a more stable, lower potential, equilibrium if agitated.

In Chapter 6 we will look at the conditions necessary for 'equilibrium' in a triaxial cell. Such equilibrium may take a significant period of time to become established particularly in unsaturated soils where internal phase pressures and strain interactions can take longer to equilibrate, though overall pressure equilibrium may apparently be satisfied.

A further thermodynamic requirement is that the properties of a system do not depend on its stress history. This is because it is necessary to consider changes from the equilibrium state as reversible. The properties are assumed to only depend on their end condition. This presents particular conceptual problems in the interpretation of thermodynamics in soils, which can exhibit significant hysteresis effects. For this reason, changes are often defined as infinitesimally small, virtual or, as in plasticity theory, as rates of change. Wroth (1973) argued for the use of increments of change as this has more direct relevance in soil mechanics and this approach will be adopted. Essentially, when dealing with changes, we will consider them to be sufficiently small so that any effects arising from non-reversibility, and deviation from the applicability of the exact differentials, can be ignored. This is a common assumption in the application of thermodynamics and not restricted to the analysis of soils. In Chapter 9 we will examine the work input into unsaturated soils in examining experimental data and the analysis will be based on small strain increments between equilibrium states. We will thus be concerned with soil systems in thermodynamic equilibrium (or meta-stable equilibrium) or in which the departure from equilibrium is negligible.

Before moving on to the four principal laws of thermodynamics, we need to define the *variables of state* employed and the meaning of *intensive* and *extensive variables*.

5.4 Variables of state

The fundamental variables of a soil are the smallest set of variables that provide a complete thermodynamic description of the soil, yet are independent and can be varied without change in the value of other fundamental variables. The fundamental variables are chosen on the basis of experience and describe the state of a soil. They are thus often termed *independent variables of state*. The following list of fundamental variables serves for a soil in the absence of external fields such as gravity:

- *Temperature* (T) is an intensive variable measured on the kelvin or absolute zero scale and controls equilibrium with respect to thermal energy change.
- *Entropy* (S) is an extensive variable and a measure of the amount of energy unavailable to do work. It is often deemed a measure of the multiplicity and disorder of a system, though such terminology can be confusing and will be avoided. Entropy is introduced in the second law of thermodynamics as the ratio of heat in a body to the absolute temperature (Q/T). It is measured in units of joule per kelvin (JK^{-1}).
- *Pressure* (p) is an intensive variable and the criterion controlling mechanical equilibrium. It is a measure of the force per unit area and has units of pascal (Pa).
- *Volume* (V) is an extensive variable that describes the spatial extent of the soil influenced by the pressure p. Change in V gives a measure of mechanical energy transfer. It is measured in units of cubic metres (m^3).
- *Chemical potential terms* (μ_i) are intensive variables that are the criteria controlling chemical transfer of matter of a component of the soil from one phase to another (such as water to water vapour) or the transformation into a different chemical compound. These are measured in units of joule per kilogram (Jkg^{-1}).
- *Mass* (M_i) is an extensive variable of mass for the component of the soil influenced by the chemical potential terms μ_i. It is measured in units of kilogram (kg).

5.5 Extensive and intensive variables

The variables (listed in Section 5.4) that depend on the quantity of matter (or size of the system) are termed *extensive* while the variables that do not depend on the quantity of matter are termed *intensive*. A clear distinction is made between the extensive and intensive variables because extensive variables possess the important property that they can be expressed as the sum of the quantities of the separate components that comprise a multi-phase material such as a mass of soil. Accordingly, the extensive variable *total volume* for an unsaturated soil is equal to the sum of the volumes of the soil particles and the water and air in the pore spaces. Similarly, both S and M_i are equal to the sum of the individual components in a soil and are extensive variables.[2] The thermodynamic potentials are also extensive variables and their meanings and significances are discussed in Section 5.7. Intensive quantities are the counterparts of extensive quantities. The intensive variables T, p and μ_i are not 'additive' and are intrinsic to a particular subsystem or component of a soil. The intensive variables are independent of the size of a system and are associated with points in a soil mass.

5.6 The laws of thermodynamics

Thermodynamics is based on a limited number of principles that are generalisations justified and supported by experimental evidence. Thermodynamics is shown to mesh well with experimentation in soils, though some of the ideas are abstract and not readily reconciled with actual soil characteristics. At present there are four recognised laws. The first law defines the internal energy and introduces the variable *enthalpy*. The significance of enthalpy as an extensive variable is used in Chapter 7 in investigating the stress conditions in unsaturated soils. The second law formulates the extensive variable *entropy* and relates to heat exchange within a system, while the third law defines the meaning of the intensive variable temperature. The fourth law is the Zeroth law and stipulates the equalisation of temperature in heat transfer. While other laws have been postulated, these are tentative and not generally agreed on by the scientific community, though intriguingly attempts have been made to develop a law relating thermodynamics to organic matter and evolution. However, this is beyond the scope of this book and we restrict ourselves to outlining the laws of direct interest in soils.

5.6.1 The Zeroth law

This law involves a simple definition of temperature equalisation. In accordance with Maxwell (1872), if two systems A and B are in thermal equilibrium with a third system C, they are in thermal equilibrium with one another. It is the temperature difference that dictates whether there is heat movement or exchange. The Zeroth law requires that for thermal equilibrium within or between systems in thermal contact, the temperatures must equilibrate.

5.6.2 The first law

The first law of thermodynamics stipulates the conservation of energy even though its form is changed. Many different definitions of the law can be found in the literature in attempts to closely define the principles involved, including the following:

- 'Energy can neither be created nor destroyed'.
- 'The total energy of a system and its surroundings must remain constant'.
- 'The energy of an isolated system is constant'.

Consider first an idealised system where work is performed on the system but no heat is generated or exchanged with the surroundings. In such a system, the work done W to the system such as a soil specimen, in taking it from an initial to a final state, is the same for all paths and only depends on the initial and final states. There must therefore be some property of the system or change in energy of the system with an initial value U_1 and final value U_2 such that:

$$W = U_2 - U_1 = \Delta U \qquad [5.1]$$

The terms U_1 and U_2 are called the internal energies of the system at the initial and final states and are dependent only on the end conditions. Suppose now that the system changes between the same initial and final states by an adiabatic process (no heat is exchanged with the surroundings) but heat is generated within the system as a result of friction and viscous effects. The work done W to the system will be different and heat Q will be generated as in compression of air in a bicycle pump. Both W and Q will now be path dependent. Taking account of the heat Q generated within the system:

$$\Delta U = Q + W \qquad [5.2]$$

Thus, the differential change in the internal energy U of a system between the same two states can be written as the sum of the work done on the system plus the heat generated. In essence, the greater the resistance to mechanical change, the greater the heat generated and the less the work done. An alternative view of Equation 5.2 is that the work done is not all transferred to internal energy in the system; some dissipates as heat.

For infinitesimal changes, Equation 5.2 can be written as:

$$dU = \delta Q + \delta W \qquad [5.3]$$

where δQ is the infinitesimal heat generated, δW is the infinitesimal work done on the system and dU is the infinitesimal change in internal energy.

Neither δQn or δW are, in general, exact differentials as they are path dependent in natural processes. Small changes in these variables are represented by δ. However, dU, or the sum $\delta Q + \delta W$, is independent of path and the change is indicated by d.

The internal energy is that part of the 'total energy' which depends on the internal state of a system. However, depending on the system being analysed, terms for elevation (or gravitational) potential and kinetic energy may also be included in a more all encompassing energy formulation. The influence of an 'external' gravitational potential is discussed in Section 5.10 and is also discussed in analysis in Chapter 6. However, for the soil systems considered in later chapters, it is unnecessary to include terms for kinetic energy. We will be concerned with the state of a soil under equilibrium conditions, or incremental changes between equilibrium conditions, and not with transient conditions involving phase movements and flows as during deformation of soils.

However, we will generalise the analysis by including a term for the influence of chemical imbalances as in Equation 5.4:

$$dU = \delta Q + \delta W + \sum \mu_i dM_i \qquad [5.4]$$

where M_i is a set of component masses and μ_i are coefficients usually termed the chemical potentials (not to be confused with the overall chemical potential $\sum \mu_i dM_i$).

The terms $\sum \mu_i dM_i$ relate to an increase of internal energy as a result of addition of matter to a system or change of matter within a system due to a chemical imbalance. Note that the term dM_i is not a path function and only relies on the initial and final conditions. This is also true of changes in the extensive variables volume V and entropy S.

It is appropriate at this juncture to define the thermodynamic meaning of work. In general terms, work can be thought of as the dispersion of mechanical energy in overcoming an inherent resistance to change. Traditionally, thermodynamics deals with pressures and the work done for a small increment of assumed reversible volume change is given by $\delta W = -pdV$, where p is the applied pressure and V is the volume through which the pressure acts ($dV < 0$ for compression and thus $\delta W > 0$). The 'driving force' behind the change is the pressure p, which both causes and controls the direction of change or displacement. However, it is important to realise that thermodynamics relates to not only pressures or isotropic stress conditions but also to the anisotropic stress conditions more relevant to soils.

5.6.3 The second law

The second law of thermodynamics introduces a further variable called *entropy*, which traditionally is taken as relating the heat absorbed by a body to the temperature. Entropy and the second law are two of the most confusing concepts in science because they are dependent not only on the interpretation of individuals but on the system being studied. Accordingly, many different definitions of the second law can be found in the literature. In essence, heat, sometimes described as energy in transit, can produce different forms of work in systems ranging from machines to plant growth, from soil mechanics to the stars. In the soil systems with which we are concerned, it is the work that tends to produce heat due to frictional effects. The relationship between heat and temperature which defines entropy is given by:

$$\delta Q \leq TdS \tag{5.5}$$

where T is the temperature and dS is the infinitesimal change in *entropy*.

It is assumed that T is constant, which may not be the case in a real system. However, in many cases, tests are carried out at a slow rate and allowance is made for the temperature to achieve equilibrium and at this point it is assumed that it is no longer necessary to worry about temperature change, while the influence of other variables is examined. However, in practice it must be ensured that the system is connected to a large reservoir at constant temperature. This is usually assumed to be the case in soil tests.

From Equation 5.5 we may say that the heat generated in a system is at most the product of the temperature and the increase in entropy, but may be less. The condition $dS = \delta Q = 0$ implies that no heat is generated and all energy is turned into work. For the condition $TdS = \delta Q$, heat is generated but none is lost from the system and the process is adiabatic. For $TdS > \delta Q$, heat is lost from the system. The inequality is consistent with natural processes where entropy is usually lost to the surroundings.

Substituting for δQ from Equation 5.5 into Equation 5.4 gives:

$$dU \leq TdS + \delta W + \sum \mu_i dM_i \tag{5.6}$$

When the equality holds, the transfer of work δW to internal energy dU is at maximum for the system. When the equality does not apply, entropy is lost from the system; this loss is irreversible and less energy is available to do work. This is also consistent with the concept of energy tending to disperse from areas of concentration and become diffused. Thus, in accordance with Equation 5.5, entropy can be depicted as a measure of how much energy is spread out at constant temperature as a result of a process. The definition of entropy has become fuzzier with time and it is now frequently referred to as a measure of the energy in a process that is unavailable to do work, and may be related to other energy losses such as light and sound. Perfect efficiency is impossible and entropy always increases in the universe and in any hypothetical isolated system within it. While entropy is reasonably well-defined in terms of heat and temperature, there is difficulty in pinning it down to a firm quantifiable definition in terms of general energy dissipation, particularly for complex systems.

It is necessary to define reversible and irreversible processes in providing further discussion on entropy and its significance to soil testing and analysis. However, those not wishing to over-complicate the issues can treat entropy as a 'black box' that allows reversible heat exchange to be expressed as an exact differential.

A process is said to be reversible if, after completion of the process, the initial states of all systems taking part in the process can be restored (by whatever means possible) without any outstanding changes in the states of other systems. The key words are 'without outstanding changes in the states of other systems'. While it may be possible to restore a soil system to its original particle configuration through some complex stress–strain path, in the process heat is developed as a result of frictional and viscous effects and there will be heat exchange with the surroundings that is not reversible. This is because while you can move heat from a hotter place to a colder place without doing work, you need to work to move heat from a colder place to a hotter place. Only if there is no heat exchange is the process reversible. This is consistent with the statement by Kelvin (1852) that 'no process is possible whose sole result is the complete conversion of heat into work'. Thus in any real system, entropy must increase. This is why perpetual motion machines are impossible and unless energy is imported to a system, increasing entropy will ensure that the system eventually grinds to a halt.

The equality in Equation 5.6 holds only for reversible processes where entropy remains constant. However, all natural physical and chemical processes proceed in the direction of increasing entropy and irreversibility, with the entropy of the universe (the system plus its surroundings) increasing to the maximum possible at which point there is equilibrium. This is where the confusing description of entropy leading to disorder arises as thermodynamics deals principally with systems that are at or move towards equilibrium.

In general terms, soil specimens tested in the triaxial cell experience irreversible thermodynamic changes in that some heat is generated and exchanged with the surroundings. In thermodynamic analysis, this is overcome by considering infinitesimal processes in which the system can be considered to closely conform to reversible ideals. In accordance with recognised precepts in thermodynamic systems, reversible processes are ones which can be considered as a series of infinitesimal changes each of which can be reversed so that the system variables have the same values as they had before the infinitesimal change took place. In Chapter 6, we will look at infinitesimal changes which decrease back to zero and which are deemed reversible. These essentially comprise virtual changes that are useful in analysing the conditions required under thermodynamic equilibrium. A reversible process must be quasi-static for which the system must essentially be in thermodynamic equilibrium. However, in Section 7.9, we examine hysteresis in soils.

This irreversibility is not just a symptom of heat generation but principally of material changes.

5.6.4 The third law

This law leads to the definition of temperature as the macro result of the kinetic energy of the molecules. The law requires that the entropy of a perfect crystal of an element is zero at the absolute zero of temperature. At the absolute zero of temperature, there is zero thermal energy or heat. Since heat is a measure of average molecular motion, zero thermal energy means that the average atom does not move at all. Since no atom can have less than zero motion, the motion of every individual atom must be zero when the average molecular motion is zero. When none of the atoms which make up a perfectly ordered crystal move at all, there can be no disorder or different states possible for the crystal.

5.7 Thermodynamic potentials

Section 5.6 defines the four laws of thermodynamics (0 to 3) and provides a precursor to understanding the significance and meaning of the thermodynamic potentials. From Equation 5.6, for a reversible process, noting that $\delta W = -p\,dV$:

$$dU = T dS - p dV + \sum \mu_i d M_i \qquad [5.7]$$

The products of the conjugate pairs on the right-hand side in Equation 5.7 give the total energy transfer dU for the condition where S, V and M_i are the independent variables of state. The temperature T, pressure p and chemical potential terms μ_i can be thought of as intensive 'forces', which drive changes in the extensive variables of entropy S, volume V and chemical mass M_i respectively. dU is an increment in the internal energy of a thermodynamic system and is the sum of the products of the intensive forces and the generalised extensive 'displacements'.

Equation 5.7 leads directly to an equation for the internal energy potential under equilibrium conditions:

$$U = TS - pV + \sum \mu_i M_i \qquad [5.8]$$

In Equation 5.8, the terms U, TS, pV and $\sum \mu_i M_i$ represent the internal energy, heat, mechanical potential and chemical potential respectively. The internal energy change dU in Equation 5.7 is the differential form of the internal energy thermodynamic potential of Equation 5.8.

The question arises as to whether it is possible to adopt the intensive variables rather than the extensive variables as independent variables of state and thus to determine different forms of thermodynamic potential. The answer is yes. However, it is important to note at this point that only one variable of each of the conjugate pairs can be chosen as an independent variable of state during an increment of change of the thermodynamic potential because one variable in the pair can be derived from the other by differentiation of the thermodynamic potential. Thus, from Equation 5.8, the intensive variables, T, p and μ_i can be derived as the partial derivatives of the internal energy with respect to the conjugate extensive variable, as in Equation 5.9a:

$$T = \partial U/\partial S \qquad p = \partial U/\partial V \qquad \mu_i = \partial U/\partial M_i \qquad [5.9a]$$

Or, the extensive variables can be treated as independent variables and the intensive variables determined as in Equation 5.9b:

$$S = \partial U/\partial T \qquad V = \partial U/\partial p \qquad M_i = \partial U/\partial \mu_i \qquad [5.9b]$$

Or, a combination of intensive and extensive variables can be considered as the independent variables of state. Accordingly, a soil can be described by T, V and a set of M_i but not T, S, V and a set of M_i. The temperature–entropy pairing provides a practical example of the importance of being able to redefine the independent variables. No laboratory instrument exists for measuring entropy, whereas devices for temperature measurement abound, allowing temperature variations to be monitored.

The independent variables of state, in particular the pV pairing, are important in soil testing. A triaxial test can be carried out by increasing the cell pressure and keeping it constant while examining the change in volume of the soil specimen as a result of the imposed pressure. This would correspond to treating the extensive variable V as the independent variable of state. However, in tests on soils, the situation is complex as there are three phases. Thus, for instance, while the total volume could change, the test could be carried out with the volume of water, as well as the volume of soil particles, maintained constant. The volume change of the soil specimen would in this case correspond to the volume change of the air. It is worth outlining the significance of the independent variables of state to some of the triaxial shearing tests assessed in later chapters:

Constant suction, fully drained test	Under controlled applied loading, the volume change is recorded; V is the independent variable of state.
Constant suction, constant pressure test	Under controlled applied loading, the axial stress and cell pressure are adjusted to maintain the mean applied stress as constant, while the volume changes are monitored; V is the independent variable of state.
Constant suction, constant volume test	The volume of the soil specimen is maintained constant by adjusting the imposed stresses; p is the independent variable of state. (In Chapter 6 it is shown that p can be taken as the mean stress as well as an isotropic pressure.)
Constant water mass tests	Under controlled applied loading, the volume change is recorded but the water mass is maintained constant; V is the independent variable of state.

The case of a saturated soil shearing test under undrained conditions is also worth mentioning. While V is the independent variable of state, because the soil particles and water are both considered incompressible, there is usually deemed to be no volume change. This is a special case.

Nature does not necessarily adhere to the restriction that only one variable of the conjugate pairs is independent at any time, with the other variable constant. In real systems, both variables in a conjugate pairing could change at the same time and true equilibrium may not be established. For instance, in the pV pairing both p and V could experience continual changes. Consider the case of an excavated clay slope. During excavation, dramatic changes to the ground stress regime will occur and the groundwater table may be drawn down. An element of soil in the slope may experience pressure and volume changes over a considerable period of time as it moves towards an equilibrium condition. However, further influences on the slope, such as those arising from the prevailing weather, have a bearing on the lack of a true equilibrium for the soil element. Successive infinitesimal

Table 5.1   The thermodynamic potentials.

Potential name	Independent variables	Potential equation	Differential potential equation
Internal energy	S, V, M_i	$U = TS - pV + \sum \mu_i M_i$	$dU = TdS - pdV + \sum \mu_i dM_i$
Enthalpy	S, p, M_i	$H = TS + \sum \mu_i M_i$	$dH = TdS + Vdp + \sum \mu_i dM_i$
Helmholtz	T, V, M_i	$A = -pV + \sum \mu_i M_i$	$dA = -SdT - pdV + \sum \mu_i dM_i$
Gibbs	T, p, M_i	$G = \sum \mu_i M_i$	$dG = -SdT + Vdp + \sum \mu_i dM_i$

changes can be used in attempting to model such conditions. For example, after allowing p to change, while maintaining V constant, V could be maintained constant while p varies. In this case the terms pdV and Vdp would be necessary to define the overall change.

Partial Legendre transformations performed on Equation 5.8 create a set of thermodynamic potentials[3]. In addition to the internal energy, the most commonly employed potentials are presented as Equation 5.10a–c:

$$\text{Enthalpy potential}\quad H = U + pV = TS + \sum \mu_i M_i \qquad [5.10a]$$

$$\text{Helmholtz potential}\quad A = U - TS = -pV + \sum \mu_i M_i \qquad [5.10b]$$

$$\text{Gibbs potential}\quad G = U - TS + pV = \sum \mu_i M_i \qquad [5.10c]$$

The incremental form of the enthalpy potential can be derived from Equation 5.10a by differentiation and substitution for dU from Equation 5.7 as follows:

$$dH = dU + pdV + Vdp = TdS - pdV + \sum \mu_i M_i + pdV + Vdp = TdS + Vdp + \sum \mu_i M_i$$

Similar manipulations are used to derive the differential forms of the Helmholtz and Gibbs potentials. Table 5.1 summarises the potentials, their differential forms and the independent variables of state.

The heat, mechanical and chemical potentials as well as the thermodynamic potentials U, H, A and G are variables that have the dimensions of energy. They are called 'potentials' because in a sense, they describe the amount of 'potential energy' in a thermodynamic system when it is subjected to certain constraints. It is possible to define other thermodynamic potentials that can be employed in describing a process. The appropriate potential depends on the test set-up and the independent variables of state[4].

It is important to recognise that thermodynamic potentials such as H, A and G are merely convenient, special cases of the general thermodynamic potential equation for U. Because of the wide applicability of thermodynamic potentials to laboratory and natural processes in different disciplines, it is not easy to generalise without appearing obscure and each problem studied requires careful individual consideration.

While the very design of an experiment dictates the appropriate form of thermodynamic potential, the potential may control different processes within the design set-up. Consider a triaxial cell. It is possible to perform a number of different tests on soil specimens within the triaxial cell from undrained tests on saturated specimens through to fully drained tests on unsaturated specimens. The enthalpy H is a form of the thermodynamic potential that can be used to describe soil equilibrium conditions in such tests. The enthalpy is minimised at equilibrium in a pressure-controlled system, such as the triaxial cell (Callen,

1965), where tests generally follow the procedure of applying an increment of loading, then holding the loading constant while the specimen adjusts to an equilibrium state, before a further increment of loading is applied[5]. The significance of the thermodynamic potentials in examining different phenomena in different triaxial test set-ups is outlined briefly below:

Internal energy	Since $dU = \delta Q + \delta W + \sum \mu_i dM_i$, for an isothermal process with $\delta Q = 0$, where there is no chemical imbalance ($\sum \mu_i M_i = 0$), dU represents the increase in internal energy as a result of the work done on the system. That is, under similar constraints it also equals the decrease in internal energy resulting from the work done by the system. Thus, for a system at constant entropy (and with no chemical imbalance), U acts as the 'potential for work'.
Enthalpy potential	In a natural system where both V and p may vary, the incremental form of Equation 5.10a is $dH = dU + pdV + Vdp$. The change in enthalpy dH can thus be described as the change in internal energy dU plus the work done pdV on the system at constant pressure. The enthalpy H represents the potential for work for a system at constant pressure.
Helmholtz potential	From Equation 5.10b, A is constant in a constant temperature, constant volume process where there is no chemical imbalance. Helmholtz potential acts as a potential for work for a system at constant temperature T (and with no chemical imbalance). It is the natural potential to describe a closed system since the variables of state are T, V and M_i. It is particularly useful in statistical mechanics.
Gibbs potential	From Equation 5.10c, G is the Gibbs chemical potential and dG is equal to the Gibbs chemical potential change for a reversible, constant temperature, constant pressure process. Phase transitions take place under these conditions (we use the full term 'Gibbs chemical potential' to distinguish it from the chemical potential terms μ_i).

The internal energy is used in Chapter 6 to examine virtual changes in variables while examining different forms of triaxial stress test, and the enthalpy and its minimisation at equilibrium is used in Chapter 7 to analyse the stress conditions in the triaxial cell under equilibrium conditions.

5.8 Thermodynamic potentials in practice

The principles of thermodynamics can be used to examine many aspects of material behaviour. This includes equilibrium and meta-stable equilibrium. Numerous examples of meta-stable equilibrium can be found in chemistry and physics, and under such conditions a material or system is not at its lowest possible thermodynamic potential and thus not at its lowest equilibrium state. Under meta-stable conditions a system is likely to change to a lower potential state following some perturbation and given sufficient time. The conditions necessary for a meta-stable state are that (i) there must be a lower potential state and (ii) mechanisms through which this meta-stable state may relax to a more stable state must be inhibited. Changes of phase between solid, liquid and gas provide the most frequent examples of meta-stable states, but the thermodynamic potential and its minimisation is not just dictated by thermal and chemical changes. The following discusses the thermodynamic potentials, equilibrium and meta-stable equilibrium and their significance in interpreting the behaviour of unsaturated soils. The discussion brings together examples of thermal, mechanical and chemical meta-stable conditions to illustrate the problem in terms of thermodynamics and allows the significance of the following to be highlighted (Murray *et al.*, 2008):

- The minimisation of the potential at equilibrium or meta-stable equilibrium.
- 'Staged'. Changes between 'equilibrium' states.
- The likelihood of abrupt energy changes at the extremes of system changes.
- The significance of a disturbing force to change from meta-stable equilibrium to a more stable equilibrium.

Later in the book the arguments are extended to describe collapse and hysteresis mechanisms in unsaturated soils, and the influence of the transition between unsaturated and saturated conditions on the critical state strength.

5.8.1 Equilibrium and meta-stable equilibrium

There are physical and chemical phenomena where analogous behaviour can be used to explain the underlying concepts. A simple mechanical analogy of a potential and its minimisation is illustrated in Figure 5.1 where a ball is shown rolling down a valley side. The ball at A has a tendency to reduce its potential energy to a minimum value, which within the defined system is at D in the valley bottom. The potential energy for a ball of given mass is dictated by its height above the valley floor. The significance of the minimisation of the thermodynamic potential for a soil is that equilibrium conditions are established and the potential of the soil system comprising the particles, water and air achieves a minimum under this equilibrium state. However, achieving an absolute minimum potential is 'easier said than done'. Conditions can arise where a meta-stable equilibrium is established where an absolute minimum potential is not achieved as can be explained by again examining Figure 5.1.

As the ball rolls down the valley side, it is possible that a ledge (position B) prevents the ball from rolling down to the valley bottom, without some other factor agitating the ball in order for it to pass over the lip of the ledge at point C, to the more stable state at D. Point B represents a position of meta-stable equilibrium as a lower equilibrium potential

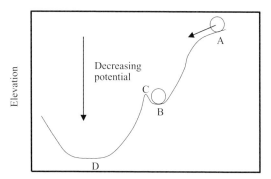

Figure 5.1   Analogy of ball rolling down valley side.

exists within the system. The lip at C represents an 'energy barrier' preventing the ball from achieving a minimum potential state.

The phase changes of water present readily appreciated meta-stable conditions. Consider the case where water is contained in a beaker and heated in a microwave. The water may not change to vapour at its boiling point of 100°C if the water is 'clean' because nucleation of the vapour phase is inhibited. When water exists at ambient pressure but at a temperature above 100°C, it is said to be in a superheated state. If this meta-stable water is agitated, possibly by the introduction of a solid substance, such as coffee, nucleation to the vapour phase occurs and the liquid suddenly boils as the superheated liquid changes to a vapour phase. Accidental superheating of a liquid is best avoided for safety reasons (Erné, 2000).

At the other extreme of freezing it is interesting to draw on the contents of a letter written by Joseph Black (professor of chemistry in Edinburgh) in 1775 to Sir John Pringle. In the letter, the writer describes some crude experiments on the freezing of water: (i) water initially boiled then cooled to room temperature and (ii) water thawed from snow to room temperature. Separate cups of boiled and unboiled water were placed outside under freezing conditions. The boiled water froze readily and the unboiled water remained fluid. However, on agitation with a toothpick, the unboiled water also froze. Drawing on earlier work by Fahrenheit, who found that boiled water placed in glass globes purged of air did not freeze at temperatures some degrees below the normal freezing point, he described how Fahrenheit found that the supercooled water suddenly froze on agitation or exposure to air. Black (1775) argued that since one effect of boiling water was to expel the air, which it naturally contained, then as soon as the water cooled it began to absorb air again over a period of time; and the air entering the boiled water provided sufficient agitation to the water to facilitate passing over the energy barrier and freezing.

The examples of meta-stable superheated and supercooled water illustrate that the creation of a new phase involves an interface, which in many situations costs energy and gives rise to an energy barrier to the formation of the new phase, analogous to the energy barrier at point C in Figure 5.1. The examples also illustrate that substances under meta-stable conditions can experience dramatic change if a mechanism for change exists. The phase changes of water are driven by heat and chemical potentials. But the thermodynamic potential of Equation 5.8 also includes a mechanical potential term (pV) and there is no reason to suppose that similar abrupt energy changes will not occur at the extremes of mechanical change in soils. Such behaviour is considered true of unsaturated soils at the extremes of near-saturation and very dry conditions.

Some caution must however be exercised as the occurrence of meta-stable conditions can represent various degrees of stability. In fact, while stable conditions may be perceived to exist, gradual changes could be taking place. Consider the nucleation of gas bubbles in water and their attachment to the sides of a container. The bubbles may attain meta-stable equilibrium and only with some other influencing factor is a lower potential achievable, for example agitation of the bubbles allows them to rise through the liquid. However, the growth and decay of gas bubbles (Keller, 1964), particularly within the void spaces of soils, is complex (Murray, 2002) and the rate of expansion and decay of the volume of free air is likely to influence the perceived equilibrium conditions. Barden and Sides (1967) concluded that in unsaturated soils there is evidence that equilibrium in terms of Henry's law may require a considerable time interval, far greater than in the absence of soil particles.

It is also interesting to note that phase transitions as well as other changes from meta-stable conditions often represent chaotic changes that are highly non-linear in their mathematics. The smooth and predictable behaviour of matter tends to break down and is of little help in understanding the transition.

5.8.2 Minimisation of the potential

The *enthalpy minimum principle*, described in more detail in Section 6.2, means that the mechanical equilibrium of a specimen in the triaxial cell is controlled by minimisation of the enthalpy. The enthalpy is shown to act as the controlling thermodynamic potential. This is true not just for isotropic loading conditions. As shown by Murray and Brown (2006) (see also Chapter 6), under anisotropic stress conditions, p in Equation 5.10a is the mean stress, which complies with the term pV being an extensive quantity.

Consistent with the minimisation of the total enthalpy of a soil system at equilibrium is the minimisation of the individual components of the total enthalpy. This is the same as saying that the stresses and pressures within the soil system will adjust to achieve a minimum energy condition. A more complete statement would be 'the stresses adjust to a minimum under the volumetric restrictions' as it is necessary to allow for meta-stable equilibrium of the soil particle structure. There could be a lower thermodynamic potential and reduced stresses and pressures associated with a redistribution of the particles. The analogy of a ball on a ledge on the valley side can again be drawn on. The ball is in meta-stable equilibrium on the 'ledge' but could achieve a lower potential if it were to roll down the slope to the valley bottom. This might correspond to collapse of a soil structure. At the ledge, the components of the soil system minimise their potential within the confines of the system.

5.8.3 Staged changes

As discussed by Mullen (2001), in the early part of the nineteenth century several researchers made the experimental observation that some aqueous solutions of inorganic salts, when cooled rapidly, first deposited crystals of a less stable form than that which normally crystallises. Ostwald (1897) attempted to generalise this sort of behaviour by propounding a 'rule of stages' which Mullen states as:

> An unstable system does not necessarily change directly into the most stable state, but into one which most closely resembles its own, i.e., into another transient state whose formation from the original is accompanied by the smallest loss of free energy.

This quote succinctly summarises the law of stages and also hints at a physical explanation for it. If the transitions to lower energy states are governed by energy barriers, then the height of these energy barriers will necessarily govern the rate at which the new states form. Hence, the state that forms from a meta-stable state is governed not by thermodynamics but by kinetics (the kinetic rate being defined as the rate of change from one state to another). In many circumstances, the energy barrier will be much smaller, and kinetics faster, for the transition to some intermediate state that more closely resembles the initial state. A good example of this is the formation of ice crystals from cold supersaturated water vapour. In an elegant experiment, Huang and Bartell (1995) cooled water vapour very rapidly in a jet expansion and at 200 K particles formed. At this low temperature, the thermodynamically most stable phase is ice; however, they demonstrated that liquid water droplets initially formed and only at some finite time later did these droplets then relax to form crystalline ice. In fact, when ice did form, it did so in a meta-stable cubic form rather than the more stable hexagonal form and only at some later stage did the cubic ice particles relax to hexagonal ice.

It is worth noting that such phase transformations can take protracted periods of time and meta-stable states can appear to be the most stable state (Murray *et al.*, 2005; Murray and Bertram, 2006).

A similar cascade through a sequence of meta-stable states may occur as soils relax: for instance, in soils that are subject to collapse or soils under post-peak shearing. The behaviour of such soils is influenced by a number of interacting factors including the interparticle stresses, the re-orientation of soil particles (or aggregates of particles) and the movement of water and air within void spaces that are changing anisotropically. The interplay between these factors leads to the formation of meta-stable states, which form more readily than the most stable state. Only with enough time and sufficient disturbance, may a more stable state be obtained; in the case of a collapsing soil this would be at greater density, and in the case of a soil experiencing post-peak shearing, this would be when a lower shear resistance is achieved.

5.9 Conjugate thermodynamic pairings

Besides the individual terms S, V and μ_i being extensive variables along with the conjugate paired terms TS, pV and $\mu_i M_i$, the thermodynamic potentials are also extensive variables and the total potential is the sum of the individual potentials within a system. This is consistent with the first law of thermodynamics which requires that energy cannot be created or destroyed and that the energy of an isolated system is constant. This implies 'packets' of energy, such as TS, pV and $\mu_i M_i$, which in an isolated system may change without overall change in the total energy, and that for a system at equilibrium the packets of energy, which can be broken down into smaller components, are additive.

The terms TS, pV and $\mu_i M_i$ are made up of the products of an intensive and extensive variable. It is useful to envisage the corollary with mathematics where an intensive variable is given a positive sign and an extensive variable a negative sign. The product of an intensive and extensive variable thus results in a negative sign or extensive variable. Similarly, the division of one extensive variable by another such as mass divided by volume to give density yields an intensive variable. The enthalpy given by $H = pV + U$ is the sum of two extensive variables and is itself an extensive variable. Thus, the total enthalpy is the sum of the enthalpies of the various components making up a soil.

The equations for the thermodynamic potentials give conjugate pairings that are measures of energy in the system. The incremental forms of the thermodynamic potentials give conjugate pairs where one of the pair is an increment of change. In this case, the pairings represent measures of the energy changes. In Chapter 9, the analysis is developed to introduce conjugate stress and strain increments under triaxial stress conditions. These represent components of work in a soil system. The analysis allows the anisotropic conditions for both the air voids and the aggregated soil packets comprising the soil particles and water to be examined.

In soils it is necessary to address the significance of the phases and their interactions. In this context it is important to appreciate the significance of *miscible* and *immiscible* interactions. Miscible interactions such as water vapour in air obey Dalton's divisional law of partial pressures. The alternative of immiscible interactions such as the adsorbed water on soil particles must be treated as having their own pressure acting through their own volume. Chapter 6 shows these ideas tie in with the ideas of extensive variables. The chapter also shows how breaking down the conjugate pairings forming the thermodynamic potential into the individual additive components for the phases and interactions within a soil system results in other conjugate pairings. This is merely summing the energy components to give the overall energy and is consistent with the first law of thermodynamics.

5.9.1 The temperature–entropy pairing

The heat term TS in Equation 5.7 applies to a soil where it is treated as an unspecified mass and not as a multi-phase material. It assumes that all phases are at the same temperature in accordance with the Zeroth law. If the soil is treated as a collection of phases and interactions, it follows directly from the fact that S is an extensive variable:

$$TS = T\sum S_i \qquad\qquad [5.11]$$

where S_i is the entropy of a phase or interaction in a soil. The differential form of the thermodynamic potential in Equation 5.8 can be modified by replacing TdS as follows:

$$TdS = T\sum dS_i \qquad\qquad [5.12]$$

In general in a soil system, heat will be generated due to applied pressure changes and soil specimen deformations. Heat exchange with external sources could also occur if a *wall* at the boundary of the system allows such interchange. Both can be accommodated in the formulation.

5.9.2 The pressure–volume pairing

The pressure p acts as the driving force producing mechanical work by change in volume V. In the more general loading case applicable to viscous fluids or plastic and elastic solids (as in soils), the pressure–volume pairing must be generalised to accommodate the stress tensor. Chapter 6 shows that the isotropic pressure p can be replaced with the mean stress in triaxial tests in the thermodynamic potential equation. In Chapter 7, this approach is developed further to examine a soil comprising distinct phases with interactions where it is shown that such a system can be viewed as comprising a dual stress regime. The differential form of the thermodynamic potential allows the work input for the different phases to be determined as well as allowing an analysis of anisotropic behaviour. But we are running ahead of ourselves. It is important to note, however, that thermodynamics deals with conjugate pairings in dealing with both equilibrium and changing conditions. Treating pressures and stresses in soils without due deference to the volumes of the phases does not follow thermodynamic principles.

The formulation does not allow for loss of mass from a system other than due to chemical imbalances as described below. For a soil specimen in a triaxial cell, drained conditions would allow the exchange of air and water with external reservoirs while maintaining the mass of the solid phase. As shown in Section 5.10, this requires some modification to the energy formulations so far discussed.

5.9.3 The chemical potential–particle number pairing

The chemical potential terms μ_i are akin to forces which when imbalance exists push an exchange of particles, either with the surroundings, or between phases inside the system. This is important in assessing the establishment of chemical equilibrium and if osmotic effects are likely to influence the soil system. For example, the pores in an unsaturated soil hold water as a phase liquid and water vapour in air. There is a possible chemical potential for the liquid pushing water molecules into the vapour (evaporation) or a reversed chemical potential for the vapour pushing vapour molecules into the liquid (condensation). Equilibrium is obtained only when these forces equilibrate. Similar arguments can be made for the interaction between phase air and dissolved air in water and the water phase and adsorption into the double layer on the surfaces of the soil particles. The chemical potentials $\sum \mu_i M_i$ can be used to allow for both energy changes as a result of chemical changes internal to a soil specimen and, assuming a suitable *wall* exists at the system boundary, for chemical imbalance, and thus mass exchange, with external sources. It is possible to express the chemical potentials in other ways. For instance, the potential for the water can be expressed as the product of $\mu_w \rho_w$ and M_w/ρ_w, where ρ_w is the density of water. The term $\mu_w \rho_w$ has units of pressure and the term M_w/ρ_w has units of volume. This alternative terminology is often employed when considering osmotic suction and ties in with the units used in describing matric suction.

5.9.4 The internal energy

On a similar basis to the foregoing, since U is an extensive variable, being the sum of other conjugate pairs which are themselves extensive variables, the total U can be equated to the sum of the internal energies of the phases and their interactions within a soil, thus:

$$U = \sum U_i \tag{5.13}$$

where U_i is the internal energy of a phase or interaction within a soil system.

5.10 Influence of a gravitational field

To take account of the influence of a gravitational field requires a modification to the thermodynamic potential equation. The gravitational or elevation potential is given by $\varphi = gz$, where g is the acceleration due to gravity and z is the elevation above a suitable datum such as the earth's surface (the analogy of the potential energy of a ball on a hillside as in Figure 5.1 is again valid). The modified equation for the thermodynamic potential can be written as:

$$U = TS - pV + \sum \mu_i M_i + \sum \varphi_i M_i \tag{5.14}$$

where φ_i is the gravitational potential and M_i is a component mass of the soil.

The differential form of the equation, where S, V and M_i are independent variables of state, is:

$$dU = TdS - pdV + \sum \mu_i dM_i + \sum \varphi_i dM_i \tag{5.15}$$

The gravitational potential comes into force when considering changes in the mass constituents of a soil, as when fluid flow from or to a soil system has occurred. In this case, the solid phase is usually considered constant and unmoving within an element of the soil. However, during consolidation or combined fluid flow and consolidation (Houlsby, 1997; Wheeler *et al.*, 2003), there will also be change in the potential of the solid phase as the soil grains move closer together. When considering infinitesimal changes in drained effective stress tests in the triaxial cell in Chapter 6, we will include terms for the gravitational potential. However, when examining equilibrium conditions we will be concerned primarily with pressure equilibrium and we can omit the terms for both the gravitational potential and the chemical potentials. As discussed in Section 5.3, for soil specimens in the triaxial cell, the pedestal on which a specimen sits counters the gravitational effect of the specimen's weight; and the specimen is considered sufficiently small that the change of gravitational field from top to bottom of the specimen can be ignored at the stress levels normally imposed on the specimen.

5.11 Concluding remarks

The following outlines the main conclusions of this chapter:

- Thermodynamics provides global principles to which soil behaviour must adhere and thus provides a basis for the understanding and interpretation of generalised soil characteristics. However, in applying the principles due consideration must be given to the heterogeneous nature of soils and their highly variable behaviour characteristics, in particular their dependence on past stress history.
- Thermodynamic equilibrium is defined as established when mechanical, thermal and chemical equilibrium has been achieved. Strictly all other forms of potential influencing a system should also be in equilibrium, but considerations of mechanical, thermal and chemical equilibrium suffice in the analysis. The total thermodynamic potential and the individual potentials governing mechanical, thermal and chemical changes must each be a minimum at equilibrium. This is true for a system whether equilibrium, dictated by an absolute minimum potential within the system, or meta-stable equilibrium, dictated by a 'local' minimisation above the lowest possible potential, is achieved. The phenomenon of 'staged' changes between meta-stable equilibrium conditions is described in the text. Under meta-stable equilibrium, a lower absolute minimum for the potential exists but the system may require some stimulus in order for the lower potential to be achieved. It is assumed in subsequent chapters that equilibrium or meta-stable equilibrium is established (or departure from such conditions is negligible), or where changes do occur these are between equilibrium states. Accordingly, the potentials associated with the phases and interactions within multi-phase materials attempt to minimise under equilibrium or meta-stable equilibrium conditions.
- When considering changing conditions, the assumption is made of system reversibility. As soils exhibit distinct deviation from reversibility it is necessary to consider small incremental changes when analysing behaviour. This is an assumption in analysis of most real materials.
- Each of the extensive variables S, V and M_i are additive in that the sum of the value for each component of a soil (the phases and their interactions) equals the total value for the soil. The thermodynamic potentials are also extensive variables. Thus, the conjugate pairings TS, pV and $\mu_i M_i$ are each the sum of the conjugate pair for each

component of a soil and can be summed to give the total thermodynamic potential given by the internal energy U. Importantly, the enthalpy given by $H = pV + U$ acts as a thermodynamic potential and is an extensive variable.

- Chapter 6 presents the minimum principles for the thermodynamic potentials. The chapter also uses the internal energy potential U to analyse virtual, infinitesimal changes in examining the fundamental considerations and assumptions in laboratory tests on soils in the triaxial cell. Chapter 7 uses the concepts associated with extensive variables and the minimisation of the thermodynamic potential at equilibrium to develop an equation describing the controlling stress regime in an unsaturated soil. The equation forms a 'cornerstone' to analysis of experimental data used in later chapters. Chapter 8 develops a theoretical approach, supported by comparison with published experimental data, for the interpretation of soil strength and volume change, which results in a three-dimensional model in dimensionless stress–volume space. Chapter 9, the last chapter, uses thermodynamic principles in a work input analysis that establishes the conjugate stresses and strains under triaxial stress conditions. This allows not only the macro-mechanical behaviour of soils under test to be examined but also allows the micro-mechanical volumetric and anisotropic stress–strain behaviour both of the aggregates and between the aggregates of fine-grained soils to be analysed.

Notes

1. Meta-stable conditions are discussed in Section 5.8. Although mechanical, thermal and chemical balance may be achieved, this is not necessarily at the lowest possible potential and a more stable, lower potential equilibrium condition may be achievable.
2. Some caution must be exercised here as it is necessary to take account of *miscible* and *immiscible* interactions between phases as discussed briefly in Section 5.9 and dealt with in greater detail in Section 7.2.
3. Callen (1965) describes Legendre transformations and the derivation of the enthalpy, Helmholtz and Gibbs potentials.
4. The independent variables deduced from the partial Legendre transformations are sometimes referred to as the natural variables of the thermodynamic potential.
5. Tests may be carried out following other procedures. This includes strain-controlled tests where a slow rate of strain may dictate that the system is never far from equilibrium conditions. The assumption of equilibrium conditions may be made if flow, deformation, drag and viscous effects can be neglected. The establishment of steady states of flow is a further form of test, but one which requires dynamic effects to be taken into account.

Chapter 6
Equilibrium Analysis and Assumptions in Triaxial Testing

6.1 Introduction

A description of triaxial testing procedures, measuring devices and equipment was presented in Chapters 2 and 3. In this chapter the fundamental considerations and assumptions in laboratory tests on soils in the triaxial cell are examined using the principles of thermodynamics and work input. First the minimum principles for the thermodynamic potential at equilibrium are examined and it is demonstrated that the total enthalpy can be used as the thermodynamic potential for a pressure-controlled system such as the triaxial cell. The internal energy and mechanical potentials making up the enthalpy must also minimise within the restrictions of a soil system. From this the stress conditions necessary for establishment of equilibrium in undrained total stress tests and both undrained and drained 'effective' stress tests in the triaxial cell are determined. Isotropic loading conditions are considered first, followed by an extension of the ideas to anisotropic loading. Generally, other than where deemed necessary, we will not make a distinction between equilibrium and meta-stable equilibrium as the principles established are equally applicable.

The first law of thermodynamics necessitates that the work equation, thus the stresses, pressures and associated strains within a soil specimen, should be determinate from the thermodynamic potential. This is undertaken and it is shown that the thermodynamic potential can be written in a similar form for both isotropic and anisotropic loading conditions, but for the latter, the mean stress replaces the isotropic pressure (Murray and Brown, 2006). This is followed by an examination of the implication to the enthalpy of adopting the axis translation technique in triaxial testing. Lastly, a discussion is presented on the significance to soil structure of a soil achieving a minimum thermodynamic potential.

6.2 The minimum principles for the potentials

The thermodynamic potential provides a set of independent variables appropriate for a particular problem. From the basic tenet of a thermodynamic potential arises the minimum principle of the thermodynamic potential at equilibrium. This will not be presented in a rigorous manner but hopefully in a way that is readily understood. Intrinsic to this is the assumption of independent variables, as discussed in Section 5.4, which allow us to examine the controlling potential under different system constraints by examination of

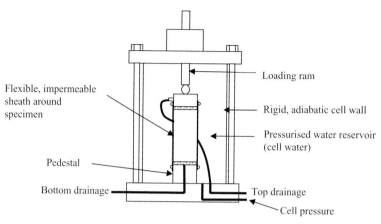

Figure 6.1   Idealised triaxial cell.

the effect of changing the variables in isolation. A comparable approach is adopted in unsaturated soils where the terminology 'independent stress state variables' is in common usage.

In accordance with Table 5.1, under equilibrium conditions the internal energy U can be written in general terms as:

$$U = TS - pV + \sum \mu_i M_i \qquad [6.1]$$

We will use this in the analysis of an idealised, isolated triaxial cell as depicted in Figure 6.1. This comprises an outer cell wall that acts as a rigid, adiabatic *wall* allowing no work, heat or chemical interaction with the surroundings. The cell contains a pressurised water *reservoir* that surrounds a soil specimen contained in a flexible sheath. For mechanical, thermal and chemical equilibrium the total energy of the system comprising the triaxial cell is a minimum and we can write the first condition for equilibrium as:

$$dU = d\,(U_m + U_r) = 0 \qquad [6.2]$$

where dU is a change in internal energy and dU_m and dU_r are changes in internal energy of the soil specimen and reservoir (water in cell) respectively.

The internal energy within the idealised triaxial cell is given by $U = U_r + U_m$ and Equation 6.2 defines zero gradient when differentiating individually against the thermal, mechanical and chemical energy components that make up the internal energy (i.e. each energy component is in equilibrium). The second condition for equilibrium is that if an infinitesimal change in $(U_m + U_r)$ was to occur, this would always be positive; thus, small changes away from the minimum condition for the individual energy components always result in an increase in the internal energy. This condition ensures that a state produced by such a change will not exhibit a lower value of U and is given by:

$$d^2U = d^2\,(U_m + U_r) > 0 \qquad [6.3]$$

This condition does not preclude meta-stable conditions where a localised minimum is achieved, though sufficient disturbance could give rise to a lower equilibrium potential.

We will be particularly interested in the *Enthalpy Minimum Principle*. This will be outlined below for the case of the idealised triaxial cell. It is necessary to make use of the independence of the variables of state, as discussed in Section 5.4, and to allow a small virtual or infinitesimal change dV_m in the volume of the specimen under constant pressure conditions. This alters the mechanical potential of the specimen but not the heat or chemical potentials. There will be an equal and opposite change $-dV_r$ of the volume of the reservoir that needs to be taken into account. From Equations 6.1 and 6.2, the first condition for equilibrium of a pressure-controlled system can be written as:

$$d(U_m + U_r) = dU_m + p_m dV_m + dU_r + p_r dV_r = dH = d(H_m + H_r) = 0 \qquad [6.4]$$

where dH is the change in enthalpy, dH_m and dH_r are the changes in enthalpy of the soil specimen and reservoir (water in cell) respectively, p_m is the pressure in the soil specimen and p_r is the pressure in the reservoir.

Accordingly, the second equilibrium condition for a pressure-controlled system is:

$$d^2 H = d^2 (H_m + H_r) > 0 \qquad [6.5]$$

Enthalpy H can thus be viewed as the thermodynamic potential governing mechanical equilibrium of a specimen in the triaxial cell under constant pressure. Equilibrium is controlled not only by minimisation of the internal energy U but also by minimisation of the enthalpy H and consequently by minimisation of the pressure potential term pV. Thus, each component of the thermodynamic potential is minimised under equilibrium conditions. This is not just true for isotropic loading conditions, as will be demonstrated in Section 6.5. Callen (1965) more rigorously analyses the foregoing *Enthalpy Minimum Principle* as well as the *Internal Energy Minimum Principle*, the *Helmholtz Potential Minimum Principle* and *the Gibbs Potential Minimum Principle*. Thus, just as internal energy acts as a potential for work for a system of constant entropy, the Helmholtz potential acts as a potential for work for a system at constant temperature and the enthalpy acts as a potential for work for a system at constant pressure.

The minimum principle for enthalpy is used later in determining the controlling stresses in an unsaturated soil and in appraising the significance of the individual components of the total enthalpy (the phase and interaction components of Section 1.4) of a soil specimen.

6.3 Isotropic loading conditions

We will consider the following cases of a soil specimen tested under isotropic loading conditions as in Figure 6.2:

- An undrained 'total' stress test;
- An undrained 'effective' stress test;
- A drained 'effective' stress test.

However, we need to refine our description of the idealised triaxial cell set-up. The cell contains the pressurised water, which acts as a thermal and volume reservoir for the system, and the flexible sheath around the specimen allows no mass exchange, including no chemical exchange, between the specimen and the reservoir. There is also no chemical imbalance with the back pressure systems. Thus, the last term $\Sigma \mu_i M_i$ can be omitted from the thermodynamic potential (Equation 6.1) as this does not represent a potential for change.

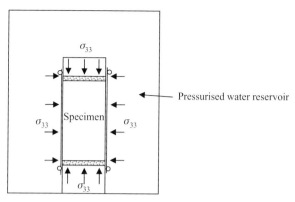

Figure 6.2 Isotropic loading conditions in triaxial cell.

6.3.1 Undrained 'total' stress test

In general, for an infinitesimal, reversible change in thermal and work energy as a result of changes dS and dV:

$$dU = \delta Q + \delta W = TdS - pdV \qquad [6.6]$$

where δQ is the heat generated in the specimen, $\delta W = -pdV$ is the work done to the specimen, dS is the increase in entropy and dV is the change in volume ($dV < 0$ for compression giving $\delta W > 0$).

Equation 6.6 reflects the situation where work is done to the specimen and is consistent with volumetric compression, and accordingly length and radius reduction, corresponding to positive strain-increments for positive compressive stresses. In the undrained test the soil is treated as a mass with no distinction as to its degree of saturation or composition and it is assumed that at equilibrium and during the infinitesimal changes considered that there is no transfer of mass or heat from or to the specimen as a result of exchange with external measurement devices. The cell reservoir (water) applies an all-round pressure p_r to the soil specimen contained in the impermeable membrane. The membrane is assumed to impart no additional loading to the specimen. The soil specimen has a responsive pressure p_m. Under the above conditions, the internal energy change for the system dU is given by Equation 6.7 (Sposito, 1981):

$$dU = dU_m + dU_r = \delta Q_m + \delta W_m + \delta Q_r + \delta W_r = T_m dS_m - p_m dV_m + T_r dS_r - p_r dV_r \qquad [6.7]$$

where $\delta Q_m = T_m dS_m$ is the change in heat of the soil, $\delta Q_r = T_r dS_r$ is the change in heat of the reservoir, $\delta W_m = p_m dV_m$ is the work done on the soil, $\delta W_r = p_r dV_r$ is the work done on the reservoir, T_m and T_r are the absolute temperatures of the soil and reservoir respectively and dS_m and dS_r are the changes in entropy of the soil and reservoir respectively.

There is no change in total entropy of the system as the cell wall acts as an adiabatic barrier preventing heat exchange with the surroundings, and thus $dS_m = -dS_r$. There is also no change in volume of the system for a rigid cell wall, and thus $dV_m = -dV_r$. In addition, assuming the establishment of thermal equilibrium within the system, $T_m = T_r$. Under these conditions Equation 6.7 can be rewritten as:

$$dU = -(p_m - p_r)dV_m \qquad [6.8]$$

Equation 6.8 essentially describes a virtual process comprising an infinitesimal change in the soil volume dV_m. For equilibrium, in accordance with Equation 6.2, $dU = 0$, as there is a requirement for the thermodynamic potential U to be a minimum. The equation confirms that under isotropic undrained loading conditions, provided the assumptions outlined are satisfied, a prime requirement in comparing equilibrium conditions in the triaxial cell with theoretical predictions is that the pressure imposed by the water in the cell p_r is balanced by the pressure exerted within the soil specimen p_m. This simple conclusion that is intuitively correct represents the condition for mechanical equilibrium, i.e. $p_m = p_r$.

6.3.2 Undrained 'effective' stress test

Similar overall triaxial conditions are assumed as for the case of an undrained 'total' stress test. However, in a detailed analysis of 'effective' stress conditions, it is necessary to account for:

- The pore air and pore water pressures within the respective volumes of air and water;
- The fluid pressure acting through the volume of solids (as discussed in Section 1.4.5);
- The phase interaction effects, such as the contractile skin.

We apologise for appearing to put the 'cart before the horse', but at this point it is necessary to draw on the conclusion of Section 8.5 that the net influence of the contractile skin and surrounding fluid pressures is equivalent to assuming that the pore water pressure acts through the solid phase, and other phase interaction effects can be ignored. While this might be perceived as applicable only at high degrees of saturation, this is later shown not to restrict the analysis. The basis of this is that the thermodynamic potential is minimised when the pressure acting through the soil particles as a result of the fluid pressures is a minimum. In an unsaturated soil the water pressure is less than the air pressure, and it is shown that the net influence of the contractile skin and the fluid pressures is to minimise the thermodynamic potential[1]. In this respect the contractile skin might be perceived as a physical manifestation of the minimisation of the thermodynamic potential.

The infinitesimal compressions of the air phase dV_a, water phase dV_w and solid phase dV_s make up the overall volume change of the specimen dV_m. It is important to take account of the compression of all three phases at this stage if a correct formulation is to be obtained. Accordingly, the internal energy equation for an undrained 'effective' stress test is given by:

$$dU = T_m dS_m - p_c' dV_m - u_a dV_a - u_w dV_w - u_w dV_s + T_r dS_r - p_r dV_r$$ [6.9]

where p_c' is an average 'effective' stress[2].

As for the case of an undrained total stress test, in the idealised triaxial cell $dS_m = -dS_r$ and $dV_m = -dV_r$, and under equilibrium conditions $T_m = T_r$ and $dU = 0$. Thus, Equation 6.9 can be rewritten as:

$$p_c' = (p_r - u_a) + (u_a - u_w) \frac{(dV_w + dV_s)}{dV_m}$$ [6.10]

The stress function that gives rise to Equation 6.10 is the requirement for mechanical equilibrium in an undrained 'effective' stress test under isotropic loading, and is given by:

$$p_c' = (p_r - u_a) + (u_a - u_w) \frac{(V_w + V_s)}{V_m}$$ [6.11]

Equation 6.11 can be written as:

$$p'_c = (p - u_a) + (u_a - u_w) v_w / v \qquad [6.11a]$$

where v is the specific volume, v_w is the specific water volume and $p = p_r$.

In Chapter 8 this equation is shown to be one of a set of three alternative representations of the stress regime in an unsaturated soil and is derived on a more rigorous basis.

6.3.3 Drained 'effective' stress test

Assumptions relating to the triaxial cell and specimen will generally be as for the undrained tests. However, for the more complex case of drained conditions, an infinitesimal exchange of air and water between the specimen and the external measuring system is allowed. It is also necessary, in order to allow recognition of the significance of the solid phase, to allow an exchange of solids, though in practice this does not occur and can be equated to zero. We will also take account of the change in gravitational potential of the matter exchanged with the measuring system. As for the case of an undrained 'effective' stress test, the fluid pressure acting through the volume of solids will be taken equal to the water pressure.

The infinitesimal volume change of the soil specimen dV_m is made up of the exchange of air dV_a, the exchange of water dV_w and the exchange of the solid phase dV_s. Accordingly, the internal energy equation for a drained 'effective' stress test is given by:

$$dU = T_m dS_m + \sum T_i dS_i - p'_c dV_m - u_a dV_a - u_w dV_w - u_w dV_s + T_r dS_r - p_r dV_r + \sum \varphi_i dM_i \qquad [6.12]$$

where $\sum T_i dS_i$ is the summed heat exchange as a result of the exchange of air, water and solids and $\sum \varphi_i dM_i$ is the summed gravitational potential change as a result of the exchange of air, water and solids.

The change of entropy of the system is given by $dS_m + \sum dS_i = -dS_r$ and the change in volume by $dV_m = dV_a + dV_w + dV_s = -dV_r$. In addition, assuming the establishment of thermal equilibrium within the system, $T_m = T_r = T_i$. Under equilibrium conditions the change in internal energy dU and the change in gravitational potential $\sum \varphi_i dM_i$ are both zero. Thus, Equation 6.13 reduces to Equation 6.11a and the requirement for mechanical equilibrium in a drained 'effective' stress test under isotropic loading is given by the same condition as in an undrained 'effective' stress test. Again this is intuitively correct as there should be only one definition of 'effective' stress.

Obviously, the case of both phase volume compression/expansion, as in the undrained 'effective' stress test, and phase exchange with the external measuring system, as in the drained 'effective' stress test, will also lead to Equation 6.11a at equilibrium.

6.4 Anisotropic loading conditions

As for isotropic loading conditions, we will consider the three basic tests on soil specimens: an undrained total stress test, an undrained 'effective' stress test and a drained 'effective' stress test. The idealised anisotropic (shear) loading conditions in a triaxial cell are illustrated in Figure 6.3.

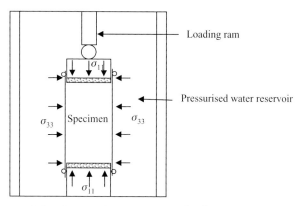

Figure 6.3   Anisotropic loading conditions in triaxial cell.

6.4.1 Undrained total stress test

Consider the case of a specimen of height h, cross-sectional area A_p and radius r (Figure 6.3) subject to a total axial stress σ_{11} and a cell pressure σ_{33} such that the mean applied stress is given by $p = (\sigma_{11} + 2\sigma_{33})/3$ and the deviator stress by $q = (\sigma_{11} - \sigma_{33})$. The double number suffix notation has been adopted to be compliant with subsequent sections of the book. The work equation takes the form $\delta W = -\sigma_{33}dV - (\sigma_{11} - \sigma_{33})\,A_p dh$.

It is also necessary to take account of the change in length of the axial loading ram in the cell as a result of the deformation of the specimen. The reservoir water pressure acts through the volume of cell water and the loading ram and the pdV term for the loading ram is included in the term $\sigma_{r,33}dV_r$ in Equation 6.13, where $\sigma_{r,33}$ is the reservoir water pressure and dV_r is the volume change of the reservoir and includes the change in volume of loading ram in the cell. However, it is necessary to include a separate term for the gravitational potential change due to the change in length of loading ram in the cell. Accordingly, for an infinitesimal transfer of thermal and work energy under similar conditions to the assumptions for isotropic undrained total stress loading, the internal energy change for the system dU is given by:

$$dU = T_m dS_m - \sigma_{m,33}dV_m - (\sigma_{m,11} - \sigma_{m,33})\,A_p dh_m + T_r dS_r - \sigma_{r,33}dV_r$$

$$- (\sigma_{r,11} - \sigma_{r,33})\,A_p dh_r + \varphi_L dM_L \qquad [6.13]$$

where $\sigma_{m,11}$ and $\sigma_{r,11}$ are the total axial stress in the soil and from the cell loading system respectively, $\sigma_{m,33}$ and $\sigma_{r,33}$ are the total radial stress in the soil and from the cell loading system respectively, dh_m and dh_r are the axial compression of the soil specimen and displacement of the cell axial loading system respectively and $\varphi_L dM_L$ is the gravitational potential change as a result of the change in length of axial loading ram in the cell (Figure 6.4).

As previously, for no change in total entropy of the system $dS_m = -dS_r$ and for no net change in volume of the system, $dV_m = -dV_r$. It is also necessary to assume compatibility of axial displacement, $dh_m = -dh_r$. In addition, $T_m = T_r$ if thermal equilibrium of the system is established and under equilibrium conditions $\varphi_L dM_L = 0$ and $dU = 0$. Under these conditions, Equation 6.13 can be written as:

$$0 = -(\sigma_{m,33} - \sigma_{r,33})\,dV_m - [(\sigma_{m,11} - \sigma_{m,33}) - (\sigma_{r,11} - \sigma_{r,33})]\,A_p dh_m \qquad [6.14]$$

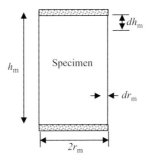

Figure 6.4 Compressive straining of specimen in triaxial cell.

Dividing throughout by the volume of the specimen V_m gives:

$$0 = (\sigma_{m,33} - \sigma_{r,33})\,(d\varepsilon_{m,11} + 2d\varepsilon_{m,33}) + [(\sigma_{m,11} - \sigma_{m,33}) - (\sigma_{r,11} - \sigma_{r,33})]\,d\varepsilon_{m,11} \qquad [6.15]$$

Rearranging and manipulating the variables gives:

$$0 = (p_m - p_r)\,d\varepsilon_{m,v} + (q_m - q_r)\,d\varepsilon_{m,q} \qquad [6.16]$$

where for stress and strain positive in compression:

$d\varepsilon_{m,v} = -dV_m/V_m = (d\varepsilon_{m,11} + 2d\varepsilon_{m,33})$

$d\varepsilon_{m,q} = 2\,(d\varepsilon_{m,11} - d\varepsilon_{m,33})/3$

$d\varepsilon_{m,11} = -dh_m/h_m$

$d\varepsilon_{m,33} = -dr_m/r_m$

dr_m is the change in radius of the soil specimen

$p_m = (\sigma_{m,11} + 2\sigma_{m,33})/3$

$q_m = (\sigma_{m,11} - \sigma_{m,33})$

$p_r = (\sigma_{r,11} + 2\sigma_{r,33})/3$

$q_m = (\sigma_{m,11} - \sigma_{m,33})$

Equilibrium analysis axially and radially gives $\sigma_{m,11} = \sigma_{r,11}$ and $\sigma_{m,33} = \sigma_{r,33}$ and thus from Equation 6.16, $p_m = p_r$ and $q_m = q_r$. However, this does not necessarily follow directly from the equation unless the influence of the mean stress, deviator stress and the conjugate strains in the work equation are treated independently. It is shown in the following that it is appropriate to treat these stress variables as independent, a situation often assumed in soil mechanics and discussed in relation to yielding in saturated soils in Section 4.5.3. Accordingly, the condition for equilibrium in an undrained total stress test is that $p_m = p_r$ and $q_m = q_r$.

6.4.2 Undrained 'effective' stress test

In this case the infinitesimal compression of the air phase dV_a, water phase dV_w and solid phase dV_s makes up the overall volume change of the specimen dV_m. Accordingly, the internal energy equation for an undrained 'effective' stress test under anisotropic (shear) loading is given by:

$$\begin{aligned} dU = {} & T_m dS_m - \sigma'_{c,33} dV_m - u_a dV_a - u_w dV_w - u_w dV_s - \left(\sigma'_{c,11} - \sigma'_{c,33}\right) A_p dh_m \\ & + T_r dS_r - \sigma_{r,33} dV_r - (\sigma_{r,11} - \sigma_{r,33})\,A_p dh_r + \varphi_L dM_L \end{aligned} \qquad [6.17]$$

where $\sigma'_{c,11}$ is the average axial 'effective' stress and $\sigma'_{c,33}$ is the average radial 'effective' stress[3].

As previously, in the idealised triaxial cell, $dS_m = -dS_r$, $dV_m = -dV_r$ and $db_m = -db_r$. Under equilibrium conditions, $T_m = T_r$ and $\varphi_L dM_L$ and $dU = 0$ and Equation 6.17 can be written as:

$$0 = -\left(\sigma'_{c,33}\sigma_{r,33}\right) dV_m - \left[\left(\sigma'_{c,11} - \sigma'_{c,33}\right) - \left(\sigma_{r,11} - \sigma_{r,33}\right)\right]$$

$$A_p db_m - u_a dV_a - u_w dV_w - u_w dV_s \qquad [6.18]$$

Dividing throughout by the volume of the specimen V and rearranging gives:

$$0 = \left(\sigma'_{c33} - \sigma_{r,33}\right)\left(d\varepsilon_{m,11} + 2d\varepsilon_{m,33}\right)$$

$$+\left[\left(\sigma'_{c11} - \sigma'_{c33}\right) - \left(\sigma_{r,11} - \sigma_{r,33}\right)\right] d\varepsilon_{m,11} + u_a d\varepsilon_{m,v} - \left(u_a - u_w\right) d\varepsilon_w \qquad [6.19]$$

where $d\varepsilon_w = -(dV_w + dV_s)/V_m$.
Rearranging and manipulating the variables gives:

$$p'_c d\varepsilon_{m,v} + q_m d\varepsilon_{m,q} = \left(p_r - u_a\right) d\varepsilon_{m,v} + \left(u_a - u_w\right) d\varepsilon_w + q_r d\varepsilon_{m,q} \qquad [6.20]$$

The stress functions governing the mechanical equilibrium requirement for an undrained 'effective' stress test under anisotropic stress conditions are thus given by $p'_c = (p_r - u_a) + (u_a - u_w)(V_w + V_s)/V_m$ and $q_m = \sigma'_{c,11} - \sigma'_{c,33} = q_r$. The equation for p'_c is in a form similar to Equation 6.11a for isotropic loading conditions but there is now an additional equation for the deviator stress under anisotropic loading.

6.4.3 Drained 'effective' stress test

The internal energy equation for a drained 'effective' stress test is given by:

$$dU = T_m dS_m + T_m dS_a + T_m dS_w + T_m dS_s - \sigma'_{c,33} dV_m - u_a dV_a - u_w dV_w - u_w dV_s$$

$$-\left(\sigma'_{c,11} - \sigma'_{c,33}\right) A_p db_m + T_r dS_r - \sigma_{r,33} dV_r - \left(\sigma_{r,11} - \sigma_{r,33}\right) A_p db_r \qquad [6.21]$$

$$+ \sum \varphi_i dM_i + \varphi_L dM_L$$

As previously, the idealised triaxial cell set-up gives $dS_m + dS_a + dS_w + dS_s = -dS_r$ and $dV_m = dV_a + dV_w + dV_s = -dV_r$ and equilibrium conditions dictate that $T_m = T_r$ and dU, $\varphi_L dM_L$ and $\sum \varphi_i dM_i = 0$. Thus, the stress function governing the mechanical equilibrium in a drained 'effective' stress test for anisotropic loading conditions is the same as that for an undrained 'effective' stress test.

Thus, Equation 6.11a gives the equilibrium stress regime for undrained or drained 'effective' stress tests on unsaturated soils in the triaxial cell for both isotropic and anisotropic (shear) loading conditions provided the pressure p under isotropic conditions is replaced with the mean stress for anisotropic conditions.

6.5 Work input and the thermodynamic potential

Compliance with the laws of thermodynamics is a required feature of any soil model if it is to be based on sound principles (Houlsby et al., 2005). In accordance with the first law of thermodynamics it is necessary that the work equation, thus the stresses, can be derived from the thermodynamic potential. In soils this is complicated by anisotropic loading conditions. Consider the case of the work input δW_m to a soil specimen under anisotropic undrained loading, as given by Equation 6.22:

$$\frac{\delta W_m}{V_m} = p_m \left(\varepsilon_{lm} + 2\varepsilon_{rm}\right) + q_m \frac{2}{3}\left(\varepsilon_{lm} - \varepsilon_{rm}\right) = -p_m \frac{dV_m}{V_m} - q_m \frac{2}{3}\left[\frac{dh_m}{h_m} - \frac{dr_m}{r_m}\right] \qquad [6.22]$$

In addition, the change in internal energy dU_m can be written as:

$$dU_m = T_m dS_m + \delta W_m \qquad [6.23]$$

Substituting for δW_m from Equation 6.22:

$$dU_m = T_m dS_m - p_m dV_m - q_m \frac{2}{3} V_m \left[\frac{dh_m}{h_m} - \frac{dr_m}{r_m}\right] \qquad [6.24]$$

The extensive thermodynamic potential U_m before the increment of work determined from Equation 6.24 is given by:

$$U_m = T_m S_m - p_m V_m \qquad [6.25]$$

Equation 6.25 is in the same form as the Euler equation (Equation 6.1) for isotropic loading without a chemical potential but p_m represents the mean stress (Murray and Brown, 2006). There is no term in the potential for the deviator stress as on integration of Equation 6.24 the deviator strain term reduces to zero. Alternatively, the appropriateness of Equation 6.25 can be demonstrated by differentiation, but it is not appropriate to merely write $dU_m = T_m dS_m - p_m dV_m$ as this is true only for isotropic loading conditions. Equation 6.25 can be written as:

$$U_m = T_m S_m - \frac{1}{3}\left(\sigma_{m,11} + 2\sigma_{m,33}\right) V_m = T_m S_m - \frac{1}{3}\left(\sigma_{m,11} - \sigma_{m,33}\right) A_p h_m - \sigma_{m,33} V_m \qquad [6.26]$$

Substituting $A_p = \pi r_m^2$ and noting that $r_m = N_m h_m$, where N_m is the ratio of radius to height of the specimen, differentiation correctly leads to:

$$dU_m = T_m dS_m - \sigma_{m,ss} dV_m - \left(\sigma_{m,11} - \sigma_{m,33}\right) A_p dh_m = T_m dS_m + \delta W_m. \qquad [6.27]$$

Thus, the thermodynamic potential given by Equation 6.25 can be written without a term for the deviator stress but leads to the correct work equation for anisotropic loading. This is important as exclusion of a term for the deviator stress from Equation 6.25 means that there is no violation of the extensive properties of the thermodynamic potential. The fact that a term for q_m does not appear in Equation 6.25 also suggests that at equilibrium it is appropriate to treat the mean stress and deviator stress as independent or uncoupled variables, as suggested in relation to Equation 6.16.

6.6 The thermodynamic potential and axis translation

The datum for pressure is often, in fact generally, taken as atmospheric (or gauge) pressure though there is no overriding thermodynamic reason for this convention. In unsaturated soils in situ, where continuous air passages are connected to the atmosphere, the air pressure in the pores is at atmospheric pressure. However, in triaxial cell testing of unsaturated soils it is generally impractical to carry out tests where the air pressure is atmospheric, or less than atmospheric, as cavitation, the disruptive nucleation of air bubbles, occurs in the tubes and measuring devices, affecting the pressure measurements. The approach adopted is to raise the air pressure above atmospheric pressure while maintaining the same difference between the air and water pressure, thus maintaining the same suction. This is the so-called axis translation technique (Hilf, 1956). Measurements are then generally reported as $(p - u_a)$ and $(u_a - u_w)$ where the datum is not atmospheric pressure but some pressure u_a above atmospheric pressure.

Where pressure or stress is the driving force governing change as in the triaxial cell, enthalpy may be considered as the thermodynamic potential. In using the axis translation technique the thermodynamic potential needs to be redefined as:

$$H = (p - u_a) V + U \qquad [6.28]$$

This is a minimum at equilibrium, which means that $(p - u_a) V$ and U are both a minimum at equilibrium. The differential form of this equation for constant $(p - u_a)$ is:

$$dH = (p - u_a) dV + dU \qquad [6.29]$$

6.7 The thermodynamic potential and an aggregated soil structure

In Chapter 1, it was concluded from the results of mercury intrusion porosimetry testing that unsaturated fine-grained soils can be viewed as having bi-modal pore size distributions with a clear division between those smaller voids comprising the intra-aggregate pore spaces and the larger inter-aggregate voids. It was also concluded that over a wide range of suctions the aggregates remain saturated with water, with air restricted to the larger inter-aggregate voids. As a precursor to the analyses in Chapter 7, and in particular the analysis of Section 7.5, an introduction is provided to the significance and creation of an aggregated soil structure. First this will be outlined in terms of the generally recognised behaviour attributable to the contractile skin in drawing particles together, and will be followed by consideration of the significance of the enthalpy minimum principle as described in Section 6.2.

Consider the case of soil particles intermixed with the fluids of water and air. Water and air fill the void spaces between the particles. Movements of the three phases are not prevented but are limited by the stresses, phase interactions and the void spaces. Air fills the larger pore spaces as it is at greater pressure than the water, and a contractile skin forms at the interface of the water and air. Surface tension means that the contractile skin is concave to the water phase and it is the minimisation of this interface phenomenon that draws the water into the smaller pore spaces. The contractile skin in turn pulls the particles together into aggregates. This results in relatively small intra-aggregate voids. This presents a logical view and reasoning for the creation of an aggregated structure in unsaturated soils. There is, however, another way of considering the issue.

At equilibrium, the thermodynamic potential given by the enthalpy $H = U + pV$ is a minimum and, as demonstrated in Section 6.2, this requires minimisation of the mechanical potential term pV. In accordance with the derivation of Equation 6.11, pV is a function of the pressure–volume terms for the three phases as well as the interactions between the phases. Thus, the mechanical potential is minimised by minimising the potentials associated with these components internal to the soil. If they are not minimised, there is an imbalance, i.e. equilibrium has not been achieved and change will occur until an equilibrium state is achieved. In this context, there are three identifiable reasons for the creation of an aggregated structure in an unsaturated soil, and which provide a basis for the interpretation of data in terms of aggregates containing all the water and soil particles. These are the differences between the air pressure and the water pressure, the contractile skin, which forms as a result of the imbalance in pressures, and the fluid pressure acting through the soil particles. These require careful consideration.

Consider the case of free water droplets in air. These would have a surface convex to the water and a pressure greater than air. For constant air pressure and a similar volume of water in an unsaturated soil, the interaction of the water and soil particles in an aggregated structure allows the water pressure to drop, and a concave interface to form with the water phase, if the water occupies the smaller intra-aggregate pores. The pressure and thus potential of the water is reduced when in close association with the soil particles. The close association also means that the area of the contractile skin is likely to be minimised and thus its potential is minimised by the water filling the smaller void spaces and the particles being drawn together into aggregates.

The pressure acting through the soil particles also influences the creation of an aggregated structure. In Chapter 1 it was established that the pressure of a fluid surrounding a particle must act through the volume of the particle. Since the air pressure is greater than the water pressure in an unsaturated soil, the component of the thermodynamic potential associated with the pressure acting through the particles is a minimum if the water pressure acts through the particles. Thus, the minimisation of the potentials associated with the water and the contractile skin leads not only to a close association of soil particles and water, but also to the requirement that the pressure acting through the particles should be minimised. The soil particles are drawn together to form aggregated packets that minimise the intra-aggregate void spaces.

However, close compaction of the soil particles in the aggregates leads to greater inter-particle stresses, which offsets the minimisation of the other components of the enthalpy potential. This can be explained by considering the analogy of the behaviour of loose sand in a container with a trap door. Opening the trap door allows the sand to escape. The sand flows from the container, reducing its potential, and does not concern itself that on falling into a lower container it might achieve a more compact condition with increased inter-particle stresses. Similarly, in an unsaturated soil, the soil particles combine with the lower pressure water minimising the potentials of the water, contractile skin and fluid pressure acting through the soil particles, even though the more compact aggregate arrangement leads to greater inter-particle stresses internal to the aggregates.

The concept of the minimisation of the thermodynamic potential and the creation of an aggregated structure in unsaturated soils will be developed further in Chapter 8.

6.8 Conclusions

The following main points emerge from the analysis:

- The extensive variable enthalpy given by $H = pV + U$ acts as a thermodynamic potential and can be used to describe the potential governing the mechanical response of soils such as in a triaxial cell.
- The enthalpy minimises at equilibrium and meta-stable equilibrium, along with the internal energy U and mechanical potential pV that make up the enthalpy.
- The equation for enthalpy can be written in a similar form for both isotropic and anisotropic (shear) loading conditions, but for the latter the mean stress replaces the isotropic pressure.

- Equilibrium stress and pressure conditions under combinations of undrained or drained and isotropic or anisotropic triaxial test conditions have been determined and shown to be logically correct.
- The enthalpy can be written as $H = (p - u_a)V + U$ when employing the axis translation approach in soil testing.
- The minimisation of the thermodynamic potential at equilibrium can be used to explain the close affinity of water and soil particles and the creation of an aggregated structure in unsaturated soil.

Notes

1. In Chapter 8 the term α is introduced which combines the influence of the contractile skin and the fluid pressure acting through the solid particles. It is the maximisation of α at a value of 1 that leads to the conclusion that the net effect is equivalent to assuming that the water pressure acts through the solid particles.
2. p'_c is more rigorously defined in Chapter 8 as the average volumetric coupling stress where its meaning and significance are explored. As previously, we use the term 'effective' in quotation marks to loosely describe the stress conditions in unsaturated soils.
3. $\sigma'_{c,11}$ and $\sigma'_{c,33}$ are more rigorously defined in Chapter 8 as the axial and radial volumetric coupling stress where their meanings and significance are explored.

Chapter 7
Enthalpy and Equilibrium Stress Conditions in Unsaturated Soils

7.1 Introduction

The prediction of soil behaviour is intrinsically linked to the need to determine the controlling stresses. The stress state in unsaturated soils under equilibrium conditions can be examined using the principles of thermodynamics, which are of general applicability to the behaviour of gases, liquids and solids. Thermodynamics allows examination of multi-phase materials and the general principles can be extended to soils without modification, though the principles need to be expressed differently for different applications. Chapters 5 and 6 provide a framework for the equilibrium thermodynamics of unsaturated soils. In particular in Section 5.7, the idea of enthalpy as an extensive thermodynamic variable, and a potential for work for a pressure-controlled system, as in the triaxial cell, was developed. This was extended in Chapter 6 to show that under anisotropic loading conditions the mean stress can replace the pressure or isotropic stress in the thermodynamic potential. Accordingly, the individual enthalpy components associated with the anisotropic stresses can be added together to give the total enthalpy. This means that in conventional triaxial cell tests the enthalpy of a soil specimen can be taken as the thermodynamic potential driving change and the enthalpy achieves a minimum at equilibrium. In fact, each component of the total enthalpy in a soil system adjusts and strives to minimise at equilibrium. An analysis using enthalpy is made in the following to yield an equation describing the controlling stress regime in an unsaturated soil (Murray, 2002; Murray and Sivakumar, 2006).

7.2 Role of enthalpy

Enthalpy is an extensive thermodynamic variable and for a material comprising a number of phases, *Principle 1* holds true under equilibrium conditions. This has particular significance for a multi-phase material such as an unsaturated soil where the phases and their interactions need to be taken into account in analysis of the controlling stresses.

Principle 1 – The total enthalpy H of a multi-phase material is the sum of the enthalpies of the individual phases and the interactions between the phases H_i.

Section 1.4 described the following interactions in unsaturated soils:

–   the water vapour in air;
–   the dissolved air in water;
–   the contractile skin at the interface between the phases of water and air;
–   the adsorbed water on the soil particles and crystalline water;
–   the fluid pressure acting through the solid phase; and
–   soil particle interactions.

The enthalpies of these interactions, as with the enthalpies of the phases, may in most cases be defined in terms of the product of a pressure and a volume term within an equation for the total enthalpy. However, for the contractile skin it is necessary to replace this with the product of the surface tension and the area of the film.

Principle 1 is applicable whether phases exhibit miscible or immiscible interactions. A miscible interaction, such as water vapour in air, can be taken as obeying Dalton's divisional law of partial pressures, whereas an immiscible interaction, such as the adsorbed water on soil particles, can be treated as having its own pressure acting through its own volume. For immiscible interactions where p_i is the total pressure arising from the miscible interactions, Dalton's divisional law states that the partial pressures p_p of the interacting phases act through the total volume V_i. Accordingly, the total pressure p_i for a miscible interaction is given by $p_i = \sum p_p$. In terms of the pressure–volume relationship within the thermodynamic potential (Equation 5.8), this can be written as $p_i V_i = \sum p_p V_i$. For immiscible phase interactions on the other hand, each phase or interaction can be represented by a term $p_i V_i$. Thus, the total pressure p for a multi-phase material with both immiscible and miscible interactions, assuming a surface encompassing a representative mass, is the sum of the pressures of the phases and their interactions factored by the relative volumes V_i/V through which the pressures act, giving $p = \sum p_i V_i / V$. Further, since the internal energy U is an extensive variable, in accordance with Equation 5.13 the total U can be equated to the sum of the internal energies of the phases and their interactions (i.e. $U = \sum U_i$) and from the definition of enthalpy:

$$H = \sum H_i = pV + U = \sum (p_i V_i + U_i) \qquad [7.1]$$

This is the mathematical representation of *Principle 1* and holds true for isotropic and anisotropic loading as the influence of the individual stresses are additive. The total enthalpy of an unsaturated soil specimen can thus be determined from the summation of the enthalpies of the individual phases and their interactions[1]. An examination of the analysis using enthalpy in Section 7.4 reveals that it is only necessary to use the mechanical potential component pV, and not include the internal energy U, but the enthalpy is a more recognisable thermodynamic potential and its use provides a closer link with thermodynamic concepts and the enthalpy minimum principle.

It is necessary in the formulation that follows to define a second principle for an operation where phases are brought together and where there is no mass loss or gain, heat exchange with the surroundings or work done in combining the phases.

Principle 2 – The sum of the enthalpies of the individual phases in isolation equals the enthalpy of the combined phases if no work is done or heat exchanged with the surroundings in combining the phases.

This principle later allows an equation to be established for aggregates comprising the water phase and the soil particles. This idealised operation is applicable where no actual process of combining the phases is undertaken but only a theoretical redistribution of the phases within a given volume is considered. We will develop *Principle 2* by initially considering the general energy equation in combining the phases, which can be written as:

$$dU = \delta Q + \delta W + \sum \mu_i M_i = TdS - pdV + \sum \mu_i M_i \qquad [7.2]$$

where $\delta Q = TdS$ is the change in heat of the soil and $\delta W = -pdV$ is the work done on the soil.

As there is no mass loss or gain or heat exchange with the surroundings, Equation 7.2 reduces to $dU = -pdV$. If the process is one where there is no volume change as the phases are combined, then no net work is done and $\delta W = dU = dH = 0$. Thus, in combining incompressible phases, equilibrium requires that:

$$dH = \sum dH_i = 0 \qquad [7.3]$$

where H_i represents the enthalpy of an individual phase or interaction.

Equation 7.3 states that there is no change in total enthalpy. Thus, the sum of the enthalpies of the individual phases in isolation equates to the sum of the enthalpies of the combined phases in accordance with *Principle 2*. As water and soil particles are usually considered incompressible, Equation 7.3 is considered to apply to the idealised unsaturated soil structure comprising aggregates of water and soil particles. The aggregates are created as a result of the difference between the surrounding inter-aggregate air and the intra-aggregate water pressures and thus arise from isotropic loading conditions.

In the following, *Principle 1* is first applied to a saturated soil to develop Terzaghi's effective stress equation and then the analysis is extended to unsaturated soils.

7.3 Enthalpy and Terzaghi's effective stress for saturated soils

In accordance with Equation 1.24, Terzaghi's effective stress equation for a saturated soil in the triaxial cell can be written in terms of the mean stress as:

$$p' = p - u_w \qquad [7.4]$$

The resistance to total stress p applied to a saturated soil specimen is provided by the sum of the water pressure u_w and the effective stress p', where p' is the result of the solid phase being present as discrete particles rather than as a coherent mass. Both p and p' can be viewed as components of enthalpy per unit total volume of the soil system. The pore water pressure u_w can also be viewed as a component of enthalpy per unit of volume of the soil as it acts through both the volume of the water phase V_w and the volume of the solid phase V_s as water surrounds the soil particles. From Equation 7.1 we may write:

$$pV + U = u_w V_w + u_w V_s + p'V + \sum U_i \qquad [7.5]$$

or

$$H = H_w + H_s \qquad [7.5a]$$

where $H = pV + U$, H_w is the enthalpy of the water phase $(= u_w V_w + U_w)$, H_s is the enthalpy of the solid phase and its interactions $(= u_w V_s + p'V + U_s)$ and $\sum U_i = U_w + U_s$.

In accordance with Equation 5.13, $U = \sum U_i$, and for a saturated soil $V = V_w + V_s$. Accordingly, Equation 7.5 reduces to Terzaghi's equation (Equation 7.4), which is not derivable unless the water pressure acting through the solid phase is taken into account as in the analysis of equilibrium on a plane in Chapter 1. It should also be noted that no terms for the components of enthalpy associated with the adsorbed water or dissolved air are present. These interactions are discussed in the following and are shown not to significantly influence Equation 7.5.

7.4 Enthalpy of unsaturated soils

A generalised equation describing the stress state of an unsaturated soil based on the enthalpy and including the coupling stress p'_c can be written in simplistic terms as:

$$pV + U = u_w V_w + H_V + H_C + H_A + u_a V_a + H_D + u_s V_s + p'_c V + \sum U_i \qquad [7.6]$$

or

$$H = H_w + H_a + H_s \qquad [7.6a]$$

where H_w is the enthalpy of the water phase and its interactions $(= u_w V_w + H_V + H_C + H_A + U_w)$; H_a is the enthalpy of the air phase and its interaction $(= u_a V_a + H_D + U_a)$; H_s is the enthalpy of the solid phase and its interaction $(= u_s V_s + p'_c V + U_s)$; V_a is the volume of the air phase; u_s is the pressure in the soil particles as a result of the pore water and pore air pressures u_w and u_a respectively; H_V, H_C, H_A and H_D are the enthalpy components as a result of the water vapour, contractile skin, adsorbed and absorbed (crystalline) water, and dissolved air respectively; and $\sum U_i = U_w + U_a + U_s$.

This formulation differs slightly from that presented by Murray (2002) as the internal energy terms are included as distinct terms for completeness, but these are shown later to cancel out. The coupling stress p'_c is shown to link the independent stress state variables and represents the average volumetric stress between particles within and between the aggregates, allowing the behaviour of an unsaturated soil to be more clearly appreciated. In accordance with previous arguments, p'_c can be viewed as a component of enthalpy per unit total volume of unsaturated soil. In the following, the individual components making up Equation 7.6 are discussed and analysed to develop a general form for the equation governing the behaviour of unsaturated soils. This expands on the discussion of the phases and their interactions given in Chapter 1.

7.4.1 Water vapour

Vapour movement is an important means of transport of water within an unsaturated soil and assists in the equalisation of water pressures. Under equilibrium conditions within a closed system, the vapour pressure is the saturated vapour pressure in air above the contractile skin. This is the case for air bubbles under equilibrium conditions in unsaturated soils. Where the air passages are connected to the atmosphere, however, the vapour pressure will be less than the saturated vapour pressure and a true equilibrium may not be established.

The partial pressures of the air and water vapour constitute a miscible interaction and are defined as the pressures they could exert if each alone occupied the volume of the mixture[2]. Dalton's divisional law states that the total pressure is equal to the sum of the partial pressures of the constituents. From Equation 7.6, this is consistent with:

$$u_a V_a + H_V = (u_a + u_v) V_a \qquad [7.7]$$

where u_v is the water vapour pressure and V_a is the volume of air and is taken to include the volume of water vapour.

Schuurman (1966), in discussing the role of air bubbles in water, stated that the compressibility of air bubbles is not influenced directly by the saturated vapour pressure, although there is an influence on u_w. However, Schuurman (1966) further showed that the influence on u_w is small and the influence of the water vapour decreases as u_a increases, as for small air bubbles. At 20°C the vapour pressure above a flat water surface, which represents likely worst conditions, is around 2.4 kPa and the volume percentage of water vapour in air is likely to be less than 1%. Thus, the contribution of the water vapour to the total enthalpy of a soil system is small and H_V in Equations 7.6 and 7.7 can in most cases be ignored.

7.4.2 Dissolved air

In general, the change in the volume of air within a soil may be a result of: (i) the interchange of air with the external surroundings, (ii) compression or expansion of the free air in the pores or (iii) air dissolving or being liberated from the pore water. At equilibrium, when water is in contact with free air, the volume of air dissolved in the water is essentially independent of air and water pressures, as can be demonstrated by examination of the ideal gas laws and Henry's law (Fredlund and Rahardjo, 1993). It is the liberation of the dissolved air under decreasing external pressure and corresponding increasing suction that leads to the formation of air bubbles. As shown by Dorsey (1940), however, dissolved air produces no significant difference between the compressibility of de-aired water and air-saturated water. Additionally, the volume of dissolved air, approximately 2% of the volume of water (as under normal temperature and pressure conditions encountered in soils), does not influence the total volume of the soil system as it is incorporated within the water without any significant change in volume. This suggests that the pressure of the air dissolved in water has little effect on the pore water pressure. Thus, the contribution of the component of enthalpy H_D in Equation 7.6 due to dissolved air, to the total enthalpy of a soil system, can in most cases be ignored.

7.4.3 Fluid pressure acting through the soil particles

In describing unsaturated soils, Barden and Sides (1970) referred to the creation of saturated aggregates comprising the soil particles and water, surrounded by air-filled voids. In Section 6.7 a discussion was presented of the reasons for the close association of the water and solid phases. A saturated aggregate arrangement of the particles provides a useful means of visualising the bi-modal structure considered to exist in unsaturated soils, with a different stress regime within and between the aggregates. Within the aggregates the particles will be tightly packed and the total volume comprises the combined volumes of the solid and water phases. The pressure u_s in Equation 7.6, contributing to the enthalpy of the particles, can be viewed as varying depending on whether the particles are on the periphery or internal to the aggregates. The air pressure u_a may be viewed as acting on any particles on the periphery of the aggregates with the water pressure u_w acting on those particles internal to the aggregates. Thus, a general expression for the component of enthalpy due to air and water pressures is given by:

$$u_s V_s = a u_w V_s \qquad [7.8]$$

where the dimensionless variable a has a minimum value of 1 as $u_a \geq u_w$. For a perfectly dry soil, u_w in Equation 7.8 should be replaced with u_a. However, the transition to such a state requires the breakdown of the large suction present at low degrees of saturation.

7.4.4 Contractile skin

The significance of the contractile skin is discussed in Section 1.4.3. For the idealised case of spherical occluded air bubbles of radius R in water, the component of enthalpy of the contractile skin is given by the surface tension multiplied by the total surface area of the bubbles. Thus, for relatively uniform size spherical bubbles, it is readily shown that the component of enthalpy associated with the contractile skin as a result of surface tension is given by the following equation:

$$\frac{3}{2} V_a (u_a - u_w) \qquad [7.9]$$

For the idealised case of occluded air bubbles, the component of enthalpy H_C in the contractile skin (Equation 7.6) is controlled by the volume of the air in accordance with Equation 7.9. However, at larger suctions where the air passages are continuous and the pore water is drawn into the aggregates, it is considered more appropriate to relate H_C to the sum of the volumes of the water and the solid phase $(V_w + V_s)$ (though it is noted that for a given volume of soil V, if V_s is known, then $(V_w + V_s)$ may be determined, and vice versa). Noting the form of Equation 7.9, a more general equation for the component of enthalpy H_C associated with the contractile skin in unsaturated soil can be written as:

$$H_C = b (V_w + V_s) (u_a - u_w) \qquad [7.10]$$

where b is a dimensionless variable influenced by the structure and size of the aggregates. For the idealised case of occluded (spherical) air bubbles, b is equal to $3 V_a / 2 (V_w + V_s)$.

7.4.5 Adsorbed double layer

Clay particles are usually considered to have a negatively charged surface. When such particles are surrounded by sufficient free water, the charge is balanced by the cations in the adsorbed double layer. For isolated particles under these conditions there would be no net component of energy due to surface charge. However, as discussed in Section 1.4.5, the surface charge comes into play when the double layers of neighbouring particles come into contact, or if there is a reduction of the cations in the double layer as would be expected under unsaturated conditions or if there is a change in pore fluid chemistry. The surface energy per unit volume is very small compared with the energy levels associated with the products of the pressures (or stresses) and volumes for the soil particle, water and air masses in Equation 7.6. The interplay of inter-particle forces is deemed to be included in the formulation of p'_c as it is for Terzaghi's effective stress in saturated soils.

Changes in the amount of crystalline water held in the inter-layer spaces of swelling clay minerals as a result of wetting and drying cause the particles to change volume. This influences the interaction between the particles and is again deemed included in the formulation of p'_c. This does not influence the enthalpy analysis based on equilibrium stress conditions, although the significance of this phenomenon with respect to soil behaviour needs further investigation. The additional enthalpy H_A in Equation 7.6 associated with the adsorbed fluid and crystalline water is considered small and is ignored, as to a large part the volume of water will be included in the volume of the water phase when measuring water contents.

7.4.6 Average volumetric coupling stress

The stress p'_c incorporates not only the influences of changes in the external applied pressures, but also the influences of changes in the degree of saturation which produce additional changes in the attractive and repulsive forces between the particles. While the principles of thermodynamics allow us to derive general relations between such variables as the imposed pressures, stresses, volumes and strains, the principles of thermodynamics also tell us which of these relations must be determined experimentally in order to completely specify all the properties of a system. This is the case with the interactions between the particles, which results in aggregation and change in behaviour as a soil becomes drier. Both p'_c for unsaturated soils and p' for saturated soils represent macro-views of the microscopic forces between particles.

7.4.7 General equation describing the state of unsaturated soils

In accordance with the foregoing discussion, Equation 7.6 can be simplified by ignoring the small effects of water vapour H_V, adsorbed water H_A and dissolved air H_D. In addition, substituting Equations 7.8 and 7.10 into Equation 7.6 gives the following general equation for unsaturated soils:

$$p = u_w n_w + b(n_w + n_s) + u_a n_a + a u_w n_s + p'_c \qquad [7.11]$$

The porosity terms n_a, n_w and n_s are given by V_a/V, V_w/V and V_s/V as defined in Section 1.2. The term $(n_w + n_s)$ represents the total volume of the aggregates per unit volume of soil.

As illustrated in Figure 7.1, the plots of u_a and u_w against p in unsaturated soils are not linear, as is often taken for saturated isotropic soils, and u_a has a value greater than u_w. This means that p'_c in the unsaturated state is less than p' would have been if the soil had remained saturated. Rearranging Equation 7.11, the following form of the equation can be obtained:

$$p'_c = (p - u_a) + \alpha (u_a - u_w)(n_w + n_s) = \overline{p} + \alpha s \frac{v_w}{v} \qquad [7.12]$$

where $v_w/v = (V_w + V_s)/V$ (ratio of specific water volume to specific volume) and is the total volume of the aggregates per unit volume of soil, and α is a dimensionless variable defined as:

$$\alpha = 1 - b - (a - 1) \frac{n_s}{(n_w + n_s)(u_a - u_w)} \frac{u_w}{} \qquad [7.13]$$

The equation for α includes the dimensionless variable a, which relates to the pressure in the soil particles from the surrounding fluid, and the dimensionless variable b, which relates to the influence of the contractile skin and the size and shape of the aggregates. It is

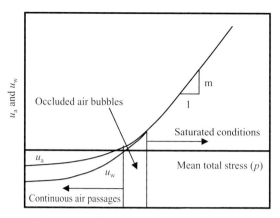

Figure 7.1   Responses of u_a and u_w due to monotonic undrained unloading.

shown in Section 7.5 how values for a and b can be estimated in exploring the significance of α.

7.5 The significance of α

Section 6.7 outlined the mechanics behind the creation of an aggregated soil structure comprising a close association of the water and soil particles. The argument was based initially on the water, at lower pressure than the air, tending to fill the smaller void spaces, with the contractile skin, concave to the air phase, acting between particles and drawing them together. This was then extended to consideration of the 'enthalpy minimum principle' and its significance in relation to the phases and their interactions. It was concluded that the net effect was the creation of an aggregated structure consistent with the conclusions of the results of mercury intrusion porosimetry (MIP) testing discussed in Chapter 1. Further analysis is given below which examines the significance of an aggregated structure and the term α. Initially an analysis will be presented which uses *Principle* 2 of Section 7.2 along with experimental data to suggest a value of $\alpha = 1$ as reasonable over a wide range of degree of saturation (Murray, 2002). This will then be supported by consideration of the thermodynamic potential to show that α equals 1 if the thermodynamic potential under equilibrium conditions is to be a minimum.

Analysis is first presented based on aggregates comprising the water phase and the soil particles. This leads to an equation relating a to b in addition to the relationship between a and b given by Equation 7.13 and allows a value of α to be determined from experimental evidence. The idealisation of unsaturated soils as comprising aggregates of water and soil particles surrounded by air voids allows a visualisation of the mathematics involved but should not be taken as restricting the use of the equations developed to a limited range of soil suctions and degrees of saturation. The primary stipulations in the analysis undertaken are that the soil particles and water are intrinsically linked by water in contact with all particles, and that at equilibrium the suction is everywhere the same. The close affinity of water and soil particles may be as water bridges at points of particle contact at low degrees of saturation, or as intra-aggregate water and water bridges between aggregates at higher degrees of saturation. While it is recognised that the idealised soil structure may not be perceived as representing the condition particularly near the extreme of a very dry soil, the equations developed are thought to apply to a wide range of conditions and soils provided these are relatively intact and do not contain open fissures. Only further experimental evidence will fully justify the range of their usage.

It was argued in Section 6.7 that there are three major reasons for the creation of an aggregated structure in terms of potentials and which provide a basis for the interpretation of experimental data on unsaturated soil. These are (i) the minimisation of the potentials associated with the water pressure, (ii) the contractile skin and (iii) the fluid pressure acting through the soil particles. It was recognised however that an exchange occurs in that while the potentials for these three components of the overall potential are minimised, there will be an increase in the inter-particle stresses within the aggregates because of closer compaction. It was argued using a 'trap door' analogy that the increase in inter-particle stresses will not prevent reduction in the other components of the overall potential but, nevertheless, the inter-particle stresses will also tend to minimise within the confines of the system. We will extend the argument of potentials to give greater justification to the conclusions.

Consider the idealised initial situation of a spherical water droplet of radius R^* surrounded by air, along with isolated soil particles of total volume V_s also surrounded by air. The water pressure within the droplet would be greater than the surrounding air pressure with the contractile skin between the water and air phases convex to the water phase. If the small soil particles were added to the droplet, the overall volume of water and particles combined would increase and the water pressure would decrease as the overall size of the droplet increased. Eventually a situation would be reached where the volume of soil particles inserted in the droplet would result in water pressure less than the surrounding air pressure and in this situation a saturated aggregate would be created. Within the aggregate the contractile meniscus between the water and air phases would now be concave to the water phase. In accordance with *Principle 2*, the sum of the individual enthalpies of the free water droplet (including that of the contractile skin, noting that the water pressure exceeds the air pressure) and the soil particles before they were combined can be equated to the total enthalpy of the saturated aggregate after they are combined (including that of the effective stress and the contractile skin, noting that now the air pressure exceeds the water pressure). This requires the assumption that the volumes of water and soil particles remain unchanged, and thus no work is done in combining the water droplet and soil particles, and that the process of aggregate creation is isothermal. Ignoring the small components of enthalpy associated with the water vapour pressure and dissolved air and using the subscripts 'i' and 'f' to denote initial and final conditions respectively:

$$H_{wi} + H_{si} = H_{wf} + H_{sf} \qquad [7.14]$$

or

$$u_w^* V_w - H_{Ci} + u_{si} V_s = u_w V_w + H_{Cf} + u_{sf} V_s + p_f' V \qquad [7.14a]$$

where

$$H_{wi} = u_w^* - H_{Ci} = u_w^* V_w - \frac{3}{2} (u_a - u_w^*) V_w$$
$$H_{si} = u_{si} V_s = u_a V_s$$
$$H_{wf} = u_w V_w + H_{Cf} = u_w V_w + A_c (u_a - u_w) V_w$$
$$H_{sf} = u_{sf} V_s + p_f' V = a u_w V_s + p_f' (V_w + V_s)$$
$$u_w^* \text{ is the water pressure in a droplet of radius } R^*$$

H_{Ci} is the component of enthalpy of the contractile skin of a spherical water droplet (determined in a way similar to Equation 7.9 for air bubbles)

H_{Cf} is the component of enthalpy of the contractile skin in an aggregate of soil particles and water

A_c is a shape factor for the contractile skin of an aggregate of soil particles and water

p'_f is the mean effective stress within a saturated aggregate due to suction alone $V = V_w + V_s$.

And in accordance with Equation 1.16:

$$u_a - u_w^* = -\frac{2T_c}{R^*} \tag{7.15}$$

For a number of isolated aggregates created by combining soil particles with water droplets, Equation 7.14a can be written as:

$$-\frac{3}{2}(u_a - u_w^*)n_w + u_w^*n_w + u_a n_s = A_c(u_a - u_w)n_w + u_w n_w + a u_w n_s + p'_f(n_w + n_s) \tag{7.16}$$

where

$$n_w = N_a \frac{4\pi}{3}(R^*)^3 \tag{7.17}$$

where N_a is the number of aggregates per unit volume of soil (the total volume of water in the aggregates must equate to the volume of water contained in the free water droplets).

In equating the enthalpies before and after combining the soil particles and water into aggregates, the increase in enthalpy associated with the inter-particle stresses must balance any reduction in potential associated with the other components of the total enthalpy. Within the aggregates the inter-particle stress is given by $p'_f = (u_a - u_w)$. This is consistent with Terzaghi's effective stress acting between the particles: $p'_f = (p - u_w)$, with $p = u_a$ as the isolated saturated aggregates are surrounded by air. Gens and Alonso (1992) and Tang and Cui (2009) argued that the effects of suction and mean net stress on the micro-mechanical volume changes of saturated aggregates are similar, and accordingly Terzaghi's effective stress applies to the aggregates. Monroy et al. (2010) from tests on London clay also suggested that the principles of effective stress apply to saturated aggregates. Substituting for p'_f in Equation 7.16 gives:

$$A_c(u_a - u_w)n_w = b(u_a - u_w)(n_w + n_s) \tag{7.18}$$

where

$$b = \frac{n_w}{n_w + n_s}\left[1 - \frac{5(u_a - u_w^*)}{2(u_a - u_w)} + \frac{(u_a - a u_w)n_s}{(u_a - u_w)n_w}\right] - 1 \tag{7.19}$$

This equation provides a relationship between the variables a and b. Substituting for b in Equation 7.13 gives:

$$\alpha = 1 + \frac{5(u_a - u_w^*)}{2(u_a - u_w)}\frac{n_w}{(n_w + n_s)} \tag{7.20}$$

where $(u_a - u_w^*)$ is negative, being the difference between the external air pressure and the water pressure inside a spherical bubble of water with the same volume as that of the water within a saturated aggregate.

For a dry soil $n_w \approx 0$ and from Equation 7.20 $\alpha \approx 1$. For a saturated or near-saturated soil, if Terzaghi's effective stress equation applies, $(n_w + n_s) = 1$ and again $\alpha = 1$. Between these two extreme cases, $\alpha \leq 1$. Murray (2002) concluded that from an analysis of a wide range of data from triaxial testing on kaolin, after Wheeler and Sivakumar (1995), that even under reasonably worst-case conditions α varied only between 0.96 and 1.00, and that there was little error in the assumption of $\alpha = 1$. The boxed section presents details of a calculation to determine α.

Example Determination of α

It is necessary to assign a size to the aggregates. These are taken to vary between silt and coarse sand size (0.01–1.00 mm) and for simplicity to be spherical, although this is not considered to detract from the conclusions reached. The smaller the aggregate, the greater is the influence on α. Silt size has been chosen as the lower limit because it has been noted that silt-size aggregates are present in clays that have been subjected to repeated extremes of wetting and drying (e.g. Chandler and Davis, 1973). The upper limit has been chosen as coarse sand size, although aggregates could be larger than this in soils where there is a low air voids content.

From a selected result of Wheeler and Sivakumar (1995), at the critical state, $n_a = 0.087$, $n_w = 0.399$, $n_s = 0.514$, $u_a = 349.75$ kPa and $u_w = 50.35$ kPa.

For a silt-size saturated aggregate: The volume V_p of a saturated aggregate $= 5.24 \times 10^{-13}$ m^3 and $N_a = (n_w + n_s)/V_p = 1.74 \times 10^{12}$ m^{-3}. From Equation 7.17, the radius of a free water droplet $R^* = 3.79 \times 10^{-5}$ m. From Equation 7.15, taking $T_c = 73 \times 10^{-6}$ kNm^{-1} at 20°C, $(u_a - u_w^*) = -3.85$ kPa. From Equation 7.20, $\alpha = 0.986$.

For coarse sand-size saturated aggregate: $V_p = 4.19 \times 10^{-9}$ m^3, $N_a = 2.18 \times 10^{-4}$ m^{-3}, $R^* = 7.59 \times 10^{-4}$ m, $(u_a - u_w^*) = -0.19$ kPa and $\alpha = 0.999$.

A range of values of a and b for the aggregates can also be determined from the experimental data as in the following boxed section. The analysis indicates that within the range of experimental data, the component of enthalpy in the soil particles as a result of imposed air and water pressures under unsaturated conditions can be taken as that under saturated conditions ($a = 1$) without any appreciable error, and the influence of the surface tension in the contractile skin is small and can be ignored ($b = 0$).

Determination of a and b

Having determined values of α, Equations 7.13 and 7.19 may be used to show that for silt-size aggregates $1 \leq a \leq 1.15$ and $0.14 \geq b \geq 0$, and for coarse sand-size aggregates $1 \leq a \leq 1.01$ and $0.0007 \geq b \geq 0$. From Equation 7.8, for the pressure in the soil particles, $a \geq 1$ and $a = 1$ under saturated conditions. In addition, from Equation 7.10 for the contractile skin, $b \geq 0$, with $b = 0$ for saturated conditions. The results indicate that there is little error in taking $a = 1$ and $b = 0$. It is not clear whether this is true under lower degrees of saturation outside the experimental range.

Combining the water phase and the soil particles has been shown to minimise the water pressure, and by comparison with experimental data for kaolin, it has been shown to minimise the fluid pressure acting through the soil particles and the influence on the thermodynamic potential of the contractile skin. This leads to the condition $\alpha = 1$. It is also necessary, however, to consider the interactions between the soil particles as the minimisation of the thermodynamic potential also requires the stresses within the soil system to adjust to minimise the 'effective' stress. For a saturated soil from Equation 7.12:

$$p'_c = p' = p - \alpha u_w \qquad [7.21]$$

As discussed previously, Terzaghi's equation holds true if α is a maximum of 1.0. This is consistent with p' being a minimum in Equation 7.21. Similarly for unsaturated conditions, Equation 7.12 can be written as Equation 7.22 if for clarity we remove the air pressure term u_a and take this as the datum pressure:

$$p'_c = p - \alpha u_w \frac{v_w}{v} \qquad [7.22]$$

p'_c is a minimum if $\alpha = 1$. There should, however, be a caveat included with the statement 'the "effective" stress achieves a minimum'. A more complete statement would be 'the "effective" stress achieves a minimum under the volumetric restrictions' as it is necessary to allow for meta-stable equilibrium of the soil particle structure as the degree of saturation reduces as discussed in Section 5.8.2.

In accordance with Equation 7.6 and ignoring the influences of the water vapour in air, dissolved air in water and adsorbed water, each of the remaining components of the total enthalpy has been shown to minimise if the water and soil particles are combined into aggregates and the condition $\alpha = 1$ is adopted. This complies with the thermodynamic principle of H being a minimum at equilibrium, as discussed in Section 5.8.2.

In Chapter 6, in the analysis of undrained and drained 'effective' stress tests in the triaxial cell, the reader was asked to accept the conjecture that the fluid pressure acting through the solid phase could be taken as equivalent to the pore water pressure and the phase interaction effects could be ignored. This is equivalent to adopting $\alpha = 1$, justification for which has now been given. It leads to similar equations governing the equilibrium stress regime in unsaturated soils in the triaxial tests based on virtual work considerations as developed in the current chapter using enthalpy as an extensive variable.

It is important to recognise that the overall stress regime in an unsaturated soil is influenced by discontinuities. Essentially the stress regime described is considered applicable where the soils remain relatively intact. However, discontinuities such as open tension cracking in very dry soil means there is no stress transmitted across the open fissure and such features have no shear strength. Any assessment of the overall behaviour of the ground must take account of such discontinuities. In stability analysis this is normally done by modelling tension cracking as distinct features outside the stress regime and strength assumed throughout the remainder of the soil mass. Further research is required on this topic.

7.6 Stress state in unsaturated soils

On the basis of $\alpha = 1$, Equation 7.12 can be written as (Murray, 2002; Murray and Sivakumar, 2005; Murray and Sivakumar, 2006):

$$p'_c = (p - u_a) + s\frac{v_w}{v} = \overline{p} + s\frac{v_w}{v} \qquad [7.23a]$$

This equation describes the mean stress condition in unsaturated soil under equilibrium conditions. By rearranging the variables the following two alternative representations of the equation can be determined:

$$p'_c = (p - u_a)\frac{v_a}{v} + (p - u_w)\frac{v_w}{v} = \bar{p}\frac{v_a}{v} + p'\frac{v_w}{v} \qquad [7.23\text{b}]$$

$$p'_c = (p - u_w) - (u_a - u_w)\frac{v_a}{v} = p' - s\frac{v_a}{v} \qquad [7.23\text{c}]$$

The equations link the stress state variables to the volumetric variables and define the average volumetric coupling stress p'_c. The stresses are not independent of the volumetric variables, as in saturated soils. The basis of much recent research has been the use of independent stress state variables. Equations 7.23a–c should not be viewed as adding complication to analysis by the inclusion of volumetric variables, but should be considered as removing a tier of uncertainty inherent when using independent stress state variables. Using the equations, more fundamental and consistent parameters are determined, and the links between the stresses and soil behaviour are brought to the fore when examining experimental data in Chapters 8 and 9.

Since enthalpy is an extensive variable, the influence of the individual stresses is additive and from Equations 7.23a–c, consistent with the analysis of Chapter 6, it is possible to write the general stress tensor equations for unsaturated soils under equilibrium conditions:

$$\sigma'_{c,ij}v = (\sigma_{ij} - u_a\delta_{ij})v + (u_a - u_w)\delta_{ij}v_w \qquad [7.24\text{a}]$$

$$\sigma'_{c,ij}v = (\sigma_{ij} - u_a\delta_{ij})v_a + (\sigma_{ij} - u_w\delta_{ij})v_w \qquad [7.24\text{b}]$$

and

$$\sigma'_{c,ij}v = (\sigma_{ij} - u_w\delta_{ij})v - (u_a - u_w)\delta_{ij}v_a \qquad [7.24\text{c}]$$

where $\sigma'_{c,ij}$ is the 'coupling stress' tensor.

For a saturated soil, Equations 7.23a–c reduce to Terzaghi's mean stress Equation 1.24, and Equations 7.24a–c reduce to Terzaghi's directional stress Equation 1.22. For a perfectly dry soil, Equations 7.23a–c and 7.24a–c do not reduce directly to Equations 1.25 and 1.23 respectively. As a soil dries, the condition $p'_c = (p - u_a) + (u_a - u_w)/v$ is approached, with the second term being a result of the high suctions associated with the very small volumes of water remaining where the particles are closest together. Only on complete breakdown of the suction does $p'_c = (p - u_a)$.

7.7 Alternative equilibrium analysis

Equilibrium of forces on a flat plane of area A_p cut through a saturated soil was considered in Section 1.4.7. Account was taken of the water pressure acting not only through the volume of the water but, because it surrounded the soil particles, also acting through the volume of the solids. For a plane cut through a saturated soil, under equilibrium conditions, the applied force σA_p acting on one side of the plane was balanced by the forces internal to the system giving Terzaghi's (1936) effective stress equation $\sigma' = \sigma - u_w$.

Consider now a similar representative plane of area A_p cut through an unsaturated soil as depicted in Figure 7.2. Equilibrium conditions are considered to prevail and the applied stress σ acting on one side of the plane is balanced by the pore air pressure u_a acting over area A_a, the pore water pressure u_w acting over area A_w and the forces associated with the soil particles. The latter can be thought of as comprising two components: (i)

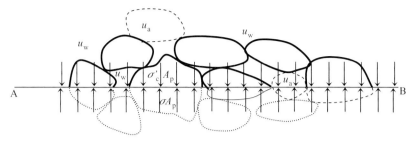

Figure 7.2   Equilibrium on plane AB.

the 'coupling' stress σ'_c as a result of interactions between the solid particles acting over area A_p, consistent with the principles of continuum mechanics, and (ii) the stress internal to the soil particles u_s acting over area A_s. The stress u_s is a result of the air and water pressures acting on the particles. Thus, from equilibrium of forces acting on the plane:

$$\sigma A = \sigma'_c A_p + u_w A_w + u_a A_a + u_s A_s \qquad [7.25]$$

Multiplying throughout by a characteristic length L and substituting for $A_p L = V$, $A_w L = V_w$, $A_a L = V_a$ and $A_s L = V_s$ gives:

$$\sigma V = \sigma'_c V + u_w V_w + u_a V_a + u_a V_a + u_s V_s \qquad [7.26]$$

A bi-modal structure exists in unsaturated soils. This can be idealised as aggregates of soil particles and water, surrounded by air voids, with different stress regimes implied as existing within and between the aggregates. The concept of saturated aggregates is thus a useful representation to illustrate the dual stress regime considered present in unsaturated soils. At high degrees of saturation the particles will be surrounded by water other than possibly at the extremities of the aggregates. Under these conditions, it can be argued that $u_s \approx u_w$. This simplification is thought to be valid over a relatively wide range of water contents. As in the foregoing discussion, the influence of other forces (as arising from the contractile skin and other phase interaction effects) can be ignored for a wide range of degrees of saturation. Thus, taking u_s as u_w and noting that $V = V_w + V_s + V_a$, Equation 7.26 can be written as:

$$\sigma'_c = (\sigma - u_a) V + (u_a - u_w)(V_w + V_s) \qquad [7.27]$$

Equation 7.27 leads directly to Equations 7.23a–c and 7.24a–c.

7.8 Graphical representation of stress state in unsaturated soils

Equations 7.23a–c and 7.24a–c can be represented graphically. As an example, Figure 7.3 provides a graphical representation of Equation 7.23b. The volumetric coupling stress p'_c can be seen as the mean stress based on the relative volumes through which the stresses and pressures of the phases in an unsaturated soil act.

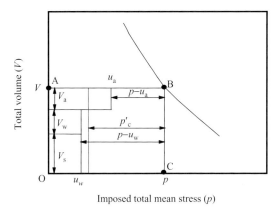

Figure 7.3   Compression plot for an unsaturated soil (after Murray, 2002). 2008 NRC Canada or its licensors. Reproduced with permission.

7.9 Stress state variables and conjugate volumetric variables

In an analysis of the behaviour of unsaturated soils, it is essential to be able to define the controlling stress regime and the volumes through which such stresses act. From such considerations, the work input due to the application of an external stress regime can be determined as in Chapter 9.

Equations 7.23a–c and 7.24a–c link the stress state variables to the conjugate volumetric variables (Murray *et al.*, 2002). The equations comply with the suggestion of Fredlund and Morgenstern (1977) that any two of the three stress state variables can be used in describing the stress state of an unsaturated soil. The equations describe three views of the dual stress regime considered to exist in an unsaturated soil as a result of a bi-modal structure. The equations do not rely on a specific soil density or particle distribution, but do rely on a close association between the soil particles and pore water and that the matric suction is everywhere the same. The equations describe the stress regime relating to the volumetric conditions.

The mean stresses in Equations 7.23a–c, and the directional stresses (or stress tensor terms) in Equations 7.24a–c, are intrinsically linked with volumetric terms that define the conjugate stress–volumetric pairings in unsaturated soils. To explain their significance, the stress tensor terms given by Equation 7.24b may be compared with the stress tensor equations 4.1 and 4.2 for saturated and perfectly dry soils respectively. In Equation 7.24b, $\sigma'_c v$ is made up of two components. The second term on the right of the equation indicates that the stress state variable given by Terzaghi's effective stress equation 4.1 acts through the aggregated regions, the volume of which is defined by the specific water volume v_w (i.e. the volume of water and solids per unit volume of solids). The first term on the right of Equation 7.24b indicates that it is necessary to account for the additional stress state variable given by Equation 4.2, for a perfectly dry soil, which acts within the remaining volume between the aggregates and is given by $v_a = (v - v_w)$ (i.e. the volume of air voids per unit volume of solids). In simple terms, Equation 7.24b for the coupling stress demonstrates that the stress state variables $(\sigma - u_w)$ and $(\sigma - u_a)$, which are in essence 'effective stresses', can be considered as acting within specific volumes of an unsaturated soil as intra-aggregate and inter-aggregate stresses respectively. An alternative view of

Equation 7.24b is provided by consideration in terms of mechanical potential. As the potential terms comprise extensive pairings, $(\sigma - u_w) v_w$ can be viewed as a component of the mechanical potential within the aggregates and $(\sigma - u_a) v_a$ as a component of the mechanical potential between the aggregates.

The simple visualisation of an unsaturated fine-grained soil as comprising saturated aggregates of soil particles and water, surrounded by air-filled voids, is useful in describing the dual stress regime and conforms to the observation by Croney et al., (1958) that compacted clays comprised aggregations of clay particles. This was confirmed by Barden and Sides (1970) from microscopic examination. Brackley (1973, 1975) proposed a model for soil structure based on the soil 'aggregates' in which the intra-aggregate voids were saturated. Tang and Cui (2009) made a similar assumption when examining the thermo-mechanical volume change behaviour of compacted expansive clays. The idealisation is supported by the MIP results reported in Chapter 1 and with the contention that the water phase has a close affinity with the soil particles, and with the requirement that the thermo-dynamic potential is a minimum under equilibrium conditions (Murray and Brown, 2006; Murray and Sivakumar, 2006). Equation 4.1 for a saturated soil and Equation 4.2 for a perfectly dry soil can be seen as special cases of Equation 7.24b, and thus of the equivalent equations 7.24a and 7.24c. They indicate that for an unsaturated soil the stress regime can be viewed, in simple terms, as complying with a saturated soil within the aggregates and with a perfectly dry soil in the remaining air void spaces between the aggregates. Within this framework the aggregates are considered to act as large 'deformable particles'. This is supported by the experimental evidence of Sivakumar (2005) who concluded that during wetting, saturated aggregates expand and deform with increasing water content closing up the inter-aggregate air voids. However, the simplified view of saturated aggregates does not sit well with the presence of water bridges between aggregates and which provide continuity of the water phase throughout a soil at high degrees of saturation. Equally on drying, the description does not sit well with the water phase being drawn back further and further into the finer intra-aggregate pore spaces. However, sight should not be lost of the fact that the term 'saturated aggregates' is a simplifying descriptive expediency and the equations merely require that the water and soil particles are intrinsically linked.

The stress state variable $s = (u_a - u_w)$ is the difference between the two effective stresses given by the other two stress state variables. Thus, Equations 7.23a–c and 7.24a–c for an unsaturated soil indicate that the stress regime can be described in terms of either two effective stresses or by one of the effective stresses and the difference between the effective stresses. However, the volumes through which these stresses act must be taken into account as only the products of the stress state variables and the conjugate volumes as in Equations 7.23a–c and 7.24a–c define additive extensive thermodynamic terms (Murray and Brown, 2006; Murray and Sivakumar, 2006). The reason the 'effective stress' equations 4.1 and 4.2 for saturated and perfectly dry soils hold true is that the volumetric terms cancel.

The overall stress–strain behaviour of unsaturated aggregated materials is controlled by two distinct internal mechanisms as a result of the bi-modal pore size distribution: interaction between the aggregates and interaction of particles within the aggregates. Sivakumar et al. (2010b) argued that for a given suction, at low confining pressures it is the interaction between the aggregates that is the dominant mechanism, but at high confining pressures the particles within the aggregates begin to play a more significant role.

Much research has been undertaken and results presented in the literature on the use of independent stress state variables. The stresses are generally considered independent of the volumes through which the stresses and pressures act. While it is shown in the foregoing analysis that this is a valid assumption in saturated soils as the volume terms cancel out in

Terzaghi's effective stress equation, in unsaturated soils any relationships that include only stresses will have coefficients with a hidden reliance on the relative volumes of the phases. In unsaturated soils, it is necessary to take account of the conjugate variables of pressure (or stress) and the volumes (or strain-increment) through which they act. Constitutive analysis which takes account only of the stress regime is akin to rowing a boat with one oar. The boat will only go round in circles.

7.10 Hysteresis, collapse and discontinuities in soil behaviour

In Section 5.3 we described equilibrium conditions and outlined the significance of stress history to soil behaviour. In this context it is necessary to talk of small changes from the equilibrium state where the deviation from non-reversibility can be ignored and the changes between equilibrium states can be described by exact differentials dependent only on end conditions. The dependence of soil behaviour on stress history is apparent in water characteristic curves obtained on wetting and drying. On de-saturation of a fine-grained soil, an aggregated structure results as air begins to fill the larger void spaces, while the water phase tends to confine itself to the smaller intra-aggregate pore spaces, although a small amount of water remains at the inter-aggregate contact points. The aggregated structure persists during subsequent wetting and drying. Only if the soil is wetted and sufficiently agitated would it be possible to restore a dispersed soil structure.

Wheeler *et al.* (2003) described the existence of *bulk water* within the water-filled voids and *meniscus water* at the inter-particle contacts around air-filled voids. The bulk water complies with the water retained within the smaller intra-aggregate pores of fine-grained soils and the meniscus water complies with the water at the points of contact between the aggregates. The suction within bulk water influences the normal and tangential forces at particle contacts, whereas the suction within meniscus water influences only the normal forces at inter-particle or inter-aggregate contacts. As discussed in Chapter 4, Wheeler *et al.* (2003) linked the irreversibility during wetting and drying, and the onset of plastic deformations at suctions below suction levels previously experienced, with hydraulic hysteresis forces arising from the bulk and meniscus water. This was argued as giving rise to not only hysteresis but also a plastic creep phenomenon during repeated wetting and drying.

In Section 5.8 a detailed description was given of the minimisation of the thermodynamic potential at equilibrium and at meta-stable equilibrium, along with the significance of staged equilibrium, the likelihood of abrupt energy changes at the extremes of system changes and the significance of a disturbing force to change from meta-stable equilibrium to a more stable equilibrium. The discussion drew on analogous behaviour observed in chemistry and physics to illustrate the concepts. It was argued that similar behavioural trends can be expected when examining unsaturated soils. This is further examined below in terms of soil collapse and hysteresis phenomena.

It is interesting to re-examine the results of a suction-controlled oedometer test on highly expansive clay reported by Alonso *et al.* (1995). The results are presented in Figure 7.4. During the first wetting path C_1, initial swelling was followed by collapse compression as the suction was progressively reduced. The plot shows significant irreversible components of compression during subsequent drying stages of the wetting–drying cycles C_2 to C_5. The phenomenon of collapse is consistent with an abrupt energy change from a meta-stable to a more stable, lower potential state. The subsequent phenomena of hysteresis

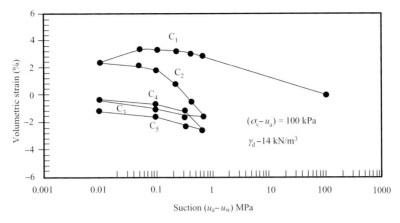

Figure 7.4   Wetting–drying cycles performed on Boom clay under oedometer conditions (after Alonso *et al.*, 1995). Reproduced by permission of Taylor and Francis Group

and plastic irreversible strains are consistent with meta-stable conditions and the soil not necessarily changing directly into the most stable state. The soil exhibits a staged transition analogous to the transient meta-stable phases observed by Ostwald (1897). Mathematical models of soils must take into consideration these meta-stable states in addition to considering the most thermodynamically favourable state. While at each point on the curves a stable state may be reached, in thermodynamic terms it is not clear that an equilibrium state with the lowest possible potential has been achieved. Hysteresis suggests that changes are not reversible and a more stable equilibrium state could exist. The existence of various stable states for similar water contents is consistent with meta-stable equilibrium conditions. Thermodynamic principles of potentials can be used to examine the significance of hysteresis.

Figure 7.5 shows a schematic plot of specific volume v against suction s to a log scale for a soil drying from slurried conditions. The plot can be compared with that of Figure

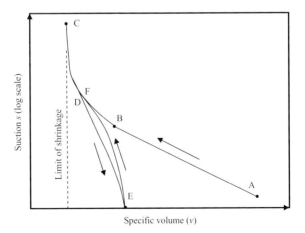

Figure 7.5   Idealised soil shrinkage characteristics.

1.9 from the results of Croney and Coleman (1960) for London clay. While saturated, and with a dispersed structure, the soil often plots as a relatively straight line A–B. On further reduction in water content and de-saturation, the volume follows a curve such as BC, which is far from linear, and suction approaches a maximum value of around 10^6 kPa. On de-saturation an aggregated structure forms as air begins to fill the larger void spaces with the water filling the smaller intra-aggregate pore spaces. The aggregated structure persists for subsequent wetting and drying and follows characteristic S-shaped paths such as CDEF. Only if the soil is wetted and agitated would it be possible to restore a dispersed soil structure. Cyclic wetting and drying tests on compacted, remoulded expansive soils have shown that there is a build-up of residual shrinkage for about three to five cycles (Subba Rao and Satyadas, 1987; Al-Homoud et al., 1995; Songyu et al., 1998 amongst others).

The wetting–drying curve for a soil is often taken as the definitive indication of non-reversibility of soils and the potential influence of soil stress history. For a given water content there is a range of suctions (or suction potentials) that could exist. It is important to note that a soil exhibits significant fabric differences, phase constituent differences, stress (pressure) differences and volume differences throughout a wetting–drying cycle. The material essentially changes from point to point. At each point, the soil minimises the thermodynamic potential but this is unlikely to be an absolute minimum potential. The soil is likely to be in a meta-stable state. Nevertheless, at each point the potentials associated with the phases and interactions within a soil will minimise.

The drier a soil becomes, the less the water pressure and the greater the suction, the influence of the contractile skin and the particle–water attraction. Sivakumar (2005) presented wetting data on compacted kaolin, which indicate that water was drawn into the saturated aggregates and on wetting did not start to fill the inter-aggregate voids until suctions were below 100 kPa. While Equations 7.23a–c and 7.24a–c for the stress conditions in an unsaturated soil are reducible to the special cases of a saturated soil and a perfectly dry soil, their applicability towards the extremes for soils with a high degree of saturation, where the air is in isolated bubbles, and for a dry soil, where the water phase is restricted to the finest intra-particle pores, needs further experimental evidence.

Toll (1990, 2003) suggested that the packets of aggregated particles act like 'large particles' that are maintained by the suction and are not easily broken down or destroyed even at shearing to the critical state. This is consistent with the suggestion of Murray (2002) that under shearing to the critical state the aggregated structure in unsaturated soils only breaks down to a more dispersed structure at low values of suction and there is a discontinuity, or abrupt energy change, between saturated and unsaturated conditions under shearing. A discontinuity in behaviour is also considered to exist on drying a soil. Arguments developed previously suggest that an unstable particle structure is likely to develop as a result of drying. This results in the water phase and contractile skin confining themselves to the smaller intra-aggregate pore spaces in order to minimise their potential. The elevated suction helps to maintain the soil particle structure. The soil particles are, as a consequence, restrained, which results in a particle structure liable to collapse settlement particularly under loading or inundation. The processes of extreme wetting and drying thus lead to latent abrupt energy changes.

Other discontinuities in behaviour occur in soils on shearing; dilation is a prime example. There has also been research into the abrupt changes associated with air entry on change from a saturated soil to an unsaturated soil. At the other extreme of dry soils, the collapse of soil structure has already been discussed. In fact, collapse has been recorded over a wide range of de-saturation. There are numerous examples in physics and chemistry of such meta-stable conditions leading to abrupt energy changes. These energy changes and the general behaviour of unsaturated soils on shearing will be examined experimentally in Chapter 9.

Despite these apparent discontinuities in behaviour, the equations for the stress regime in unsaturated soils are still applicable under equilibrium and meta-stable equilibrium conditions.

7.11 Conclusions

A number of important conclusions have been reached in this chapter:

- Using enthalpy as the thermodynamic potential in a pressure-controlled test (triaxial cell), three forms of an equation governing the stress regime under equilibrium conditions have been developed. These are presented as mean stress equations (Equations 7.23a–c) and directional stress equations (Equations 7.24a–c) and indicate that any two of the three stress state variables can be used to describe the stress regime in an unsaturated soil.
- The use of the three independent stress state variables leads to models that need portrayal in three-dimensional stress space. A major advantage of the coupling stress is that it allows the shear strength data presented in Chapter 8 to be unified on two-dimensional plots.
- The importance of the conjugate pairings of stress state variables and volumetric variables is described, which are consistent with the principles of thermodynamics. The equations do not rely on a specific soil density or particle distribution other than that the soil particles and the water are intrinsically linked and that the matric suction is everywhere the same.
- Evidence is presented to suggest that the equations are applicable under equilibrium (lowest thermodynamic potential) and meta-stable equilibrium conditions. It is tentatively suggested that they apply over a wide range of degrees of saturation although further experimental evidence is required particularly at the drier end of this spectrum.
- The significance of hysteresis and stress history is described in terms of equilibrium and meta-stable equilibrium as a precursor to the experimental evidence in the following chapters.

Notes

1. Amenu et al. (2005) and Chen and Kumar (2004) use enthalpy summation equations to examine the interaction of the soil moisture profile and subsurface temperature profile with land–atmosphere energy fluxes due to weather variations over time.
2. This is analogous to the condition in a saturated soil where the water pressure acts through the total volume of the soil comprising the water phase and solid phase, as discussed in Chapter 1.

Chapter 8
Shear Strength and Compression Characteristics of Unsaturated Soils

8.1 Introduction

The strength and deformation characteristics of unsaturated soils are complex. Material behaviour is influenced by changes under drying, wetting, loading and unloading, as well as the 'old devil' time. Yet predictive approaches need to be developed if the performance of engineering structures comprising unsaturated soils, or the interaction of engineering structures with the ground in an unsaturated state, is to be adequately assessed. In this chapter we propose methods of interpreting soil strength and volume change, which form a three-dimensional model in dimensionless stress–volume space. We will first deal with the strength of unsaturated soils, in particular, the critical state strength, and will then examine the volume change behaviour (Murray, 2002; Murray *et al.*, 2002). Relationships will be established and validated by comparison with published experimental data.

Attempts to improve predictive methods must take account of the stress regime controlling the physics of the material behaviour. The formulation describing the stress regime in unsaturated soils under equilibrium conditions developed in the preceding chapter will be used. The stress equations (Equations 7.23a–c and 7.24a–c) confirm that any two of the three stress state variables can be used to describe the stress state in unsaturated soils, although functions of these variables can also be considered. An equation for the deviator stress q at peak strength or at the critical state must reflect the duality of the stress regime implicit in this statement. The duality of the stress regime is a consequence of the three interacting phases and results from the bi-modal structure exhibited by fine-grained soils and comprising aggregates of soil and water surrounded by air voids. A relationship is presented for q that incorporates the 'coupling' stress p'_c which links the dual stress regime to the specific volume v and the specific water volume v_w. The strength of unsaturated soils defined in terms of q is thus not independent of phase volumes as is usually considered to be the case in saturated soils.

Published data on unsaturated kaolin, a lateritic gravel, a residual soil, bentonite enriched sand (BES) and a rock powder are used to validate the relationship for q. Consistent trends are indicated for all five materials. The strength equation based on the coupling stress provides a compelling basis for understanding and predicting large-strain strengths of a wide range of unsaturated soils.

Subsequent examination of the volume change characteristics also utilises the equations for the stress regime in unsaturated soils developed in the preceding chapter. Published experimental results for kaolin are used to examine the *iso-ncl* and *csl* for unsaturated soils and to link them in three-dimensional stress–volume space using the dimensionless variables $\bar{p}/s$, q/s and v_w/v. Consistent but complex trends are indicated by the model.

8.2 Shear strength and critical state characteristics of unsaturated soils

We will first examine the critical state strength and critical state volumetric characteristics of kaolin and determine an equation for the strength using the coupling stress p'_c defined in Chapter 7. The equation will then be used to examine published shear strength data for other materials before drawing generalised conclusions on material behaviour.

8.2.1 Kaolin (Sivakumar, 1993; Wheeler and Sivakumar, 1995, 2000; Sivakumar, 2005; Sivakumar et al., 2010a)

The results of shearing tests in the triaxial cell on lightly and heavily statically compressed specimens and lightly dynamically compacted specimens of speswhite kaolin have been reported by Sivakumar (1993) and Wheeler and Sivakumar (1995, 2000). The kaolin had liquid and plastic limits of 70 and 34% respectively. Aggregates of particles were prepared using a 1.12-mm sieve prior to specimen preparation (as discussed in Chapter 3). The one-dimensionally compressed and compacted specimens were prepared at different water contents and were subsequently subjected to isotropic consolidation in the triaxial cell, prior to shearing to critical state along various stress paths. The shearing stages were carried out at constant suctions of 100, 200 or 300 kPa using the axis translation technique. For brevity, these tests will be referred to as 1d-cs tests, i.e. one-dimensionally compressed, constant suction shearing tests.

Figure 8.1 presents plots of the conjugate variables v and p'_c (from Equation 7.23a–c) at the critical state with the stress plotted to a log scale. The nature of the initial one-dimensional compression or compaction of the specimens does not appear to influence the conditions at critical state. The results portray a logical, sequential trend with the plots for constant suction s converging on the $s = 0$ line for increasing stress level as the significance of s reduces. The results of Maâtouk *et al.* (1995) amongst others support such a trend in behaviour.

Wheeler and Sivakumar (1995) produced critical state plots of v against $\ln(p - u_a)$ for different suctions (see Figure 8.2) and recognised the apparent anomalous position of the line for $s = 100$ kPa lying significantly below the line for $s = 0$. They suggested the possibility of a discontinuity in behaviour at the transition between unsaturated and saturated conditions. Wheeler and Sivakumar (2000) examined additional data for kaolin and reported similar behaviour.

In accordance with Equation 7.23a, v plotted against $(p - u_a)$ does not take account of the duality of the controlling stress regime. The apparent anomalous change in behaviour between unsaturated and saturated conditions for kaolin is considered consistent with the development and influence of a dual stress regime and the difference in behaviour between a dispersed and aggregated structure. The use of p'_c takes account of the duality of the stress regime and Figure 8.1 for kaolin does not exhibit the apparent inconsistency that is observed in plotting v against $(p - u_a)$ in Figure 8.2.

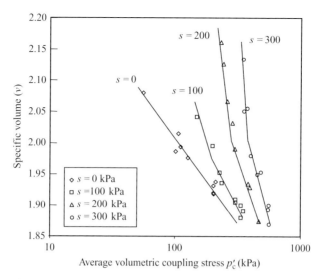

Figure 8.1   Specific volume v against p_c' at critical state from 1d-cs test results of Sivakumar (1993) and Wheeler and Sivakumar (1995, 2000) for kaolin.

Figure 8.3 also presents results from the shearing stages of the triaxial tests on kaolin. The deviator stress q at critical state has been plotted against p_c'. All lines are tentatively interpreted as straight and parallel within the range of data presented. The question that must be asked is why do the plots for greater suctions lie below the line $s = 0$? The primary reason is that within an unsaturated soil, there is preferential shearing between (not through) the more highly stressed aggregates. Preferential shearing tends to reduce

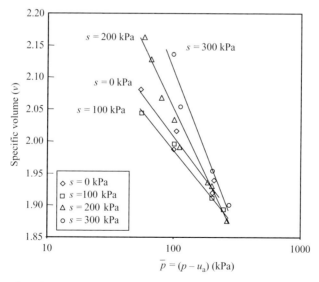

Figure 8.2   Specific volume v against $\overline{p}$ at critical state from 1d-cs test results of Wheeler and Sivakumar (1995) for unsaturated kaolin.

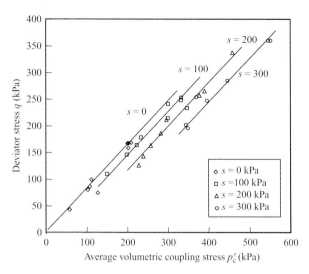

Figure 8.3   Deviator stress q against p'_c at critical state from 1d-cs test results of Sivakumar (1993) and Wheeler and Sivakumar (1995, 2000) for unsaturated kaolin.

shearing resistance. The conjugate variables of Equation 7.23b indicate that the stress between the aggregates is given by $(p - u_a)$, but a greater stress $(p - u_w)$ acts within the aggregates, leading to greater shearing resistance. Following the 'path of least resistance' is a theme that runs through science and includes the shearing of unsaturated soils.

In Figure 8.4, the critical state results for q and p'_c for kaolin have been normalised with respect to s. The data lie close to a unique line that can be written as:

$$\frac{q}{s} = M_a \left[\frac{p'_c}{s} \right] + C \qquad [8.1]$$

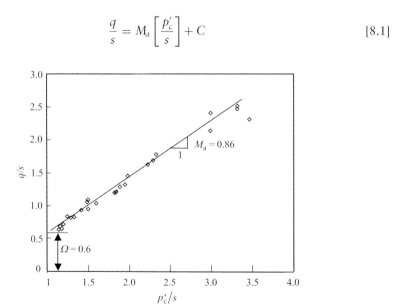

Figure 8.4 q/s against p'_c/s at critical state from 1d-cs test results of Sivakumar (1993) and Wheeler and Sivakumar (1995, 2000) for unsaturated kaolin.

where M_a is the slope of the critical state line based on the normalised axes and C is the intercept of the line at $p'_c/s = 0$.

The equation can be rearranged in the following form:

$$\frac{q}{s} = M_a \left[\frac{p'_c}{s} - 1 \right] + \Omega \qquad [8.2]$$

where Ω is the intercept on the q/s axis at $p'_c/s = 1$.

Substituting for p'_c from Equation 7.23a gives:

$$q = M_a \overline{p} + M_b s \qquad [8.3]$$

where

$$M_b = M_a \left[\frac{v_w}{v} - 1 \right] + \Omega \qquad [8.4]$$

While Equation 8.3 is in a form similar to that of Equation 4.41 suggested by Toll (1990), the interpretation and values of the total stress ratio parameter M_a and suction stress ratio parameter M_b are fundamentally different. The analysis permits an alternative method of analysing the experimental data without the assumption of a smooth transition from unsaturated to saturated conditions. The approach thus allows the results to 'speak for themselves'.

The value of M_a from Figure 8.4 is constant at 0.86. This is slightly greater than $M = 0.82$ for a saturated soil ($s = 0$) in Figure 8.3, determined in accordance with Equation 4.22. This difference, though small, is consistent with differences in soil fabric in unsaturated and saturated soils.

In Figure 8.4, the intercept on the q/s axis is given by $\Omega = 0.6$. This is also the intercept value for other materials tested to critical state and discussed in subsequent sections. The significance of this is appraised in Section 8.2.6. Taking $\Omega = 0.6$, Figure 8.5 presents values for the suction stress ratio M_b from Equation 8.4 plotted against v_w/v. M_b decreases linearly with decreasing v_w/v at an inclination of M_a. It decreases as the degree of saturation decreases, that is, with increasing values of suction. This is consistent with an increasing proportion of the shearing taking place between (not through) the aggregates

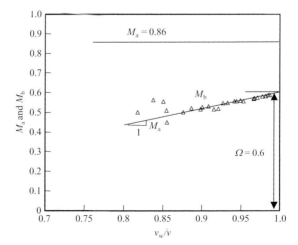

Figure 8.5 Stress ratios M_a and M_b plotted against v_w/v from 1d-cs tests of Sivakumar (1993) and Wheeler and Sivakumar (1995, 2000) for unsaturated kaolin.

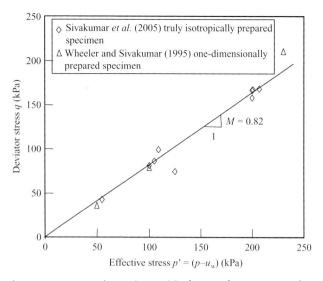

Figure 8.6 Deviator stress q against p' at critical state from test results of Wheeler and Sivakumar (1995) and Sivakumar (2005) for saturated kaolin.

as the soil becomes drier (Leroueil, 1997). As saturation is approached, Figure 8.5 shows that M_b tends towards the condition $M_b = \Omega = 0.6$.

Most published data on unsaturated soils, as with the 1d-cs results of Sivakumar (1993) and Wheeler and Sivakumar (1995, 2000) on kaolin, are on initially anisotropically compacted or compressed soil specimens. Sivakumar (2005) and Sivakumar et al. (2010a) presented data on truly isotropically prepared specimens of kaolin. As for the tests reported by Sivakumar (1993) and Wheeler and Sivakumar (1995, 2000), the aggregated kaolin was graded using a 1.12-mm sieve prior to specimen preparation. Critical state strength results for saturated specimens are presented in Figure 8.6 and results for unsaturated specimens in Figure 8.7. The authors reported results of tests on unsaturated

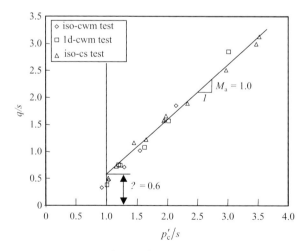

Figure 8.7 q/s against p_c'/s at critical state from iso-cs, iso-cwm and 1d-cwm test results of Sivakumar (2005) and Sivakumar et al. (2010a) for unsaturated kaolin.

specimens where the suction was maintained constant by the axis translation technique during shearing, and the results of constant water mass tests where the suction was allowed to vary and was recorded by a thermocouple psychrometer. The critical state results on the normally compressed unsaturated specimens are presented in Figure 8.7 as normalised plots of q/s against p'_c/s, along with some additional results from constant water mass tests on specimens initially one-dimensionally compressed in layers than subject to isotropic compression prior to shearing. The range of suctions in the suction-controlled tests was up to 300 kPa and in the constant water mass tests was between 400 and 1000 kPa. The tests reported by Sivakumar (2005) and Sivakumar *et al.* (2010a) are given the following abbreviations:

- Constant suction tests on truly isotropically prepared specimens – designated iso-cs tests;
- Constant water mass tests on truly isotropically prepared specimens – designated iso-cwm tests;
- Constant water mass tests on specimens initially one-dimensionally compressed in layers – designated 1d-cwm tests.

The data for isotropically prepared specimens and one-dimensionally prepared specimens iso-cs, iso-cwm and 1d-cwm in Figure 8.7 can be reasonably represented by a single critical state strength line. The value of $M_a = 1.00$ is notably greater than the value of $M_a = 0.86$ obtained for kaolin based on the 1d-cs test results of Sivakumar (1993) and Wheeler and Sivakumar (1995, 2000) as shown in Figure 8.4. There is no material characteristic difference to account for the greater value of M_a, and the value of M for saturated kaolin ($s = 0$) is close to that from the results of Sivakumar (1993) and Wheeler and Sivakumar (1995) as shown in Figure 8.6. The difference in the unsaturated kaolin test data is discussed in Section 8.2.7 in terms of the form of the aggregated soil structure and the influence of test procedure.

For unsaturated specimens, the trend line of Figure 8.7 has been drawn through a value of $\Omega = 0.6$ as for Figure 8.4, though points close to $p'_c = 1$ appear to lie a little below this point. This will be discussed in Section 8.2.6 in terms of the fissured nature of these specimens at elevated suctions.

Figure 8.8 presents the values of M_b, based on $\Omega = 0.6$, but omitting the results considered significantly influenced by fissuring. The data correspond to those of Figure 8.7, and again indicate a consistent trend with results close to a line drawn at an inclination of M_a in accordance with Equation 8.4. The steeper plot in Figure 8.8 compared to that in Figure 8.5 indicates proportionally less shearing taking place between the aggregates in the former figure.

8.2.2 Kiunyu gravel (Toll, 1990)

The lateritic Kiunyu gravel used in these tests had a clay fraction of 8–9% and fines content (clay and silt) of approximately 15%. The specimens were prepared by either static or drop hammer compaction in layers. The triaxial tests were carried out with the water contents and pore air pressures kept constant. Shearing was continued until the specimens were at or reasonably close to critical state with recorded suctions in the range 0–537 kPa. The original test data have been re-evaluated and these are presented in Figures 8.9–8.11. The only results not included are those where s was zero or very close to zero. These tests are designated 1d-cwm tests as defined in Section 8.2.1.

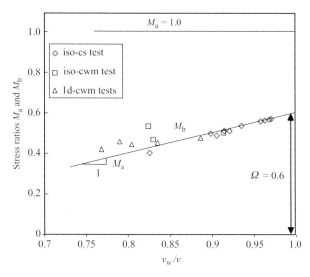

Figure 8.8 Stress ratios M_a and M_b plotted against v_w/v from iso-cs, iso-cwm and 1d-cwm test results of Sivakumar (2005) and Sivakumar *et al.* (2010a) for unsaturated kaolin.

Figure 8.9 shows q at critical state plotted against p'_c. Similar to the results for kaolin shown in Figure 8.3, end-of-test results for the unsaturated lateritic gravel specimens plot below the saturated strength envelope. This is again consistent with the conclusion expressed earlier that there is preferential shearing through the inter-aggregate void spaces. Figure 8.9 does not show plots for constant suctions below the saturated envelope as suctions varied during the tests.

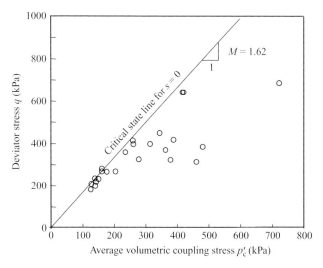

Figure 8.9 Deviator stress q against p'_c at critical state from 1d-cwm test results of Toll (1990) for unsaturated Kiunyu gravel.

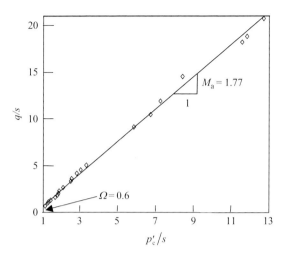

Figure 8.10 q/s against p'_c/s at critical state from 1d-cwm test results of Toll (1990) for unsaturated Kiunyu gravel (after Murray, 2002). 2008 NRC Canada or its licensors. Reproduced with permission.

If the values of q and p'_c shown in Figure 8.9 are normalised with respect to s, the results for all suctions and different initial compaction techniques lie close to a unique line as shown in Figure 8.10. As for kaolin, M_a is again constant and larger than the value of M for saturated material ($s = 0$); for the lateritic gravel $M_a = 1.77$ and $M = 1.62$. Also in Figure 8.10, the intercept $\Omega = 0.6$ is the same as for kaolin in Figures 8.4 and 8.7.

Taking $\Omega = 0.6$, Equation 8.4 can be used to determine values of M_b. These are plotted against v_w/v in Figure 8.11. An important difference from the behaviour of kaolin shown in Figures 8.5 and 8.8 is that the value of M_b drops rapidly to zero at $v_w/v = 0.66$ because of the large value of M_a for Kiunyu gravel, which dictates the inclination of the plot for

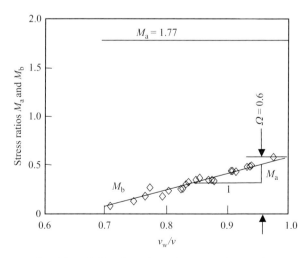

Figure 8.11 Stress ratios M_a and M_b plotted against v_w/v from 1d-cwm test results of Toll (1990) for unsaturated Kiunyu gravel (after Murray, 2002).

M_b. The condition $M_b = 0$ is where the remaining water phase has been drawn back into the finer pores within the aggregates and shearing is wholly concentrated through the regions between the aggregates. Under these conditions, the suction no longer has a direct influence on q other than in maintaining the aggregates.

8.2.3 Jurong soil (Toll and Ong, 2003)

The residual Jurong soil used in these tests had liquid and plastic limits of 36 and 22% respectively, and test specimens were prepared by static compaction in layers. A series of constant water mass triaxial tests was carried out on unsaturated specimens with the matric suction controlled by the axis translation technique. The matric suction varied between 170 and 400 kPa. These tests are designated 1d-cwm tests.

Figures 8.12 and 8.13 indicate a consistent interpretation of the data for unsaturated Jurong soil with M_a constant and M_b decreasing with decreasing v_w/v (increasing suction) at an inclination of M_a. The value of M_a is approximately 1.27, which is again greater than $M = 1.23$ for the saturated soil as quoted by the authors. As with the foregoing results, the intercept Ω on the q/s axis at $p'_c/s = 1.0$ appears close to 0.6.

8.2.4 Bentonite-enriched sand (Stewart et al. 2001)

These triaxial tests were carried out on specimens of saturated Sherburn sand and unsaturated mixtures of bentonite and Sherburn sand. The sand was uniformly grained with minimal fines. The bentonite (sodium-montmorillonite) had a liquid limit of 354% and a plastic limit of 27%. The bentonite-enriched sand (BES) contained 10% by dry weight of bentonite. The sand specimens were prepared by pouring, and the BES specimens by heavy compaction in layers. The tests on the sand were conventional drained triaxial compression tests. The tests on the BES were consolidated triaxial compression tests run with

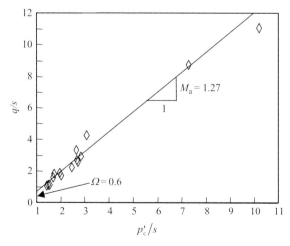

Figure 8.12 q/s against p'_c/s at critical state from 1d-cwm test results of Toll and Ong (2003) for unsaturated Jurong soil.

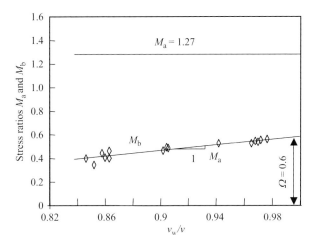

Figure 8.13 Stress ratios M_a and M_b plotted against v_w/v from 1d-cwm test results of Toll and Ong (2003) for unsaturated Jurong soil.

constant water mass. The suctions in the BES specimens were between 68 and 185 kPa at failure. Failure in this case was defined as peak deviator stress and so may not correspond to the critical state condition. The tests on unsaturated BES are designated 1d-cwm tests.

Figure 8.14 shows q plotted against p_c' at failure for both the saturated sand and for the unsaturated BES. All results appear to lie close to a unique line. The value of M for this line is approximately 1.45. The plot supports the conclusion of Stewart *et al.* (2001) that the inclusion of bentonite had no influence on the shear strength, which was controlled

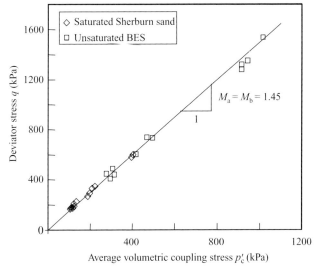

Figure 8.14 Deviator stress q against p_c' from drained tests on Sherburn sand and 1d-cwm tests on unsaturated BES of Stewart *et al.* (2001).

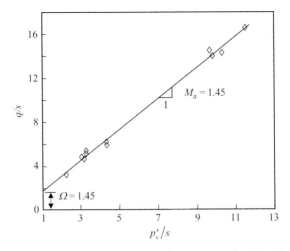

Figure 8.15 q/s against p'_c/s from 1d-cwm tests of Stewart *et al.* (2001) for unsaturated BES.

by the frictional properties of the sand grains. The results reported by Blatz and Graham (2003) on a compacted 50–50 mixture of quartz sand and sodium-rich bentonite support the general behaviour reported by Stewart *et al.* (2001).

Normalisation of q and p'_c with respect to s for the peak deviator stress results for the unsaturated BES in Figure 8.15 again leads to the results following a unique line for all suctions. While this finding leads to a formulation (Equations 8.3 and 8.4) similar to that for kaolin, Kiunyu gravel and the Jurong soil tested to critical state, for the BES $M_a = \Omega = M = 1.45$. This differs from the relationship between M_a and M for the other three materials. Using Equation 8.4 and taking $\Omega = 1.45$, Figure 8.16 shows values of the suction stress ratio M_b plotted against v_w/v. The decrease in M_b with decreasing v_w/v is

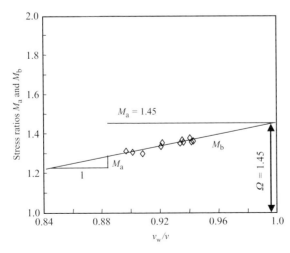

Figure 8.16 Stress ratios M_a and M_b plotted against v_w/v from 1d-cwm test results of Stewart *et al.* (2001) for unsaturated BES.

again linear and at an inclination of M_a, consistent with the critical state data analysed. However, as saturation is approached, the plot of M_b against v_w/v tends towards the condition $M_b = M_a = \Omega = 1.45$.

8.2.5 Trois-Rivières silt (Maâtouk et al. 1995)

These data appear intermediate in character between (i) those for kaolin, Kiunyu gravel and Jurong soil and (ii) those for the BES. The material tested comprised a rock powder (16% sand, 66% silt and 18% clay) eroded and deposited along the St. Lawrence River, Quebec. The plasticity index was 7% and the soil was described as not significantly expansive. Specimens were formed by initially lightly tamping the material into a mould. Triaxial shearing tests were reported for saturated and unsaturated soil specimens that were compressed quasi-isotropically in the triaxial cell prior to shearing. Consolidated undrained shearing tests were performed on the saturated specimens and consolidated drained tests performed on unsaturated specimens. The tests on unsaturated specimens employed the axis translation technique with suctions held constant during shearing at values between 80 and 600 kPa. Failure was defined as the critical state. It is unclear whether to designate the tests on unsaturated Trois-Rivières silt as 1d-cs tests or iso-cs tests.

Figure 8.17 shows q plotted against p'_c at critical state for both the saturated and the unsaturated specimens. As for the other soils examined other than the BES, the results for the unsaturated silt plot below the saturated strength envelope. The slope of the critical state strength envelope for the saturated specimens is given by $M = 1.63$.

Normalising q and p'_c with respect to s, the critical state deviator stress results in Figure 8.18 lie close to a unique line for all suctions. A line through the data at $M_a = 1.63$ with an intercept of $\Omega = 0.6$ on the q/s axis at $p'_c/s = 1$ reasonably follows the data points. This suggests $M \approx M_a = 1.63$, which differs from the results for kaolin, Kiunyu gravel and Jurong soil where $M_a > M$, but complies with the behaviour of BES. However, the

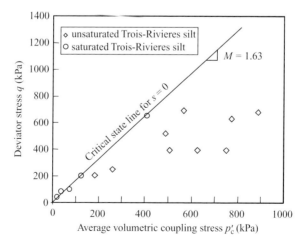

Figure 8.17 Deviator stress q against p'_c at critical state from the results of Maâtouk *et al.* (1995) for Trois-Rivières silt.

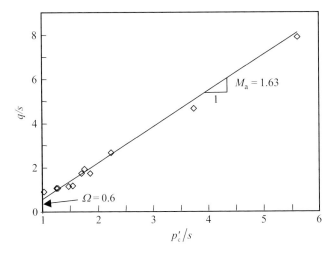

Figure 8.18 q/s against p'_c/s at critical state from 1d-cs test results of Maâtouk *et al.* (1995) for unsaturated Trois-Rivières silt.

value of $\Omega = 0.6$ complies with the results for kaolin, Kiunyu gravel and Jurong soil, but differs from the behaviour of BES.

Using Equation 8.4, Figure 8.19 shows values of the suction stress ratio M_b plotted against v_w/v. The decrease in M_b with decreasing v_w/v is at an inclination of M_a as for all materials examined. As for Kiunyu gravel, the value of M_b drops rapidly and approaches zero at an intercept of $v_w/v = 0.63$ where shearing can be expected to become concentrated between the aggregates.

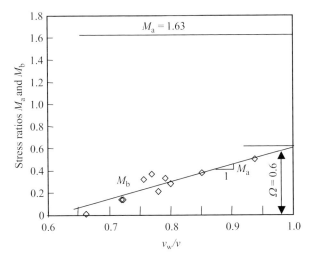

Figure 8.19 Stress ratios M_a and M_b against v_w/v from 1d-cs test results of Maâtouk *et al.* (1995) for unsaturated Trois-Rivières silt.

8.2.6 Significance of Ω

The significance of Ω and the reason for it having a value close to 0.6 at critical states in the shearing tests on kaolin, Kiunyu gravel, Jurong soil and Trois-Rivières silt are examined. Further investigation is needed to confirm whether $\Omega = 0.6$ is applicable to other soils. It is important to recognise that linear regression gives values of Ω slightly different for each of the materials, but the critical state results examined suggest that $\Omega = 0.6$ provides a reasonable approximation for the materials. On this basis, when the critical state shear strength results are extrapolated to the condition $p'_c = s$, then $q = 0.6s$. It should also be noted that M_b tends to Ω as saturation is approached. This is attributable to the aggregated structure rather than the nature of the fine particles. However, to create an aggregated structure, sufficient fines must be present and there must be a predisposition for the fines to aggregate.

8.2.6.1 Components of q

Rearranging Equation 8.3 gives Equation 8.5, which can be used to explain the significance of the different components making up the shear strength and the form of the experimental relationship of q/s plotted against p'_c/s.

$$qv = M_a \bar{p} v + s M_a v_w - s (M_a - \Omega) v \qquad [8.5]$$

It is informative to examine each of the terms in the equation separately using the idea of conjugate pairings originally introduced in Chapter 5:

- qv is the conjugate pairing for the total deviator stress q acting through the total soil volume represented by v.

  This is made up of:

- $M_a \bar{p} v$, which is the conjugate pairing $M_a \bar{p}$ and v for the deviator stress component due to $\bar{p}$ acting through the total soil volume given by v.
- $s M_a v_w$, which is the conjugate pairing $s M_a$ and v_w for the deviator stress component due to the addition of suction s acting through the combined volume of the water and solids represented by v_w.
- $-s (M_a - \Omega) v$, which is the conjugate pairing $-s (M_a - \Omega)$ and v. This must account for all other factors that influence the shearing resistance, in particular (i) the reduction in total deviator stress due to preferential shearing through regions where suction has less influence, and (ii) the difference that may potentially exist between shearing resistance within and between aggregates due to the variation in soil fabric. This is consistent with the first two terms on the right of Equation 8.5 being in direct proportion to the volumes through which the net stress $\bar{p}$ and suction s act, and both being controlled by the stress ratio parameter M_a.

The test results of Stewart *et al.* (2001) for BES (10% bentonite) do not indicate an influence from soil particle aggregation with the plot of q/s against p'_c/s for unsaturated BES following a similar plot to that of q against p' for saturated BES (Figures 8.14 and 8.15). In these tests $\Omega = M_a$ and the last term in Equation 8.5, which includes the influences of both preferential shearing and the difference between inter- and intra-aggregate soil structure, is zero. $\Omega = M_a$ suggests a relatively uniform soil structure with no net influence from aggregate formation. For this condition $q = M_a (\bar{p} + s v_w/v)$ and the deviator stress components are directly related to the volumes through which $\bar{p}$ and s act. It is a little surprising that the results of Toll (1990) appear to indicate aggregation of particles and a value of $\Omega = 0.6$ as the Kiunyu gravel tested had a clay content of only 8–9%. However, this material was more widely graded than the BES, which contained 90% uniform sand. This suggests that the clay mineralogy and the grading of a soil influence the development of aggregates and the behaviour at critical state.

The plots of M_b against v_w/v for the soils examined indicate that as the soil dries M_b decreases to zero and as p'_c/s approaches 1 all shearing can be expected to be between the aggregates with q given by:

$$q = M_a \bar{p}$$ [8.6]

Under these conditions, from Equation 8.5, for the condition $\Omega < M_a$, where aggregation is inferred as influencing soil behaviour:

$$s\,M_a \frac{v^*_w}{v^*} - s\,M_a + s\Omega = 0$$ [8.7]

And from Equation 8.7, noting that for a dry soil $V_w = 0$ (and $v^*_w = 1.0$):

$$\Omega = M_a \left[1 - \frac{v^*_w}{v^*}\right] = M_a \left[\frac{v^*_a}{v^*}\right] = M_a n^* = M_a \left[1 - \frac{\rho^*_d}{\rho_w G_s}\right]$$ [8.8]

where the superscript $*$ denotes the conditions for a dry aggregated soil with only a relict water phase internal to the aggregates, ρ^*_d is the density of a dry aggregated material and G_s is the specific gravity of the soil particles.

The intercept Ω at $p'_c/s = 1$ is determined from extrapolation of the critical state strength line for an unsaturated soil. At this extreme, the soil can be expected to be subject to a very high suction and in a loose condition because of shearing to the critical state. Measurements of dry aggregated speswhite kaolin in a loose state indicate a dry density of around 0.80 Mg/m^3. With $G_s = 2.65$, Equation 8.8 gives a value of $v^*_a/v^* = 0.70$, and with $M_a = 0.86$ (determined from Figure 8.4) it yields a value of $\Omega = 0.60$, which is in agreement with the experimental value.

The fact that $\Omega = 0.60$ is a reasonable approximation for the four soils tested to critical state, and for which it is believed that aggregation of particles influenced the results, suggests that $\Omega = M_a n^*$ may be relatively constant for a wide class of materials. Further experimental evidence is required to draw firm conclusions.

8.2.6.2 Influence of a fissured structure

Figure 8.20 presents plots of p'_c/s against v_a/v and n at the critical state for speswhite kaolin based on the 1d-cs results of Sivakumar (1993) and Wheeler and Sivakumar (1995, 2000). The drier a soil, the closer v_a/v must be to n. The plots are reasonably consistent with convergence at $n^* = v^*_a/v^* = 0.70$, as determined for dry aggregated soil in the foregoing using Equation 8.8. Extrapolating the plot for M_b in Figure 8.5 to the condition $M_b = 0$, where shearing is wholly between the aggregates, gives a value of $v_w/v = 0.30$. This again corresponds to $v_a/v = v^*_a/v^* = 0.70$.

The other soils that have been examined generally comply with the above interpretation. However, a note of caution is required in extrapolation of the data for p'_c/s against v_a/v and n to the condition $p'_c/s = 1$. Both the volumetric variables and the variable p'_c, which includes the volumetric term v_w/v, are based on overall measurements of soil volume terms and are thus influenced by any open fissures in the soil specimens. As a consequence, p'_c/s values less than 1 (not possible for an intact material) can be recorded for fissured soils. Fissures mask the behavioural trend that would be obtained for an intact material. In fact, fissuring is likely to be present in soils with values of p'_c/s a little greater than 1,

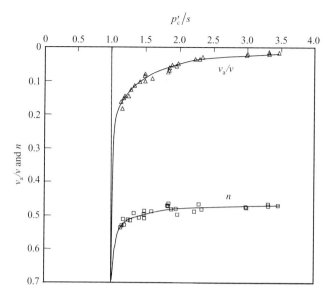

Figure 8.20 p'_c/s against v_a/v and n at critical state from 1d-cs test results of Sivakumar (1993) and Wheeler and Sivakumar (1995, 2000) for unsaturated kaolin.

particularly where confining pressures are low and suction is elevated. This is evident in Figure 8.21 for kaolin from the results of Sivakumar (2005) and Sivakumar *et al.* (2010a) where the specimens were subject to low confining pressures ($\leq$100 kPa) with elevated suctions of around 800–1000 kPa. The strength relationship established assuming a constant M_a applies to intact volumes of unsaturated soil, or at least requires the fissures to be essentially closed. The shear strength of fissured soils where p'_c/s is less than or close

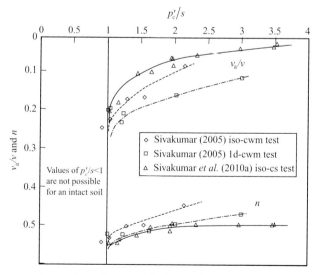

Figure 8.21 p'_c/s against v_a/v and n at critical state from iso-cs, iso-cwm and 1d-cwm test results of Sivakumar (2005) and Sivakumar *et al.* (2010a) for unsaturated kaolin.

to 1 can be expected to deviate from the relationships established for intact materials and leads to values deviating from the extrapolated value of $\Omega = 0.60$.

8.2.7 Influence of soil fabric

A number of factors influence the soil fabric and the results of shearing tests. These are discussed in the following.

8.2.7.1 Particle aggregation and preferential shearing

The materials that have been examined indicate a decrease in M_b with decreasing v_w/v. This is consistent with the water phase being drawn back into the finer pore spaces and the suction having less influence on the shear resistance (Toll, 1990). When $M_b = 0$, the suction plays no part in the shear resistance other than in maintaining any aggregation of particles. While this is true as a general statement, the test results for M_a and Ω for kaolin, Kiunyu gravel and Jurong soil contrast with those for BES, while the results for Trois-Rivières silt exhibit intermediate behaviour characteristics. The results for BES are considered not to have been influenced by aggregation of particles. The different behaviour characteristics suggest a significantly different soil structure.

Other than for the BES tests, $M_a \geq M$ and $\Omega = 0.60 < M$. This is indicative of aggregates of particles acting as larger particles. The experimental evidence indicates that once created the aggregates give rise to a relatively constant M_a that appears unaffected by change in suction. It is generally accepted that larger soil particles produce greater frictional resistance under shearing. In an unsaturated fine-grained soil comprising a significant percentage of clay and silt, it is not surprising therefore that when aggregates are present, M_a can exceed M if the suction is sufficiently high to maintain the aggregates under the application of shear stresses (Toll, 1990).

For saturated BES with 5–10% bentonite, Mollins *et al.* (1999) concluded that the uniformly grained sand matrix was primarily responsible for supporting the applied stresses and the bentonite made a negligible contribution to the strength. Stewart *et al.* (2001) suggested this was also true for the unsaturated BES tests with 10% bentonite. They concluded that the sand grains became coated with bentonite 'gel' but that the sand grains were brought into mechanical contact when the BES was compacted. The sand grains thus controlled the strength of the BES whether the material was saturated or unsaturated. This is considered to have resulted in $M_a = \Omega = M = 1.45$ for the BES. In accordance with Equation 8.3, the shearing resistance is made up of that attributable to the net stress and that attributable to the suction. The equality $M = M_a$ for the BES suggests that the component of shearing resistance attributable to the net stress was unaffected by particle aggregation, while the equality $\Omega = M_a$ suggests there was no net influence from preferential shearing or soil fabric variations. This is consistent with the uniformly grained sand in the BES controlling the shear resistance in a relatively homogeneous material.

The Trois-Rivières silt results exhibit strength characteristics intermediate between those of the BES and the other materials examined. For the Trois-Rivières silt and the BES, $M \approx M_a$ and the relatively low percentage of fines is considered to have resulted in the coarser soil fraction controlling the influence due to the net stress. However, for the Trois-Rivières silt as well as for kaolin, Kiunyu gravel and Jurong soil, $\Omega = 0.6 \langle M_a$ indicates a reduction in shearing resistance due to the presence of aggregates and, in particular, the influence of preferential shearing.

8.2.7.2 Specimen preparation and test procedure

Different preparation procedures produce different soil fabrics. However, Wheeler and Sivakumar (2000) concluded from tests on kaolin that differences in fabric caused by different one-dimensional compaction procedures could be erased by shearing to the critical state, and any remaining influence of soil fabric was likely to be related only to the suction and the initial compaction water content. This does not tell the whole story. The constant suction critical state strength results 1d-cs in Figure 8.4 for kaolin appear to confirm that the initial structure, whether from one-dimensional light or heavy static compression or light dynamic compaction, has no influence on the strength envelope. The results of Figure 8.7, again for kaolin, also appear to lie on a unique critical state strength line, but this line indicates greater deviator stress than in Figure 8.4. As there was no measurable property change in the kaolin, it is necessary to consider carefully the differences in the methods of specimen preparation and the test procedures adopted in the test series in order to appreciate the reasons for the difference. It is argued that this is attributable to the aggregation of the fine particles and the behaviour of the aggregates during specimen preparation and shearing.

A unique critical state strength line for saturated soils in $p' : q : v$ space (see Section 4.5) is based on the premise that any original soil structure due to specimen preparation history is destroyed during shearing. This appears justified for a wide class of materials. However, this requires that in the shearing tests there is no influence from change in particle size or shape due to specimen preparation procedure or shearing action. In unsaturated soils, aggregation of particles influences the shearing characteristics. In the constant suction tests 1d-cs on kaolin, differences in the initial soil structure as a result of different preparation procedures were erased, though the suction-controlled aggregates influenced the strength characteristics. In these tests, an example of which is analysed in Section 9.9, the action of increasing deviator stress resulted in the specimens taking up water. Mercury intrusion porosimetry (MIP) analysis, as discussed in Section 1.7, indicates that water added to an unsaturated specimen is preferentially absorbed by the aggregates, allowing them to expand and deform in a manner influenced by the imposed stresses and the available surrounding inter-aggregate void spaces. Such expansion along with the shearing action can be perceived as eradicating any initial differences in soil structure.

The results for kaolin examined indicate differences in critical state behavioural characteristics depending on the specimen preparation procedure and test conditions. In particular, the results in Figure 8.21 for the constant suction tests on truly isotropically prepared specimens iso-cs; constant water mass tests on truly isotropically prepared specimens iso-cwm; and constant water mass tests on specimens initially one-dimensionally compressed in layers 1d-cwm show marked differences. The dimensionless volumetric terms v_a/v and n at critical state plotted against p_c'/s follow consistent but different trend lines. Thus, although all the results appear to plot on the same strength envelope q/s against p_c'/s at critical state (Figure 8.7), they do not lie on the same critical state line in $p_c'/s : q/s : v_w/v$ space in the three-dimensional model developed in Section 8.5.

The shearing tests 1d-cs on one-dimensionally prepared specimens (Figures 8.4, 8.5 and 8.20) and the shearing tests iso-cs on truly isotropically prepared specimens (Figures 8.7, 8.8 and 8.21) were carried out under constant suction conditions but indicate different strength and volumetric conditions at critical state. The differences in test results are considered attributable to differences in soil fabrics as a result of the initial specimen preparation. Thus, while Wheeler and Sivakumar (2000) concluded that any influence from soil fabric due to different one-dimensional compression procedures was destroyed on shearing to the critical state, specimens prepared under truly isotropic conditions are inferred as leading to a different soil structure at critical state. The initial one-dimensional compression in the 1d-cs tests will have resulted in a heterogeneous soil fabric and distortion and possible breakage of aggregates arguably at odds with the specimens iso-cs prepared by isotropic compression. Subsequent shearing at constant suction entailed uptake of water and expansion of the aggregates into the surrounding macro-pores, which can be expected to be different under the one-dimensional and isotropic preparation procedures, and will have resulted in further changes in the size and form of the aggregates. It is these likely differences in aggregate shape and size at critical state that are considered the reason for the differences between the critical state results for 1d-cs and iso-cs.

The test series iso-cs and iso-cwm on truly isotropically prepared specimens on the other hand (Figure 8.21) differ only in that the iso-cs results relate to shearing tests carried out under constant suction and the iso-cwm results to shearing under constant water mass. This leads to the conclusion that the differences in the volumetric conditions at critical state are dictated by the shearing procedure. Constant water mass tests are likely to maintain the aggregates more rigidly than constant suction tests where shearing was associated with water uptake and expansion and deformation of the aggregates. Again it is argued that the nature of the aggregates at critical state is the primary reason for the differences in test results.

The results for test series iso-cwm and 1d-cwm shown in Figure 8.21 relate to shearing tests carried out under constant water mass conditions but the iso-cwm tests were on specimens initially prepared under truly isotropic conditions and the 1d-cwm tests on specimens under one-dimensional compression. In both series of tests the control of the water mass is considered to have resulted in relatively rigid aggregates. It is the difference in specimen preparation that is considered to have resulted in the different volumetric test results. The heterogeneous soil fabric and distortion and possible breakage of aggregates under one-dimensional compression are likely to be at odds with the aggregate structure resulting from isotropic preparation.

The discussion is developed further in Sections 8.4 and 8.5, but it appears evident from the test data on kaolin that the critical state is non-unique. This contrasts with the evidence for saturated soils. The critical state in unsaturated soils appears to be influenced by the size and rigidity of the aggregates, which is influenced by the test specimen preparation (whether one-dimensional or isotropic) and the test procedure (whether constant suction or constant water mass). Other materials may exhibit further differences in shear behaviour and a wider range of critical state strength data.

8.2.7.3 Transitional behaviour

A further important influence of soil structure is the transition between unsaturated and saturated soils. Toll (1990) and Toll and Ong (2003) along with other researchers have assumed a smooth transition from unsaturated to saturated behaviour in analysing experimental shear strength data. However, the foregoing analysis does not require this as an assumption and direct comparison with the experimental data shows consistent trends with a discontinuity between unsaturated and saturated behaviour where aggregation of particles significantly influences the soil strength. No obvious breakdown of the aggregate structure during shearing is indicated for kaolin even for suctions below 100 kPa. Monroy *et al.* (2010) noted that the transition from an aggregate to a matrix structure in unsaturated London clay took place as the suction in wetting tests was reduced from 40 kPa to 0 kPa. It might be reasonably argued that if the suction in an unsaturated soil is sufficient to maintain the integrity of individual aggregates when the soil is at critical state, then it is likely to hold everywhere within a soil mass and at all lower stress levels. Only at relatively low suctions when the integrity of the aggregates can no longer be maintained under shearing will they break down. In this case, the aggregates in the zones of shearing can be expected to break down. If this is so, it may be unreasonable to expect a smooth transition from unsaturated to saturated behaviour. This needs further experimental investigation. The lack of a smooth transition from unsaturated to saturated behaviour adds some complexity to strength analysis but as shown by Murray and Sivakumar (2004) and discussed in Section 8.6, the analysis that leads to this conclusion justifies a reduction in the number of variables and parameters necessary to describe the critical state of unsaturated soils.

8.3 Equivalent strength parameters

It is possible to relate the critical state stress ratios M_a and M_b to the equivalent ϕ' and c' parameters as below:

$$M_a = \frac{6 \sin \phi'}{3 - \sin \phi'} \text{ and } M_b = \frac{c'}{s} \left[\frac{6 \cos \phi'}{3 - \sin \phi'} \right] \qquad [8.9a]$$

$$\phi' = \sin^{-1} \left[\frac{3 M_a}{6 + M_a} \right] \text{ and } c' = M_b \left[\frac{3 - \sin \phi'}{6 \cos \phi'} \right] s \qquad [8.9b]$$

For constant M_a, ϕ' is also constant and independent of s. However, while M_b decreases with decreasing degree of saturation and increasing s, the apparent cohesion parameter c' incorporates s in its definition and increases with increasing suction. Cohesion implies tensile strength, which increases with increasing suction. At the extreme of a perfectly dry soil, both M_b and c' must be zero, indicative of a discontinuity in behaviour between unsaturated conditions and perfectly dry conditions. Determination of ϕ' and c' from M_a, M_b and s, using Equations 8.9a and 8.9b, allows conventional ϕ', c' stability analysis calculations to be carried out.

8.4 Compression and critical state characteristics of unsaturated kaolin

This section looks at the compression characteristics of one-dimensionally and isotropically prepared kaolin specimens prior to shearing, and the volumetric conditions at critical state.

8.4.1 Isotropic compression of initially one-dimensionally prepared specimens

Figure 8.22 presents results from isotropic, constant suction compression tests of initially one-dimensionally prepared specimens of kaolin. The tests are after Wheeler and Sivakumar (1995) and were carried out in the triaxial cell prior to shearing under axially increasing loads at constant suction. The tests correspond to those reported in Section 8.2.1. Figure 8.22 indicates consistent reproducible results and a clear linearisation of the plots of $\ln p'_c$ against specific volume v as the compression pressure increases. The normally compressed straight-line portions of the plots are given by:

$$v = N_t - \lambda_t \ln p'_c = N_t - \lambda_t \ln [\bar{p} + s v_w / v] \qquad [8.10]$$

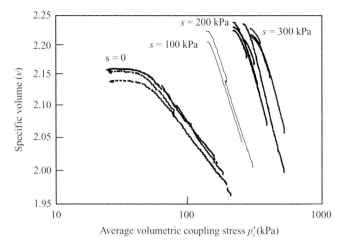

Figure 8.22 Specific volume v against p'_c during ramped compression at constant suction from the results of Wheeler and Sivakumar (1995) for kaolin (after Murray, 2002). 2008 NRC Canada or its licensors. Reproduced with permission.

where Equation 7.23a has been used to substitute for p'_c, N_t is the extrapolated value of v at $p'_c = 1.0$ kPa and λ_t is the slope of the straight-line portion of the compression plots.

The values of λ_t and N_t plotted in Figure 8.23 have been reassessed from those published by Murray (2002). The data suggest it reasonable to interpret a constant λ_t of 0.313 for $s \geq 100$ kPa. On this basis, N_t cannot also be constant and must be a function of s, though the results suggest a reasonable approximation of around 3.9 for $s \geq 100$ kPa. The corresponding values of $\lambda_t = 0.120$ and $N_t = 2.61$ for a saturated soil ($s = 0$) are notably less and indicate convergence of the plots for unsaturated and saturated soils as p'_c increases. This interpretation of the data is consistent with that from the shearing

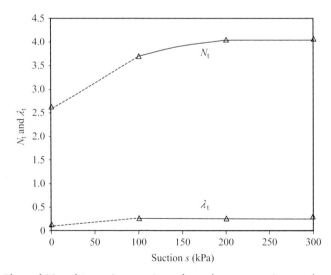

Figure 8.23 Plots of N_t and λ_t against suction s from the compression results of Wheeler and Sivakumar (1995) for kaolin.

tests and indicates a marked difference in behaviour between that of an unsaturated soil with an aggregated structure and a saturated soil with a more dispersed structure. The aggregated structure was not broken down though inevitably modified during the isotropic compression tests.

8.4.2 Critical state volumetric conditions for initially one-dimensionally prepared specimens

Figure 8.24 presents the critical state data of Sivakumar (1993) and Wheeler and Sivakumar (1995, 2000) for kaolin and corresponds to the data of Figure 8.1 but as plots of p'_c against v_w/v. Again the plots indicate convergence of the critical state results for constant suction on the saturated line as p'_c increases, but they also show a distinct change in behaviour represented by an inflection in the plots for unsaturated conditions. Convergence towards the condition for a saturated soil occurs as v (or the volume of the soil) reduces and v_w/v (or the relative volume of the aggregates) increases. In other words, the aggregates comprising the soil particles and water come closer together and the air voids reduce. The change in behaviour either side of the inflection is inferred as being a result of the greater interaction of the aggregates due to reducing air voids. This adds complexity to the interpretation of the critical state results and probably to the number of parameters necessary to define the critical state. The change occurs at $v \approx 2.0$ for all the suction values (i.e. when the volume of voids equals the volume of solids). This was the approximate value of v at the inflections in the critical state data for unsaturated kaolin presented in Figure 8.1.

8.4.3 Analysis of critical state data and compression data

Figure 8.25 (after Murray, 2002) presents the results of both the compression data of Figure 8.22 and the critical state data of Figure 8.24 for unsaturated kaolin. The data

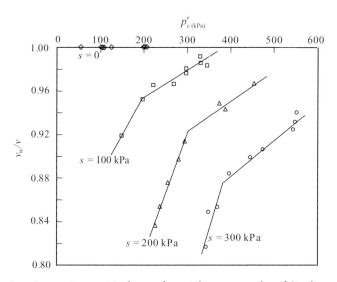

Figure 8.24 p'_c against v_w/v at critical state from 1d-cs test results of Sivakumar (1993) and Wheeler and Sivakumar (1995, 2000) for unsaturated kaolin.

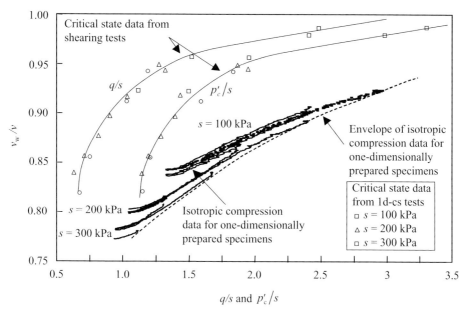

Figure 8.25 Plots of p'_c/s and q/s against v_w/v from the 1d-cs test results of Sivakumar (1993) and Wheeler and Sivakumar (1995) for unsaturated kaolin (after Murray, 2002). 2008 NRC Canada or its licensors. Reproduced with permission.

are presented in terms of the dimensionless stress ratios q/s and p'_c/s plotted against the dimensionless volumetric variable v_w/v based on Equation 7.23a, which describes the stress regime in an unsaturated soil, and Equation 8.3, which describes the critical state strength of an unsaturated soil. As discussed in Section 4.6, the Barcelona Basic Model (BBM) describes soil behaviour, including yielding, in $q : \overline{p} : s$ stress space and does not include any volumetric terms. However, the relative volumes of the phases in an unsaturated soil are considered important when describing behaviour patterns. In saturated soil this is achieved in the Cam Clay models by the inclusion of the specific volume v. The use of the normalised stresses q/s and p'_c/s along with the dimensionless volumetric term v_w/v in unsaturated soils allows the behaviour characteristics to be more fully appreciated.

The plots of Figure 8.25 show a high degree of unification of the critical state data for different suctions. The plots also indicate that the isotropic compression results for initially one-dimensionally prepared specimens tend towards a single curved line, identified in Figure 8.25 as an envelope of isotropic compression. This is shown in Figure 8.26, as discussed in the following, to correspond to the isotropic normal compression line (*iso-ncl*). Consistent with the argument of Wheeler and Sivakumar (2000) from theoretical considerations, specimens initially one-dimensionally compacted and then subsequently isotropically loaded at a given value of suction should gradually converge on the truly isotropic normal compression line (*iso-ncl*) for an isotropically prepared specimen at the same suction. They explain this in terms of an elasto-plastic anisotropic model involving a rotated yield surface as discussed in Section 4.7.5. Though Wheeler and Sivakumar (2000) employed $q : \overline{p}$ to describe the yield surface, which differs from the dimensionless parameters of Figure 8.25, the basic principles are considered applicable.

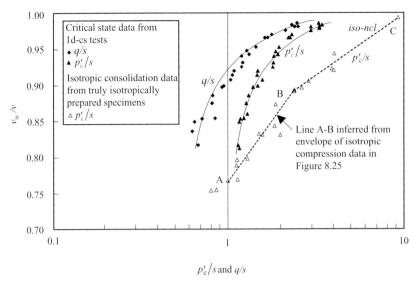

Figure 8.26   In p_c'/s and ln q/s against v_w/v at critical state from 1d-cs test results of Sivakumar (1993) and Wheeler and Sivakumar (1995, 2000) and the isotropic compression data for truly isotropically prepared specimens of Sivakumar (2005) and Sivakumar *et al.* (2010b) for unsaturated kaolin.

The critical state data from initially one-dimensionally compressed specimens and the envelope for isotropic compression in Figure 8.25 have been plotted in Figure 8.26 as dimensionless stress terms p_c'/s and q/s to a logarithmic scale against v_w/v. Also included are the compression results from truly isotropically prepared specimens of kaolin after Sivakumar (2005) and Sivakumar *et al.* (2010b). The isotropic experimental results can be reasonably represented by a bi-linear relationship, with the plot for results between A and B (values of p_c'/s of 1–2.4) being significantly steeper than the plot between B and C (values of p_c'/s between 2.4 and 10). As indicated, the plot between A and B corresponds to the inferred *iso-ncl* of Figure 8.25. Equation 8.11 can be used to represent the *iso-ncl*.

$$v_w/v = \Gamma_c + \lambda_c \ln\left[p_c'/s\right] \qquad [8.11]$$

where Γ_c is the intercept of the *iso-ncl* on the v_w/v axis at $p_c'/s = 1$ and λ_c is the gradient of the *iso-ncl*.

A value for the intercept $\Gamma_c = 0.767$ along with a gradient $\lambda_c = 0.143$ are applicable to the *iso-ncl* between A and B, and values of $\Gamma_c = 0.827$ and $\lambda_c = 0.0746$ to the *iso-ncl* between B and C. A combination of imposed stress and suction given by p_c'/s controls the isotropic volume characteristics. The inflection point B in the *iso-ncl* is at an approximate value of v between 2.0 and 2.1. This differs only slightly from the inflection value of $v = 2.0$ determined from the critical state data for initially one-dimensionally compressed specimens of Figures 8.1 and 8.24. The steeper section of the *iso-ncl* A–B in Figure 8.26 (i.e. where the change in v_w/v for a given change in $\ln\left(p_c'/s\right)$ is greatest) is the region where v is greatest and the experimental data give values between 2.02 and 2.23. The flatter section B–C corresponds to lesser values of v between 1.85 and 2.10. The data suggest that between A and B where p_c'/s is less than 2.4, the suction is sufficiently large in relation to the imposed isotropic stress to maintain a relatively loose soil structure.

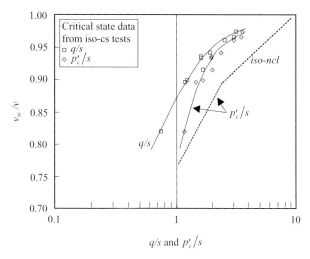

q/s and p'_c/s

Figure 8.27 In p'_c/s and ln q/s against v_w/v from the critical state results for constant suction tests on isotropically prepared specimens (iso-cs) of Sivakumar (2005) and Sivakumar *et al.* (2010a) for unsaturated kaolin.

However, increasing imposed isotropic stress under the same suction (increasing p'_c/s) results in relatively large compression changes and closure of the inter-aggregate voids, resulting in greater interaction between aggregates. Continued increase in isotropic loading at constant suction into the region B–C where $p'_c/s > 2.4$ and the soil is more compact results in a change in behaviour with reduced compression for a given change in p'_c/s.

It is interesting to note that as discussed in Section 8.2.6, p'_c/s cannot be less than 1.0 for an intact material and for increasing suction and decreasing p'_c/s the critical state lines in Figures 8.25 and 8.26 become asymptotic to a value of $p'_c/s = 1$. The plots also converges on the *iso-ncl* line as the relative influence of the suction increases.

Figure 8.27 presents the iso-cs critical state results from Sivakumar (2005) and Sivakumar *et al.* (2010a). Again, p'_c/s and q/s to a logarithmic scale have been plotted against v_w/v. Though behaviour trends similar to those of the initially one-dimensionally prepared specimens of Figure 8.26 are indicated, the critical state data do not follow the same lines. As argued in Section 8.2.7, it appears that the initial preparation of the specimens, whether one-dimensionally compressed or isotropically compressed, influences the conditions at critical state. It is thought that the primary influence is the nature of the aggregates with more rotund aggregates likely to be present for the specimens prepared initially under the truly isotropic preparation regime.

Figure 8.28 presents a limited amount of data for iso-cwm tests on kaolin specimens. Though Figure 8.7 suggests it reasonable to include the critical state strength data along with that for constant suction tests as they appear to lie close to the same strength line, the results of Figure 8.28 suggest that there is an influence on critical state volumetric conditions from the type of test carried out. Constant water mass tests will tend to preserve the aggregate shape and size more readily than constant suction tests.

Figure 8.29 shows that the 1d-cwm results of p'_c/s against v_w/v at critical state plot outside the *iso-ncl* (opposite side to the critical state results for the 1d-cs, iso-cs and iso-cwm tests). This does not follow the general rule for saturated soils. In unsaturated soils the influence of aggregation means that the rule can no longer be relied upon. The *iso-ncl* is based on aggregates initially sieved to 1.12 mm size. A different *iso-ncl* would be expected

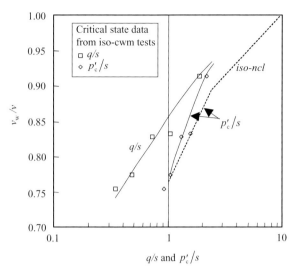

Figure 8.28 ln p'_c/s and ln q/s against v_w/v from the critical state constant water mass results on isotropically prepared specimens (iso-cwm) of Sivakumar (2005) and Sivakumar *et al.* (2010a) for unsaturated kaolin.

if the sieved aggregate size was varied. The data for q/s, p'_c/s and v_w/v are considered influenced by the aggregate size and shape at critical state.

8.5 Modelling of unsaturated kaolin

The plots of p'_c/s against v_w/v and q/s against v_w/v at critical state of Figures 8.26 to 8.29 are projections of the critical state lines *csl* in three-dimensional $p'_c/s : q/s : v_w/v$ space. Figure 8.30 combines the *csl* plots from one-dimensionally prepared specimens sheared under constant suction (1d-cs tests) from Figure 8.26, with the critical state strength line q/s against p'_c/s of Figure 8.4 ($M = 0.86$ and $\Omega = 0.60$). Also shown in Figure 8.30 is the *iso-ncl*. An advantage over the BBM discussed in Chapter 4 is that all stress and volumetric data can be viewed in three-dimensional space and there is no need to revert to hyperspace (four-dimensional space) as in the model proposed by Wheeler and Sivakumar (1995) also discussed in Chapter 4.

Figure 8.31 shows the *csl* and *iso-ncl* along with the shearing stages of four tests carried out and reported in Figure 8.4. Plot A is for a drained shear test on a specimen with suction maintained at 300 kPa, plot B for a specimen maintained at constant v and a suction of 100 kPa, plot C for a fully drained specimen at a suction of 100 kPa and plot D for a specimen at constant $\bar{p}$ and suction of 200 kPa.

Figure 8.32 presents the critical state lines (*csl*) for the iso-cs, iso-cwm and 1d-cwm tests of Sivakumar (2005) and Sivakumar *et al.* (2010a) for kaolin in $p'_c/s : q/s : v_w/v$ space along with the 1d-cs results of Wheeler and Sivakumar (1995). The *csl* are markedly different for the different tests though showing a limited range of projected strength in the $p'_c/s : q/s$ plane.

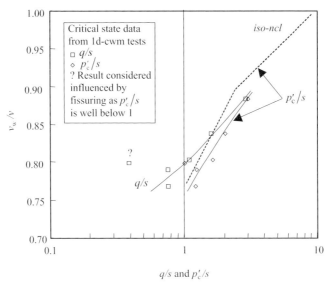

Figure 8.29 $\ln p_c'/s$ and $\ln q/s$ against v_w/v from the critical state results for constant water mass tests on one-dimensionally prepared specimens (1d-cwm) of Sivakumar (2005) and Sivakumar *et al.* (2010a) for unsaturated kaolin.

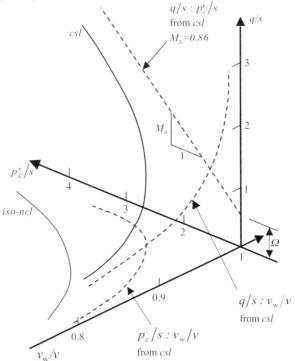

Figure 8.30 *csl* from the results of Sivakumar (1993) and Wheeler and Sivakumar (1995, 2000) and *iso-ncl* from the results of Sivakumar (2005) and Sivakumar *et al.* (2010b) in $p_c'/s : q/s : v_w/v$ space for unsaturated kaolin.

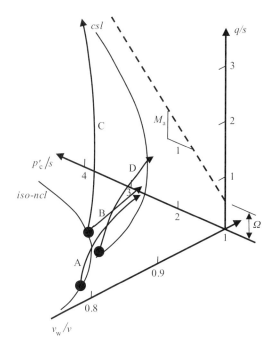

Figure 8.31 Shearing stress paths in $p'_c/s : q/s : v_w/v$ space from the results of Wheeler and Sivakumar (1995) for unsaturated kaolin.

8.6 Structure, variables and parameters

Sivakumar *et al.* (2010b) concluded from MIP tests on kaolin that an aggregated structure was maintained on wetting up to saturated conditions. This is important in that the behaviour of a suction-controlled aggregate structure can be expected to differ from that of a dispersed particle structure. The BBM and comparable constitutive models such as proposed by Toll (1990) and Wheeler and Sivakumar (1995) assume there is a smooth transition in shear strength results between unsaturated and saturated soils. This relies on progressive breakdown in the aggregated structure associated with unsaturated conditions in fine-grained soils to the more dispersed condition associated with saturated soils. However, as discussed in Section 8.2.7, an alternative philosophy is advocated, and it is argued that if the suction in an unsaturated soil is sufficient to maintain the structure at critical state, it is likely that it will hold everywhere. Only at relatively low suctions when the structure can no longer be maintained under shearing will it break down. A smooth transition may thus not occur. In accordance with Section 5.8, which discusses thermo-dynamic potentials, at low suctions the aggregated structure may represent a meta-stable condition. This structure is maintained unless disturbing forces, such as during shearing, facilitate the 'energy barrier' between an aggregate structure and a dispersed structure being overcome. Shearing appears to result in breakdown of the structure to a dispersed and lower potential structure only at low values of suction.

The degree of saturation S_r is frequently used to describe the volumetric variables in an unsaturated soil. It is argued that v_w/v is a more appropriate volumetric variable. The significance of v_w/v is that it represents the volume of the aggregates per unit volume

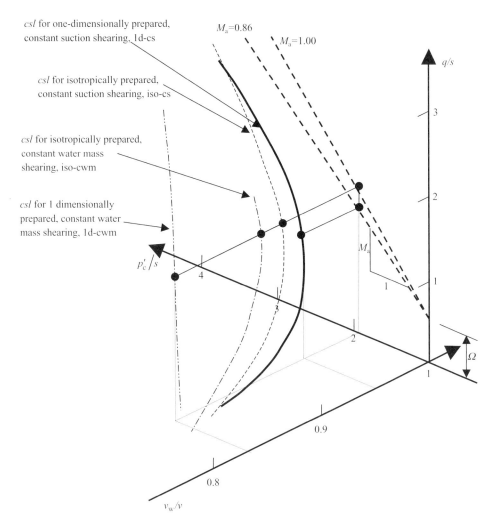

Figure 8.32 Critical state lines from the 1d-cs results of Sivakumar (1993) and Wheeler and Sivakumar (1995, 2000) and the iso-cs, iso-cwm and 1d-cwm test results of Sivakumar (2005) and Sivakumar *et al.* (2010a) in $p'_c/s : q/s : v_w/v$ space for unsaturated kaolin.

of soil, and the foregoing analysis describes the critical state strength using the variables q, $\bar{p}$, s and v_w/v. Two parameters also emerge from the analysis for critical state strength: M_a and Ω.

While the discontinuity in behaviour between an unsaturated soil and a saturated soil requires further investigation, importantly, M_a appears to be constant for a given soil, other than where an open fissured structure develops. It is also not necessary to include M_b as a parameter as this is shown in Equation 8.4 to be a function of the other variables and parameters, i.e. $M_b = f(M_a, v_w/v, \Omega)$. It may be possible to further reduce the parameters to 1 if, as for kaolin, Kiunyu gravel, Jurong soil and Trois-Rivières silt, Ω can be shown to be relatively constant at 0.6 for other soils. In accordance with Equations 8.3 and 8.4, this means that only one or two carefully controlled tests may be necessary to characterise the critical state strength of an unsaturated soil. In attempting to extend this to specific site conditions it would be necessary to measure or assess $(p - u_a)$, s and v_w/v values at critical state within the soil mass under analysis.

Equation 8.10 presents an equation for isotropic compression of specimens of kaolin initially one-dimensionally compressed. The equation is presented in terms of the variables $\bar{p}$, s, v and v_w (or v_w/v). Two parameters also emerge from this analysis: λ_t and N_t.

The dimensionless Equation 8.11 is shown to represent the *iso-ncl* using the variables $\bar{p}$, s and v_w/v and the parameters λ_c and N_c. The critical state strength and *iso-ncl* parameters are thus represented by a gradient and an intercept term.

A warning is issued at this point that the foregoing interpretation of soil compression behaviour is principally based on the results of tests on kaolin. It must be recognised that other soils, particularly more plastic clays and materials with significant proportions of sand and gravel, may exhibit variations in behaviour characteristics. In particular, there is some evidence to suggest that graded materials exhibit a change in behaviour characteristics depending on stress levels and the degree of compaction, leading to greater or lesser interaction of the coarser fraction and thus its relative importance to soil behaviour.

The aim of research into unsaturated soils is to provide working guidelines and a methodology for practising engineers to be able to predict the strength and deformation characteristics under field conditions. To this end the approach must be as simple as possible, but robust in order to provide the predictive accuracy required. Acceptable simplification can be justified only in the light of detailed research. While it is tempting to advocate the wielding of Occam's razor[1] in choosing between the approach described in the foregoing discussion and those of Chapter 4, this must be tempered by considerations of the current general shortcomings in the modelling of unsaturated soil behaviour. However, the approach outlined above has a theoretical basis and relies on a limited number of variables and parameters to describe the critical state strength and compression behaviour as well as providing a clear explanation of experimental data.

8.7 Conclusions

The shear strength and compression behaviour in laboratory triaxial stress testing have been analysed using the coupling stress variable p'_c that links the stress state variables to the volumes of the phases. The following main conclusions have been reached:

- Using p'_c and published experimental triaxial shearing data for the critical state strength of kaolin, a lateritic gravel (Kiunyu gravel), a residual soil (Jurong soil) and rock powder (Trois-Rivières silt) and the peak strength of BES, an equation (Equation 8.3) for the large-strain deviator stress q has been developed and analysed. This equation is in a form similar to that proposed by Toll (1990) from consideration of the independent stress state variables and includes the net stress ratio M_a and suction stress ratio M_b that give the components of deviator stress associated with the stress state variables $\bar{p}$ and s respectively. However, the interpretation of the stress ratios, which reflect the duality of the stress regime considered present in unsaturated soils, differs significantly from that of Toll (1990).
- The significance of M_a and M_b to the shear strength of unsaturated soils is described in terms of the soil fabric and preferential shearing within and between particles and aggregates. Consistent trends are indicated for all materials analysed, indicating the

equation for deviator stress as a reliable basis for the understanding and prediction of the strength of a wide class of unsaturated soils.

- The results demonstrate that M_a for an unsaturated soil is greater than or equal to M for the soil in a saturated state. The inequality $M_a > M$ exists for soils where the aggregates have a significant influence on the strength, and the equality $M_a = M$ exists for soils where a relatively low percentage of fines means the coarser soil fraction controls the influence due to the net stress.

- All materials examined have indicated a decrease in M_b with decreasing v_w/v (increasing s) at a slope of M_a consistent with the water phase being drawn back into the finer pore spaces.

- Preferential shearing between aggregates is indicated by $\Omega = 0.60 < M_a$ with Ω appearing to be a constant for the materials sheared to the critical state. An equation for Ω in terms of basic soil properties is presented. However, for BES where the uniformly grained sand controlled the shearing resistance, and for which it is unclear whether critical state conditions were established, $\Omega = M_a = M$ and no reduction in shearing resistance due to preferential shearing is indicated suggesting a relatively homogeneous material.

- Critical state data and isotropic compression data on kaolin, examined using the volumetric coupling stress p'_c, have shown consistent trends. The test data indicate well-defined critical state lines (*csl*) and isotropic normal compression lines (*iso-ncl*). The *iso-ncl* plotted as $\ln\left(p'_c/s\right)$ against v_w/v is interpreted as bi-linear. It is also shown that the critical states and isotropic compression results can be presented as three-dimensional plots in $p'_c/s : q/s : v_w/v$ space. The use of the dimensionless variables allows the interaction of the stresses and volumetric terms to be more fully appreciated. It is shown that the *csl* is not unique but is influenced by the aggregate fabric and changes under preparation and shearing though for a given preparation procedure and test conditions it appears well defined.

- There is some evidence to suggest a distinct change in characteristics in the *csl* and *iso-ncl* for kaolin at a value of v of 2.0–2.1 (i.e. close to the condition when the volume of voids equals the volume of solids).

- While the analysis is shown to limit the number of variables and parameters necessary to describe strength and normal compression characteristics, there is evidence of a discontinuity in behaviour between unsaturated and saturated fine-grained soils as the aggregated structure is maintained in an unsaturated soil even at relatively low values of suction.

- No attempt has been made to define yielding in the test results. The BBM and other models examined in Chapter 4 assume elastic behaviour at stresses below the previous stress level and the onset of plastic behaviour when the yield surface is encountered. This is a simplification of actual conditions and on a micro-mechanical level a degree of yielding and plastic behaviour can be expected at stress levels below the stress levels previously experienced. In Chapter 9 the micro-mechanical characteristics of unsaturated soil is examined and provides insight into the definition of yielding, strain and the dispersion of energy due to the application of loading to a soil specimen.

Note

1. Also spelled Ockham's razor after the fourteenth-century English Franciscan friar William of Ockham. Occam's razor, or the principle *lex parsimoniae* (the law of succinctness), recommends

that the theory which is simplest and introduces the fewer assumptions amongst conflicting theories should be chosen. Essentially this relies on the unproven principle that while the universe is under no obligation to make life as easy as possible for researchers to understand, there is no requirement to make life more complex than it has to be. Experience tells us that solutions to problems tend to be simpler than first imagined.

Chapter 9
Work Input, Conjugate Variables and Load-Deformation Behaviour of Unsaturated Soils

9.1 Introduction

The stress regime governing the behaviour of unsaturated soils is defined by the three forms of the equation for the directional stresses, Equations 7.24a–c. These equations, along with Equations 7.23a–c for the average volumetric coupling stress under triaxial stress conditions, are used in this chapter to determine three forms of a work input equation. Sections 9.2–9.6 detail this analysis of the conjugate stresses and strain-increments. Importantly, the analysis allows the volumetric and anisotropic stress–strain behaviour both of the aggregates and between the aggregates of fine-grained soils to be examined. The conjugate variables summarised in Table 9.1 and discussed in Section 9.7 are intuitively correct and sight of this should not be lost in following the analysis. The work conjugate stress and strain-increment variables provide a theoretical basis, centred on the thermodynamic principles developed in Chapters 5 and 6, for analysing triaxial experimental data.

Section 9.8 addresses the meanings of stress and strain in analysing experimental data and, in particular, the strain continuity requirements and the need to balance the work input with the energy dispersion internal to the soil. In Section 9.9 the use of the conjugate stress and strain-increments is illustrated by the analysis of the shear behaviour of specimens of kaolin. The influence of the bi-modal structure on the shearing resistance of unsaturated soils is clearly demonstrated.

9.2 Work input under triaxial stress conditions

First we will look at the work input at the two extremes of soil conditions – a saturated soil and a perfectly dry soil – for a specimen under triaxial stress conditions. The specimen is subjected to imposed stresses and experiences a small axial strain-increment $d\varepsilon_{11}$ and a radial strain-increment $d\varepsilon_{33}$. For a saturated soil specimen, the incremental work input per unit volume $\delta W'$ (Schofield and Wroth, 1968) can be written as:

$$\delta W' = p'd\varepsilon_v + qd\varepsilon_q = \sigma'_{11}d\varepsilon_{11} + 2\sigma'_{33}d\varepsilon_{33} \qquad [9.1a]$$

For a perfectly dry soil with no influence from suction, the incremental work input per unit volume can be written as:

$$\delta W' = \overline{p}d\varepsilon_v + qd\varepsilon_q = \overline{\sigma}_{11}d\varepsilon_{11} + 2\overline{\sigma}_{33}d\varepsilon_{33} \tag{9.1b}$$

Equations 9.1a and 9.1b for incremental work input can be abbreviated to Equation 9.1c, which is written in terms of the decoupled volumetric and deviatoric work components:

$$\delta W' = \delta W_v' + \delta W_q' \tag{9.1c}$$

where $p' = (p - u_w) = \left(\sigma_{11}' + 2\sigma_{33}'\right)/3$ is the mean effective stress for a saturated soil, $\overline{p} = (p - u_a)\left(\overline{\sigma}_{11} + 2\overline{\sigma}_{33}\right)/3$ is the mean effective stress for a perfectly dry soil, $q = \sigma_{11}' - \sigma_{33}' = \overline{\sigma}_{11} - \overline{\sigma}_{33}$ is the deviator stress, $d\varepsilon_v = -dv/v = d\varepsilon_{11} + 2d\varepsilon_{33}$ is the volumetric strain-increment, $d\varepsilon_q = 2\left(d\varepsilon_{11} - d\varepsilon_{33}\right)/3$ is the deviator strain-increment, $\delta W_v' = p'd\varepsilon_v$ or $\overline{p}d\varepsilon_v$ is the volumetric work input per unit volume due to p' or $\overline{p}$ respectively and $\delta W_q' = qd\varepsilon_q$ is the deviatoric work input per unit volume due to q.

In the analysis, volumetric compression and length and radius reduction are taken as positive strain-increments, consistent with compressive stresses being positive and that work is done to the specimen[1]. Equations 9.1a and 9.1b are applicable for a saturated soil and a perfectly dry soil respectively, but are not applicable for the more general class of a partially saturated soil in which a dual stress regime controls material behaviour.

Equations 7.23a–c, and accordingly Equations 7.24a–c, give alternative but equivalent views of the dual stress regime within an unsaturated soil. The correct interpretation of σ_c' is essential in determining the work input equation for unsaturated soils. It is not correct to simply multiply σ_c' by an appropriate strain-increment as for the effective stresses in saturated or perfectly dry soils. This would lead to an anomalous situation of the strain-increments for the aggregates and air voids merely being in proportion to their relative volumes. This would clearly be incorrect as the changes in relative volumes are dependent on test drainage conditions and differences in phase compressibility. For an unsaturated soil, if the differences in the strain-increment responses both of the aggregates and between the aggregates are to be correctly determined, it is essential to segregate the stresses and the work conjugate strain-increments based on a dual stress regime and bi-modal structure. The bi-model structure comprises the aggregates, formed from the soil particles and water, along with the surrounding air voids.

For unsaturated soils it is first necessary to define the stresses applicable under axially symmetrical loading conditions. Under such conditions the average volumetric coupling stress p_c' is given by:

$$p_c' = \left(\frac{\sigma_{c,11}' + 2\sigma_{c,33}'}{3}\right) \tag{9.2}$$

where $\sigma_{c,11}'$ is the axial coupling stress and $\sigma_{c,33}'$ is the radial coupling stress.

To demonstrate the derivation of the work input equation, Equation 7.23a and accordingly Equation 7.24a will be used as these contain the most commonly employed stress state variables. Equation 7.23a is rewritten below as Equation 9.3:

$$p_c' = \overline{p} + s\frac{v_w}{v} \tag{9.3}$$

where $\overline{p} = (p - u_a)$ is the mean net stress (here $p = (\sigma_{11} + 2\sigma_{33})/3$ is the mean total stress).

Consistent with thermodynamic principles, the work input components are additive, and under triaxial stress conditions the following equation applies to unsaturated soils:

$$\delta W_u' = \delta W_{vu}' + \delta W_{qu}' \qquad [9.4]$$

where $\delta W_u'$ is the total incremental work input per unit volume in an unsaturated soil, $\delta W_{vu}'$ is the incremental volumetric work input per unit volume due to the stress state variables and $\delta W_{qu}'$ is the incremental deviatoric work input per unit volume due to the components of the deviator stress.

Equation 9.4 decouples the volumetric and deviatoric work components for unsaturated soils and may be compared with Equation 9.1c for saturated and perfectly dry soils. However, for unsaturated soils the duality of the stress regime must be taken into account in determining $\delta W_{vu}'$ and $\delta W_{qu}'$.

9.2.1 Work input $\delta W_{vu}'$

Within an unsaturated fine-grained soil there will be both air-filled voids and aggregates. The application of an external stress regime will, in general, result in straining of both the air and water voids and will influence the solid particle fabric. Whichever of Equations 7.24a–c is used to describe the stress regime, and thus considered to control the deformations in an unsaturated soil, it is important to correctly account for the regions through which the stress state variables act in defining the work conjugate strain-increments. In accordance with Equation 7.24a, the net stress $\bar{\sigma}_{ij}$ is conjugate to the specific volume v, which represents the total volume of the soil, and is thus work conjugate to the total strain-increment in the direction of the stress. Similarly, under triaxial test conditions, in accordance with Equation 9.3, $\bar{p}$ is conjugate to the specific volume v, which represents the total volume of the soil, and is thus work conjugate to the volumetric strain-increment $d\varepsilon_v$. On the same basis, in Equation 9.3 the suction s is conjugate to the specific water volume v_w, which represents the total volume of the aggregates and is work conjugate to the volumetric strain-increment of the aggregates $d\varepsilon_w$. Thus, in accordance with Equations 9.3 and 9.4, under triaxial test conditions, the component of work input $\delta W_{vu}'$ can be written as:

$$\delta W_{vu}' = \bar{p}d\varepsilon_v + sd\varepsilon_w \qquad [9.5]$$

where $d\varepsilon_w = -dv_w/v = d\varepsilon_{w.11} + 2d\varepsilon_{w.33}$ is the volumetric strain-increment of the aggregates, $d\varepsilon_{w.11}$ is the overall axial strain-increment of the aggregates and $d\varepsilon_{w.33}$ is the overall radial strain-increment of the aggregates.

For a saturated soil $v_w = v$ and $d\varepsilon_w = d\varepsilon_v$, and $\delta W_{vu}'$ correctly reduces to $\delta W_v'$ of Equation 9.1c. For a perfectly dry soil the strain-increment $d\varepsilon_w = 0$ and again $\delta W_{vu}'$ correctly reduces to $\delta W_v'$.

Equation 9.5 describes the overall incremental volumetric straining of an unsaturated soil specimen and is made up of the volumetric straining of the aggregates and the volumetric straining of the air voids. The definition of $d\varepsilon_w$ indicates that the volumetric straining of the aggregates can be further divided into axial and radial components. This is also true of the air voids.

The first term on the right of Equation 9.5 is the work input per unit volume due to $\bar{p}$ acting through the whole soil mass and the volumetric straining of the soil $d\varepsilon_v$. The second term relates to the additional work input per unit volume due to the suction s acting within the more highly stressed aggregates and the volumetric straining of the aggregates $d\varepsilon_w$.

9.2.2 Work input $\delta W'_{qu}$

For a saturated soil at the critical state (Schofield and Wroth, 1968) the deviator stress can be shown by experimental data to be closely represented by:

$$q = Mp' \qquad [9.6]$$

where M is a stress ratio being the slope of the critical state line in $p' - q$ space for a saturated soil. This is illustrated in Figure 9.1 for tests on saturated kaolin. For the general case of stresses other than at the critical state, Schofield and Wroth (1968) suggested the use of $q = \eta p'$, which is also illustrated in Figure 9.1. On this basis, the work input Equation 9.1a for a saturated soil can be written as:

$$\delta W' = p'\left(d\varepsilon_v + \eta d\varepsilon_q\right) \qquad [9.7]$$

where η is the mobilised stress ratio q/p'.

A comparable approach can be adopted for unsaturated soils. However, in such soils the deviator stress has two components. These depend on which two of the three stress state variables are considered to represent the stress regime. This is consistent with the shearing resistance being different through and between the aggregates and is thus influenced by the dual stress regime. The significance of the stress state variables to the deviator stress q for an unsaturated soil is demonstrated by the following equation for the critical state strength of unsaturated soils as developed in Chapter 8 and which uses the stress state

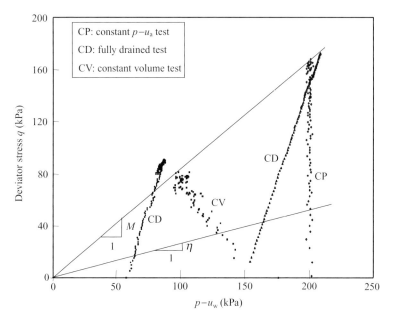

Figure 9.1   Deviator stress q against mean effective stress $p' = (p - u_w)$ from the results of Sivakumar (1993) and Wheeler and Sivakumar (1995) for saturated kaolin ($s = 0$).

variables $\bar{p}$ and s.

$$q = M_a \bar{p} + M_b s \qquad [9.8]$$

where M_a is the total stress ratio and M_b is the suction stress ratio.

For the general case of unsaturated soils under conditions other than at the critical state, adopting a similar approach to that for saturated soils, the deviator stress can be written as:

$$q = \eta_a \bar{p} + \eta_b s \qquad [9.9]$$

where η_a and η_b are the mobilised stress ratios with values at critical state of M_a and M_b respectively.

The first term on the right of Equation 9.9, i.e. $\eta_a \bar{p}$, represents the component of q as a result of $\bar{p}$ acting through the whole soil mass and is thus work conjugate, under triaxial loading conditions, to the overall deviator strain-increment $d\varepsilon_q$. Thus, the component of distortional work input is given by $\eta_a \bar{p} d\varepsilon_q$. The second term $\eta_b s$ represents the additional component of q within the more highly stressed aggregates, where the additional stress state variable s acts. In the triaxial cell the work conjugate deviator strain-increment for the aggregates is $d\varepsilon_{qw}$. Thus, the additional distortional work input in the aggregates is $\eta_b s d\varepsilon_{qw}$. Accordingly, the work input per unit volume resulting from the deviator stress is given by:

$$dW'_{qu} = \eta_a \bar{p} d\varepsilon_q + \eta_b s d\varepsilon_{qw} \qquad [9.10]$$

where $d\varepsilon_{qw} = 2(d\varepsilon_{w.11} - d\varepsilon_{w.33})/3$ is the deviator strain-increment for the aggregates.

For a saturated soil, $v_w = v$ and $d\varepsilon_{qw} = d\varepsilon_q$. The work increment $\delta W'_{qu}$ given by Equation 9.10 reduces smoothly to $\delta W'_q$ of Equation 9.1c if η_a and η_b tend smoothly to η. At the other extreme, for a perfectly dry soil the deviator strain-increment for the aggregates $d\varepsilon_{qw} = 0$, and $\delta W'_{qu}$ reduces smoothly to $\delta W'_q$ if η_a tends smoothly to η. The transitions from unsaturated conditions to saturated conditions and unsaturated to perfectly dry conditions are discussed in the following.

9.2.3 Total work input $\delta W'_u$

Substituting Equations 9.5 and 9.10 for $\delta W'_{vu}$ and $\delta W'_{qu}$ respectively into Equation 9.4 gives the following equation for the total incremental work input (per unit volume) into an unsaturated soil under triaxial stress conditions:

$$\delta W'_u = \bar{p}(d\varepsilon_v + \eta_a d\varepsilon_q) + s(d\varepsilon_w + \eta_b d\varepsilon_{qw}) \qquad [9.11a]$$

Following similar lines of argument, the following two alternative equations for the work input can be determined consistent with the use of any two of the three stress state variables:

$$\delta W'_u = \bar{p}(d\varepsilon_a + (\eta_a - \eta_b)d\varepsilon_{qa}) + p'(d\varepsilon_w + \eta_b d\varepsilon_{qw}) \qquad [9.11b]$$

or

$$\delta W'_u = p'(d\varepsilon_v + \eta_a d\varepsilon_q) - s(d\varepsilon_a + (\eta_a - \eta_b)d\varepsilon_{qa}) \qquad [9.11c]$$

where $d\varepsilon_a = -\delta v_a/v = d\varepsilon_{a.11} + 2d\varepsilon_{a.33}$ is the volumetric strain-increment for the air voids, $d\varepsilon_{qa} = 2(d\varepsilon_{a.11} - d\varepsilon_{a.33})/3$ is the deviator strain-increment for the air voids, $d\varepsilon_{a.11}$ is the axial strain-increment for the air voids and $d\varepsilon_{a,33}$ is the radial strain-increment for the air voids.

Equations 9.11a–c are considered logical and consistent with the bi-modal structure present in unsaturated soils. Only two of the three stress state variables are necessary to define the incremental work input along with the deviator stress which is represented by the mobilised stress ratios η_a and η_b.

9.3 Components of the deviator stress

Equation 9.9 can be written in terms of the two mobilised components of deviator stress in three compatible ways:

$$q = q_a + q_b \qquad [9.12a]$$

or, based on the alternative views of the dual stress regime:

$$q = q_c + q_d \qquad [9.12b]$$

or

$$q = q_e + q_f \qquad [9.12c]$$

where q_a to q_f are components of q and are dependent on the stress regime considered to represent an unsaturated soil and given by Equations 7.23a–c:

- $q_a = \eta_a \bar{p}$ is the component as a result of $\bar{p}$ acting through the whole soil mass (Equation 7.23a).
- $q_b = \eta_b s$ is the additional component as a result of s acting through the aggregates (Equation 7.23a).
- $q_c = (\eta_a - \eta_b) \bar{p}$ is the component as a result of $\bar{p}$ acting between the aggregates (Equation 7.23b).
- $q_d = \eta_b p'$ is the component as a result of p' acting within the aggregates (Equation 7.23b).
- $q_e = \eta_a p'$ is the component as a result of p' acting through the whole soil mass (Equation 7.23c).
- $q_f = (\eta_a - \eta_b) s$ is the reduction component as a result of s not acting through the whole soil mass (Equation 7.23c).

Irrespective of the two stress state variables chosen to describe the stress regime, the components of deviator stress in an unsaturated soil can be written in terms of the two mobilised stress ratios η_a and η_b. The following paragraphs show that the alternative definitions lead to the correct determination of the deviator strain-increments.

9.4 Work input to unsaturated soils

Internal to a soil specimen, the work input results in energy dissipation as plastic, friction straining, and free energy[2] as elastic, recoverable straining. While this is discussed further in Section 9.8.4, it is important to realise that the incremental work input results in both elastic and plastic strains, which are combined in the strain-increment terms.

It is instructive to rewrite Equations 9.11a–c by substitution for η_a and η_b using the definitions of Section 9.3. Equation 9.11a can be written as:

$$dW_u' = \overline{p}d\varepsilon_v + q_a d\varepsilon_q + sd\varepsilon_w + q_b d\varepsilon_{qw} \qquad [9.13a]$$

The first two terms on the right of the equation are the components of the work input due to a mean net stress $\overline{p}$ and a deviator stress q_a respectively acting through the whole soil mass. The second two additive terms are the additional work input due to the suction s and deviator stress q_b acting through the aggregates. Thus, the work input and the associated mechanical energy that is transferred to the soil internally can be divided into two components consistent with the bi-modal structure and dual stress regime.

In a similar manner, Equation 9.11b can be written as:

$$\delta W_u' = \overline{p}d\varepsilon_a + q_c d\varepsilon_{qa} + p'd\varepsilon_w + q_d d\varepsilon_{qw} \qquad [9.13b]$$

The first two terms on the right of the equation are the components of the work input due to a mean net stress $\overline{p}$ and a deviator stress q_c acting between the aggregates. The second two terms are the additional work input due to the mean effective stress p' and deviator stress q_d acting through the aggregates. The form of the equation can be directly compared to Equations 9.1a and 9.1b, which apply to saturated and perfectly dry soils respectively. The first two terms on the right in Equation 9.13b correspond to a dry soil, where the aggregates are treated as large particles, and the second two terms to a saturated soil, and apply to the aggregates.

Adopting an approach similar to the above, Equation 9.11c can be rewritten as:

$$\delta W_u' = p'd\varepsilon_v + q_e d\varepsilon_q - sd\varepsilon_a - q_f d\varepsilon_{qa} \qquad [9.13c]$$

In this third form of the work input equation, the first two terms on the right-hand side are the components of the work input due to a mean effective stress p' and deviator stress q_e acting through the whole soil mass. The second two terms are the subtractive work input components due to the suction s and deviator stress q_f not acting through the volume of the air voids.

It is possible to further rewrite the work input equations in terms of the directional stresses, noting that the mean net and effective stresses $\overline{p}$ and p' can be written as $\overline{p} = \left(\overline{\sigma}_{11} + 2\overline{\sigma}_{33} \right)/3$ and $p' = \left(\sigma_{11}' + 2\sigma_{33}' \right)/3$ respectively, and writing the volumetric and deviator strains in terms of the directional strains. We will be particularly concerned with the form of Equation 9.13b, which can be written as:

$$\delta W_u' = \overline{\sigma}_{11} d\varepsilon_{a.11} + 2\overline{\sigma}_{33} d\varepsilon_{a.33} + \sigma_{11}' d\varepsilon_{w.11} + 2\sigma_{33}' d\varepsilon_{w.33} \qquad [9.14]$$

The first two terms on the right of the equation are the components of the work input between the aggregates and the second two terms are the additional work input to the aggregates. The form of the equation can again be directly compared with Equations 9.1a and 9.1b, which apply to saturated and perfectly dry soils respectively. The first two terms in Equation 9.14 correspond to a dry soil, where the aggregates are treated as large particles, and the second two terms to a saturated soil.

The incremental work input components result in both elastic and plastic straining of the bi-modal structure. The distinction between elastic and plastic strains is usually based on changes of slope of stress–strain plots or recoverable strains on unloading back to original conditions. It is important to realise, however, that the strains internal to a soil must reflect the total energy changes necessary to satisfy the work input equations as well as satisfying overall continuity requirements.

9.5 Analysis of the mobilised stress ratios

In saturated soils it is normal to take the mobilised stress ratio η as defined by a line passing through the origin as illustrated in Figure 9.1. The mobilised q is thus defined by the polar coordinate parameter η and stress state variable p'. Similarly for unsaturated soils, the mobilised stress ratio η_a is taken as defined by a line passing through the intercept on the p_c'/s axis given by the strength envelope, defined by the stress ratio M_a, as in Figure 9.2. Thus, the intercept ω on the q/s axis made by the mobilised stress ratio line η_a (at $p_c'/s = 1.0$) is given by:

$$\omega = \Omega \frac{\eta_a}{M_a} \qquad [9.15]$$

The equation for the line η_a is given by:

$$\frac{q}{s} = \eta_a \left[\frac{p_c'}{s} - 1 \right] + \omega \qquad [9.16]$$

Substituting for ω from Equation 9.15 and p_c' from Equation 9.3 and rearranging gives:

$$\eta_a = \frac{q}{\bar{p} + s \left(\frac{v_w}{v} + \frac{\Omega}{M_a} - 1 \right)} \qquad [9.17]$$

and from Equations 9.9 and 9.13:

$$\eta_b = \eta_a \left(\frac{v_w}{v} + \frac{\Omega}{M_a} - 1 \right) = \eta_a \left(\frac{\Omega}{M_a} - \frac{v_a}{v} \right) \qquad [9.18]$$

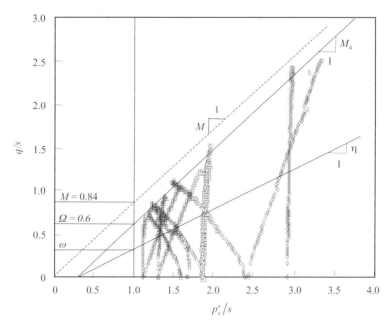

Figure 9.2 q/s against p_c'/s at critical state from 1d-cs test results for unsaturated kaolin, and q against p' for saturated kaolin from the results of Sivakumar (1993) and Wheeler and Sivakumar (1995, 2000).

Equations 9.17 and 9.18 allow the determination of η_a and η_b from triaxial experimental data.

9.6 Continuity relationships between strain-increments

It is necessary to define the relationships between the various strain-increment variables in order to allow analysis of experimental data. For volumetric compatibility, $dv = dv_w + dv_a$. With $d\varepsilon_v = -dv/v$, the following volumetric strain-increment equation holds true only if $d\varepsilon_w = -dv_w/v$ and $d\varepsilon_a = -dv_a/v$, i.e. the strain-increments for the aggregates and the air voids are normalised with respect to v as previously defined:

$$d\varepsilon_v = d\varepsilon_w + d\varepsilon_a \qquad [9.19]$$

In addition, substituting in Equation 9.19 using the definitions of $d\varepsilon_w$ in Equation 9.5 and $d\varepsilon_a$ in Equations 9.11a–c, and comparing with the definition of $d\varepsilon_v$ in Equations 9.1a–b, it is necessary for axial and radial deformation continuity that:

$$d\varepsilon_{11} = d\varepsilon_{a,11} + d\varepsilon_{w,11} \qquad [9.20]$$

and

$$d\varepsilon_{33} = d\varepsilon_{a,33} + d\varepsilon_{w,33} \qquad [9.21]$$

Equations 9.11a–c and 9.12a–c show that $d\varepsilon_q$ is a function of $d\varepsilon_{qw}$, $d\varepsilon_{qa}$ and the components of deviator stress:

$$d\varepsilon_q = \left[\frac{q_c d\varepsilon_{qa} + (q_d - q_b) d\varepsilon_{qw}}{q_a} \right] \qquad [9.22]$$

Thus, using Equations 9.22 and 9.12a–b it is readily shown that:

$$d\varepsilon_q = d\varepsilon_{qa} + \frac{\eta_b}{\eta_a} \left[d\varepsilon_{qw} - d\varepsilon_{qa} \right] \qquad [9.23]$$

As a soil becomes drier the deviator strain-increment $d\varepsilon_q$ should tend towards the deviator strain-increment for the air voids $d\varepsilon_{qa}$. For this to happen, η_b should tend towards zero, which complies with the experimental evidence in Chapter 8. However, the transition may not be smooth and at very high suctions a discontinuity in volumetric behaviour, and thus in the work input, can be experienced. Similarly, as a soil approaches saturation, $d\varepsilon_q$ should tend towards $d\varepsilon_{qw}$. Equations 9.10 and 9.19 suggest that for a smooth transition of the work input to saturated conditions, η_a and η_b should both tend smoothly to η. However, experimental data at the critical state suggest that the transition from unsaturated to saturated conditions is not smooth (Murray, 2002). At critical state, $\eta_b = M_b$, and Equation 9.3 indicates that M_b tends towards Ω as the degree of saturation increases, which from experimental evidence is less than M_a (η_a at critical state). The experimental evidence thus suggests that the transition from unsaturated to saturated conditions may not be smooth and is taken as indicative of the breakdown of the aggregate structure, leading to relatively abrupt changes in work input (Murray et al., 2008). Experimental observations of discontinuities in behaviour, as discussed in Section 5.8, are attributable to meta-stable conditions.

The overall deviator strain-increment is correctly determined as the sum of the deviator strain-increments for the air voids and the aggregates (Equation 9.24), lending strong support to the validity of the analysis. This can be shown by substituting from the definition for the deviator strain-increment for the air voids $d\varepsilon_{qa}$ from Equation 9.11a–c, into Equations 9.22 and 9.23, noting the definition of the deviator strain-increment for the aggregates $d\varepsilon_{qw}$ in Equation 9.10. Accordingly we may write:

$$d\varepsilon_q = d\varepsilon_{qw} + d\varepsilon_{qa} \qquad [9.24]$$

Thus, from Equations 9.21 and 9.22, and the definitions of the mobilised stress ratios:

$$d\varepsilon_{qa} = d\varepsilon_q \left[\frac{\eta_b - \eta_a}{2\eta_b - \eta_a} \right] \qquad [9.25]$$

$$d\varepsilon_{qw} = d\varepsilon_q \left[\frac{\eta_b}{2\eta_b - \eta_a} \right] \qquad [9.26]$$

and

$$d\varepsilon_{qa} = d\varepsilon_{qw} \left[\frac{\eta_b - \eta_a}{\eta_b} \right] \qquad [9.27]$$

Equations 9.25 and 9.26 indicate that $d\varepsilon_{qa}$ and $d\varepsilon_{qw}$ go to infinity when $2\eta_b = \eta_a$, as the denominators in the equations go to zero. This is an important observation in analysing the experimental data in later sections as it corresponds to discontinuities in material behaviour under shearing. On either side of the discontinuities the values of $d\varepsilon_{qa}$ and $d\varepsilon_{qw}$ change sign.

Other important relations between the directional strain-increments and volumetric and deviator strain-increments for the aggregates and air voids can be established. From the definition of $d\varepsilon_w$ in Equation 9.5, substituting for $d\varepsilon_{w,11}$ into the definition for $d\varepsilon_{qw}$ in Equation 9.10 and rearranging gives:

$$d\varepsilon_{w,33} = \frac{1}{3} d\varepsilon_w - \frac{1}{2} d\varepsilon_{qw} \qquad [9.28]$$

Substituting for $d\varepsilon_{w,33}$ from Equation 9.28 into the definition of $d\varepsilon_{qw}$ from Equation 9.10 and rearranging gives:

$$d\varepsilon_{w,11} = \frac{1}{3} d\varepsilon_w + d\varepsilon_{qw} \qquad [9.29]$$

On a similar basis, from the definition of $d\varepsilon_a$ in Equations 9.11a–c, substituting for $d\varepsilon_{a,11}$ into the definition for $d\varepsilon_{qa}$ in the same equations and rearranging gives:

$$d\varepsilon_{a,33} = \frac{1}{3} d\varepsilon_a - \frac{1}{2} d\varepsilon_{qa} \qquad [9.30]$$

Substituting for $d\varepsilon_{a,33}$ from Equation 9.30 into the definition of $d\varepsilon_{qa}$ from Equations 9.11a–c and rearranging gives:

$$d\varepsilon_{a,11} = \frac{1}{3} d\varepsilon_a + d\varepsilon_{qa} \qquad [9.31]$$

As Equations 9.28–9.31 are functions of $d\varepsilon_{qw}$ or $d\varepsilon_{qa}$ given by Equations 9.25 and 9.26 respectively, $d\varepsilon_{w,11}$, $d\varepsilon_{w,33}$, $d\varepsilon_{a,11}$ and $d\varepsilon_{a,33}$ go to infinity when $2\eta_b = \eta_a$ as do $d\varepsilon_{qw}$ and $d\varepsilon_{qa}$. On either side of the discontinuities the values of $d\varepsilon_{w,11}$, $d\varepsilon_{w,33}$, $d\varepsilon_{a,11}$ and $d\varepsilon_{a,33}$ change sign. The discontinuities are described as 'singularities' as there are abrupt energy changes associated with changes between compressive and expansive behaviour.

In accordance with the definition of the total deviator strain-increment $d\varepsilon_q$ given by Equations 9.1a–c, if a specimen experiences length reduction and radial expansion, $d\varepsilon_q$ is positive. If a specimen experiences radius reduction and length expansion, $d\varepsilon_q$ is negative. The definitions of shear strain-increment as well as volumetric strain-increment $d\varepsilon_v$ and the associated directional strain-increments $d\varepsilon_{11}$ and $d\varepsilon_{33}$ are sufficient to satisfy continuity and work input requirements based on overall specimen deformation. However, as discussed in Section 9.8.5, the deviator strains as well as the total and directional strain-increments for the air voids and the water–soil particle aggregates are internal strains and must satisfy the work/energy dispersion requirements at this structural level as well as satisfying the overall continuity requirements. The analysis of the work input does not restrict the strain-increments for the air voids or the aggregates other than that the summed incremental changes must equate to the external volumetric and directional strain-increment measurements for a soil specimen.

9.7 Stress state variables and conjugate volumetric and strain-increment variables

Table 9.1 presents the stress state variables, stress conjugate volumetric variables and work conjugate strain-increment variables appropriate in constitutive modelling under triaxial test conditions. This is based on Equations 7.23a–c and 7.24a–c and the foregoing analysis.

Table 9.1 Conjugate variables.

Condition	Stress variable	Stress conjugate volumetric variable	Work conjugate strain-increment variables
1	$\bar{p} = (p - u_a)$	v	$d\varepsilon_v$
	s	v_w	$d\varepsilon_w, d\varepsilon_{w.11}, d\varepsilon_{w.33}$
	q_a	v	$d\varepsilon_q$
	q_b	v_w	$d\varepsilon_{qw}$
	$\bar{\sigma}_{11} = (\sigma_{11} - u_a)$	v	$d\varepsilon_{11}$
	$\bar{\sigma}_{33} = (\sigma_{33} - u_a)$	v	$d\varepsilon_{33}$
2	$\bar{p} = (p - u_a)$	v_a	$d\varepsilon_a$
	$p' = (p - u_w)$	v_w	$d\varepsilon_w$
	q_c	v_a	$d\varepsilon_{qa}$
	q_d	v_w	$d\varepsilon_{qw}$
	$\bar{\sigma}_{11} = (\sigma_{11} - u_a)$	v_a	$d\varepsilon_{a.11}$
	$\bar{\sigma}_{33} = (\sigma_{33} - u_a)$	v_a	$d\varepsilon_{a.33}$
	$\sigma'_{11} = (\sigma_{11} - u_w)$	v_w	$d\varepsilon_{w.11}$
	$\sigma'_{33} = (\sigma_{33} - u_w)$	v_w	$d\varepsilon_{w.33}$
3	$p' = (p - u_w)$	v	$d\varepsilon_v$
	$-s$	v_a	$d\varepsilon_a, d\varepsilon_{a.11}, d\varepsilon_{a.33}$
	q_e	v	$d\varepsilon_q$
	$-q_f$	v_a	$d\varepsilon_{qa}$
	$\sigma'_{11} = (\sigma_{11} - u_w)$	v	$d\varepsilon_{11}$
	$\sigma'_{33} = (\sigma_{33} - u_w)$	v	$d\varepsilon_{33}$

Conditions 1, 2 and 3 in Table 9.1 describe the three different ways of looking at the dual stress regime and work input equations for an unsaturated soil. Condition 1 relates to the view of net stress $\bar{p}$ acting through a soil with the addition of the suction s within the more highly stressed aggregates. Condition 2 relates to $\bar{p}$ acting between the aggregates, which act as large particles, with the effective stress p' acting within the aggregates. Condition 3 relates to an effective stress p' acting throughout a soil with a reduction for the suction s, which does not act through the volume of the air voids.

Chapter 8 demonstrated the use of the stress state variables and conjugate volumetric variables in developing an understanding of strength and compression data for unsaturated soils (Murray, 2002; Murray *et al.*, 2002). In particular, the significance of Equation 8.3 for shear strength is demonstrated. The stress state variables, deviator stresses and work conjugate strain-increment variables in Table 9.1 are used in the following to examine the volumetric and anisotropic behaviour both of the aggregates and between the aggregates in unsaturated soils during triaxial shearing.

9.8 The meaning and interpretation of stresses and strains

In understanding the significance of stresses and strains in unsaturated soils and before embarking on a description of the correct interpretation of experimental data, it is important to first outline the assumptions on which the analysis of the stress–strain behaviour of unsaturated soils is based. The discussion that follows highlights the significance of the directional strain-increments $d\varepsilon_{w,11}$, $d\varepsilon_{w,33}$, $d\varepsilon_{a,11}$ and $d\varepsilon_{a,33}$ and the conjugate stresses for unsaturated soil specimens in the triaxial cell. The strain-increments are described in terms of the summed components of the free energy associated with elastic straining and the energy dissipated as plastic, friction straining that an unsaturated soil experiences.

9.8.1 Principal assumptions

It is assumed in the isothermal work input analysis that there is coincidence of the principle axes of stress, strain and strain-increment. While real soils may deviate somewhat from this assumption, the coincidence of the axes is an assumption frequently made and does not invalidate general conclusions of material behaviour based on this premise. In addition, the thermodynamic requirement of reversibility requires that strain-increments are infinitesimally small. While this is unrealistic in undertaking soil tests, small incremental changes between equilibrium states, as adopted in the tests analysed in the following, comply reasonably with this requirement. In the tests, small imposed stress-increments were held constant until there was no further measurable change to a soil specimen. During the shearing stages up to the critical states, each test analysed was carried out by imposing around 200 small incremental changes between equilibrium conditions.

It is also important to recognise the general assumption in triaxial cell testing and analysis that specimens experience uniform stress and strain distributions, though this is not strictly the case. A soil specimen under test is generally considered to represent a point or small element of soil, and an intrinsic assumption is that both axial and radial strains are uniform throughout the specimen. This assumption becomes less valid as the magnitude of strain increases. A manifestation of the non-uniform straining is the development of radial barrelling by many soil specimens in response to a non-uniform internal stress distribution. Experimental measurements and analysis generally assume that specimens deform uniformly as right-sided cylinders and thus relate to mean conditions.

9.8.2 Continuity, work and energy dispersion requirements

In saturated soils, Terzaghi's effective stress has been shown to be a valid, usable interpretation of the controlling stress. This is, however, a simplification of the actual interactions between soil particles (or their adsorbed double layers). Similarly, the average volumetric coupling stress as defined by Equations 7.23a–c and 7.24a–c is a simplification of the actual interactive stresses within an unsaturated soil. These effective stresses represent the net interactive stresses and are simplifications of the actual stress conditions internal to a soil. Similar comments may be applied to the strains. The externally measured strains represent the net effects of the movements and deformations of the phases in a soil. Nevertheless, there are two important principles that the strain-increments, and accordingly the total strains, must satisfy and it is worth re-emphasising these here (Schofield and Wroth, 1968; Fredlund and Rahardjo, 1993; Houlsby, 1997):

- They must comply with the continuity requirements of continuum mechanics.
- They must satisfy thermodynamic work and energy dispersion requirements.

These requirements apply equally to the overall behaviour of a soil specimen, based on external measurements, and the behaviour of the phases internal to a soil specimen. It is the implications of these requirements to the interpretation of experimental data that is discussed. We will use the terms strain and strain-increment somewhat interchangeably as the comments and conclusions apply equally to both.

9.8.3 Macro-mechanical specimen straining in the triaxial cell

In Section 4.3, body distortion was defined as comprising the directional strains and the shear strains, though it was recognised that on a more general level total strains may also include body displacement and body rotation (Schofield and Wroth, 1968). Nevertheless, in triaxial cell testing, the strains associated with body distortion, determined from overall strain measurements of a specimen, are sufficient and adequate to describe the external changes. It is common practice to describe the distortional straining of a specimen under test in terms of the volumetric strain and deviator strain, while also recording the change in water volume and, accordingly, the change in air volume of the specimen. Thus, the following strain-increment variables are commonly recorded: $d\varepsilon_{11}, d\varepsilon_{33}, d\varepsilon_{v}, d\varepsilon_{q}, d\varepsilon_{w}$ and $d\varepsilon_{a}$. The externally measured directional strain-increments $d\varepsilon_{11}$ and $d\varepsilon_{33}$ are shown to be sufficient to define the volumetric and deviator strain increments $d\varepsilon_{v}$ and $d\varepsilon_{q}$ respectively and to satisfy the overall continuity requirements. The externally measured strain-increment variables in conjunction with their conjugate stress state variables, as summarised in Table 9.1, have also been shown in the foregoing analysis to be compliant with the work input requirements. The definitions of the strain-increments, and accordingly the total strains, thus satisfy the governing continuity and work principles, and serve perfectly well in normal practice of measuring overall macro-mechanical specimen deformations. However, the meaning of strain at the aggregate and inter-aggregate micro-mechanical level in an unsaturated soil is more complex.

9.8.4 Elastic and plastic straining, free energy and frozen elastic energy

Elastic straining in soils is usually defined as occurring within the current yield surface, with plastic straining occurring when the yield surface is reached. The assessment of yield is frequently based on change of curvature of stress–strain plots. However, the distinction between elastic and plastic behaviour presents a knotty problem. The applicability of the theory of 'thermomechanics' to constitutive modelling of soils as described by Collins and Muhunthan (2003), Collins (2005) and Collins *et al.* (2007) leads to insight into the elasto-plastic behaviour. This relies on the premise that both cohesionless granular materials and fine-grained materials (so-called cohesive soils) exhibit similar characteristics. Historically, a major difference has been the understanding that fine-grained soils exhibit cohesion while granular soils do not. More recent research has led to a redefining of the cohesion intercept of the strength envelope, as exhibited by fine-grained soils at peak strength, as the manifestation of the degree of interlocking of the soil particles as expressed by Taylor (1948) for granular materials. Schofield (2005, 2006) has argued that the peak strength of both granular and cohesive deposits is the sum of the critical state strength and interlocking (or the rate of dilation). In accordance with this interpretation, the strength of fine-grained soils does not rely on the surface chemistry of bonds between clay particles. Unsaturated fine-grained soils with an aggregate structure can thus be perceived to comply closely with a granular material, and interlocking plays a major role in the strength and deformation behaviour, including dilation under shearing action.

The basic energy equation for work input based on the Cam Clay plasticity model of Roscoe *et al.* (1963) is given by:

$$\delta W = \delta W^e + \delta W^p \qquad [9.32]$$

where the work input δW is made up of the recoverable elastic work δW^e and the irrecoverable plastic work δW^p.

It is interesting to compare Equation 9.32 with Equation 5.3 and to tentatively equate the elastic stored work δW^e with the internal energy dU, and the plastic dissipated work (or energy) δW^p with the heat term δQ, noting that entropy change given by $dS = \delta Q/T$ is a quantity that never decreases and is a measure of the amount of energy change unavailable to do work. Equating δW^p to δQ assumes that all heat is turned into mechanical work. However, there are subtle differences between the work (or energy) input dissipated plastically, and deemed irrecoverable on unloading back to original conditions, and the definition of entropy change as a measure of the energy dissipated and unavailable to do work. As discussed in Section 5.6.3, there is difficulty in fully reconciling entropy in terms of general energy dissipation.

Elastic and plastic straining, and the ideas of energy dissipation and stored plastic work (or frozen elastic energy), have been discussed by Collins (2005). By definition, plastic work is irrecoverable and the energy is dissipated, while elastic work is recoverable on unloading. However, problems arise in interpreting experimental data at the macro-mechanical or continuum level. On unloading to original imposed stress conditions, the heterogeneous nature of deformations, and the resulting stress changes at the micro-mechanical level, mean that not all elastic straining at the micro-mechanical level is likely to be recovered. This gives rise to 'frozen plastic energy' when considering macro-mechanical behaviour.

At the micro-mechanical level this elastic energy component is not recovered on external unloading, and thus not detected as external elastic straining. This straining is included within the plastic component of strain measured externally. Reversed plastic loading, or possibly time, is required to free the frozen energy. There is growing experimental evidence to support these contentions which are manifest in hysteresis and may also be linked to viscous (time-dependent) behaviour in soils. Collins (2005) expressed this mathematically as:

$$\delta W = \delta W^c + \delta W^p = d\Psi^c + d\Psi^p + \delta\Phi \qquad [9.33]$$

where δW is the increment of applied work, δW^c is the increment of applied elastic work, δW^p is the increment of applied plastic work, $d\Psi^c$ is the increment of elastic work not frozen, $d\Psi^p$ is the increment of elastic work frozen as applied plastic work and $\delta\Phi$ is the increment of plastic work dissipated and is always positive.

Accordingly, the total plastic work based on external measurements is given by,

$$\delta W^p = d\Psi^p + \delta\Phi \qquad [9.34]$$

In the tests analysed in Section 9.9, the analysis is restricted to appraisal of loading during the shear stages of triaxial tests and we will generally concern ourselves with the summed elastic and plastic strains. However, the foregoing discussion emphasises the significance of the frictional dissipative strains as a result of the sliding and rolling contact of neighbouring particles or aggregates, and the free energy strains that are not dissipative and are recoverable. In the following, as in the foregoing discussion, we will use the term 'dispersed' energy to represent the sum of the stored elastic energy and the dissipated plastic energy at the micro-mechanical level.

9.8.5 Micro-mechanical specimen straining in the triaxial cell

External measurements of overall specimen distortion define the *macro-mechanical* straining and satisfy the overall continuity and work (or mechanical energy input) criteria. Internal to a soil specimen, the stresses and strains associated with the aggregates must also satisfy the compatibility and energy requirements but the situation is complex. The internal straining constitutes the *micro-mechanical* behaviour and may be defined as the behaviour of composite or heterogeneous materials on the level of the individual phases that constitute the material. While we will consider the water and solid phases combined into aggregates in unsaturated fine-grained soils, and not consider these phases separately, the definition serves to differentiate between the overall macro-mechanical distortional straining of a specimen and the internal strains both of the aggregates and between the aggregates.

There are obvious benefits in determining the magnitudes of the relevant micro-mechanical strains as these represent the changes in internal structure and allow an appraisal of the significance of phenomenon such as anisotropy, yielding, compression and dilation. The use of the conjugate stress and strain-increment variables for the air voids and aggregates allows insight into such behaviour.

9.8.5.1 Micro-mechanical strain components

It is possible to identify a number of strain components internal to a soil specimen that give rise to energy dispersion resulting from the application of an external load. The primary identifiable sources are:

- Displacement of particles, aggregates, water and air;
- Rotation of particles and aggregates;
- Distortion of phases and aggregates (including compression/expansion and shear deformation);
- Breakage of particles and aggregates;
- Changes and interactions of the adsorbed double layer, the contractile skin, dissolved air and water vapour in air.

Visualisation of tests on soil specimens and reference to published work allow us to describe how these internal strain components, along with any air and water exchanges with external reservoirs, as in triaxial testing, interact to give the net specimen macro-mechanical distortion determined from external overall measurements.

First consider the case of a specimen of saturated gravel under triaxial undrained loading. Application of the loading would lead to deformation of the specimen, which may be perceived to be primarily a result of *displacement* straining of the gravel and the water internal to the specimen. However, Alramahi and Alshibli (2006) noted from X-ray tomography measurements of spherical plastic particles under triaxial loading conditions that *rotation* straining also played a major role in the shearing resistance and must, accordingly, be taken into account in work and energy dispersion analysis. Cui and O'Sullivan (2006) discussed the use of the distinct element method (DEM) and direct shear box tests on uniform steel spheres. They noted the occurrence of sphere rotation and displacement and confirmed the conclusion of Masson and Martinez (2001) from DEM analysis that particle rotation is a significant indicator of strain localisation. For continuity, the externally measured straining of the specimen must equate in an axial and radial direction with the net effects of displacement and rotation straining as well as other possible strains at a micro-mechanical level. Bolton *et al.* (2008) noted the influence of energy dispersion due to rearrangement (displacement and rotation) of assemblies of bonded balls in computer simulations using DEM analysis, but also identified the energy stored as elastic strain energy at points of contact between balls. While the elastic *compression* straining of solid particles and water is usually ignored in volume change analysis, in terms of energy dispersion it is unclear how significant and what role such compression plays. Particle breakage, which may be described as *degradation straining*, may also occur and result in energy dispersion.

Consider now an unsaturated soil where aggregates of fine soil particles and water exist and these are surrounded by air-filled voids. Under undrained loading there will be *displacement*, *rotation* and *distortion* straining along with possible *degradation* straining of the aggregates. There will also be *displacement* and *distortion* straining of the air phase. The distortional straining of the aggregates may be perceived as comprising both compressive and expansive directional straining and shear straining under constant volume conditions if the soil particles and water phases are considered incompressible. In this respect the aggregates can be thought of as deformable large particles. While we will be concerned only with overall straining of the aggregates, it is important to note that internal to the aggregates, energy would be dispersed as a result of displacement and rotation (ignoring compression) at the particle level. There could also be an influence on external strain measurements from breakage of soil particles as well as degradation and reformation of the aggregates. Further influences would accrue from air dissolving or being liberated from the pore water, as well as changes in water vapour in the air. Other sources of mechanical energy dispersion associated with the adsorbed double layer and the contractile skin may also occur.

In the foregoing undrained tests, the externally measured strains, whether directional, volumetric or deviatoric, must be the net effects of the displacement, rotation, distortion and degradation straining, plus other possible strain effects, internal to the specimen. If drainage of either or both the water and air phases is allowed, external strain measurements would also include the effects of the phase mass changes.

The energy dispersion requirement as a result of external work input is not satisfied by merely considering the compressive and expansive changes of the aggregates and air voids. It is essential to understand that at the micro-mechanical level the straining is multifaceted and requires the meaning of the generalised strain-increment variables to be interpreted in terms of energy dispersion as plastic dissipation of energy and elastic recoverable energy. While some of the displacement, rotation, compression and distortion straining may be recoverable on removal of external loading, in any particular test much is likely to lead to energy dissipation as a result of overcoming frictional resistance.

There are identifiable mechanical changes internal to a soil specimen that give rise to energy dispersion but do not contribute to the net strain-increments determined from external measurements. For example, energy may be dispersed by rotation of aggregates, but rotation may not give rise to externally measurable strains. Similarly, displacement of aggregates, particles, water or air in one direction may to a greater or lesser degree be balanced by movement elsewhere in the specimen in an opposite direction, giving rise to a reduced net external effect, though energy is dispersed by the movements in both directions. Similarly, both directional straining and shear straining of aggregates in unsaturated soils in opposing directions may occur internal to a specimen, with only the net effects measurable external to the specimen.

9.8.5.2 Micro-mechanical strains as a measure of energy dispersion

While the individual directional strain-increments $d\varepsilon_{11}$ and $d\varepsilon_{33}$ are adequate in describing overall macro-mechanical soil specimen behaviour, the directional strain-increments $d\varepsilon_{w,11}, d\varepsilon_{w,33}, d\varepsilon_{a,11}$ and $d\varepsilon_{a,33}$ relate to the micro-mechanical behaviour and energy dispersion in or between the aggregates. Nevertheless, to satisfy continuity, the sums of the changes associated with the aggregates and air voids in both an axial and radial direction must equate to the overall macro-mechanical strain-increments $d\varepsilon_{11}$ and $d\varepsilon_{33}$ experienced by the specimen. This equality is expressed in Equations 9.20 and 9.21. Thus, the summed internal strains must satisfy the overall volumetric and deviatoric straining of a specimen. Examination of the equations for the strain-increments $d\varepsilon_{w,11}, d\varepsilon_{w,33}, d\varepsilon_{a,11}$ and $d\varepsilon_{a,33}$ shows that they are linked to the mobilised stress ratios η_a and η_b, indicating their dependence on the dispersion of the mechanical energy as a result of the work input.

The outcome of the argument, which is confirmed by examination of laboratory test results, is that the magnitudes of the strain-increments $d\varepsilon_{w,11}, d\varepsilon_{w,33}, d\varepsilon_{a,11}$ and $d\varepsilon_{a,33}$ can be expected to be significantly greater than might be assessed by merely examining the magnitudes of the externally measured directional strain-increments. This requires that the strain-increments $d\varepsilon_{w,11}$ and $d\varepsilon_{a,11}$ are of opposite sign, as their sum must equate to the net strain-increment $d\varepsilon_{11}$ measured external to the soil specimen. A similar conclusion can be drawn for the strain-increments $d\varepsilon_{w,33}$ and $d\varepsilon_{a,33}$, the summed influence of which must equate to the strain-increment $d\varepsilon_{33}$.

It is possible to gain further insight into the behavioural trends of the strain-increments by simple conjecture. In a conventional triaxial test where the axial load is increased and the cell pressure is maintained constant, radial expansive straining $-d\varepsilon_{33}$ occurs and is in a direction opposite to the imposed cell pressure. This leads to apparent negative energy dispersion. Negative energy dispersion is also indicated when analysing strains on a micro-mechanical level. Consider the case of saturated aggregates in a specimen tested under undrained triaxial conditions. Assuming no change of volume of the solid particles and water under application of loading, and a positive axial strain-increment $d\varepsilon_{w,11}$ for the aggregates, there would be expected to be negative expansive radial straining of the aggregates $-d\varepsilon_{w,33}$ though the magnitudes of $d\varepsilon_{w,11}$ and $-d\varepsilon_{w,33}$ will be influenced by other energy dispersion effects. Thus, when the strain-increments $d\varepsilon_{w,11}$ and $-d\varepsilon_{w,33}$ are plotted against the conjugate stress variables σ'_{11} and σ'_{33} respectively in accordance with Table 9.1 Case 2, positive energy dispersion is indicated for the aggregates in an axial direction and negative energy dispersion in a radial direction.

Dilation also gives rise to apparent negative energy dispersion as a soil expands against imposed loading. Alramahi and Alshibli (2006) have noted the importance of dilation in assessing the behaviour of granular soils under triaxial test conditions. This is relevant to unsaturated fine-grained soils as they develop aggregates of particles that behave in many respects as a granular material, though the aggregates are likely to exhibit a greater degree of deformability than individual granular particles. The greater the suction and the stiffer the aggregates, the less the influence of the inter-aggregate bonding (Wheeler *et al.*, 2003). Consequently, with increasing desiccation the behaviour of an unsaturated fine-grained soil resembles more closely that of a true granular material.

Negative energy dispersion is a manifestation of macro-mechanical measurements of stress or pressure and is a phenomenon that arises in analyses such as turbulent fluid flow. The significance of negative energy dispersion is discussed further in Section 9.8.6, but it is important to appreciate that this is a necessary outcome of the continuity and work/energy dispersion requirements that must be satisfied. As noted by Atkinson (1993), during an increment of straining the work done must be invariant. In other words, the net work done must be independent of the choice of parameters.

The foregoing arguments suggest two primary energy dispersion mechanisms as likely to occur in unsaturated soils subject to conventional axial compression tests in a triaxial cell. These are defined by positive and negative directional strain-increments as either $+d\varepsilon_{w,11}$, $-d\varepsilon_{w,33}$, $-d\varepsilon_{a,11}$ and $+d\varepsilon_{a,33}$ or $-d\varepsilon_{w,11}$, $+d\varepsilon_{w,33}$, $+d\varepsilon_{a,11}$ and $-d\varepsilon_{a,33}$. While other test protocols may dictate other relationships between the strain-increments, these two dispersion mechanisms are exhibited in the tests on specimens of kaolin examined in Section 9.9.

What does the foregoing discussion mean?
- The conjugate variables allow analysis of the net stress–strain behaviour of the aggregate structure of an unsaturated soil specimen in a triaxial cell based on fundamental theoretical principles.
- The straining at the micro-mechanical level of the aggregate structure is in terms of energy dispersion and not just body distortion as at the macro-mechanical level of overall specimen deformation.

- While it is possible to identify components of energy dispersion that contribute to the various strains, it is difficult to definitively correlate all energy components with specific strains and the stress state variables making up the dual stress regime. However, general conclusions can be drawn as discussed in Section 9.9.
- The analysis does not allow the behaviour within the aggregates to be analysed at a particle level but only the net deformation behaviour of the aggregates and the air voids between the aggregates. Accordingly, the analysis does not simplify down to the interparticle analysis of a saturated soil.
- The analysis allows the dispersion of energy of the aggregates and between the aggregates in terms of volumetric straining, deviatoric straining as well as directional straining to be determined.

9.8.6 Significance of apparent negative energy dispersion

Consider the case of a soil specimen (no consideration is given to the nature of the soil) in a triaxial cell, with the specimen initially under equilibrium conditions. Let there be a change in the axial loading while maintaining the cell pressure constant, and let the new total applied axial stress be given by σ_{11} and the total radial stress be given by σ_{33}. As illustrated in Figure 9.3, following the application of loading, the total stresses internal to the soil specimen are not in equilibrium with the applied stresses and are given by $\overline{\overline{\sigma}}_{11}$ axially and $\overline{\overline{\sigma}}_{33}$ radially. It is the differences between the total stresses external and internal to the specimen that give rise to the strains. The increment of work input per unit volume $\delta W'$ (or energy dispersion) due to the stress differences is given by:

$$\delta W' = \int \left(\sigma_{11} - \overline{\overline{\sigma}}_{11}\right) d\varepsilon_{11} + 2 \int \left(\sigma_{33} - \overline{\overline{\sigma}}_{33}\right) d\varepsilon_{33} \tag{9.35}$$

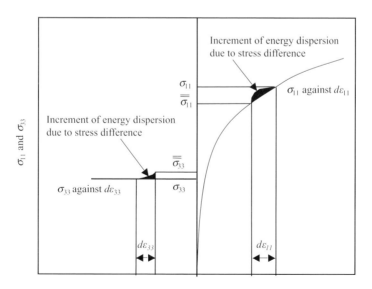

Figure 9.3   Incremental energy dispersion under triaxial conditions.

If $(\sigma_{11} - \bar{\bar{\sigma}}_{11})$ is positive (i.e. $\sigma_{11} > \bar{\bar{\sigma}}_{11}$), $d\varepsilon_{11}$ will be positive indicating compression in an axial direction. If $(\sigma_{33} - \bar{\bar{\sigma}}_{33})$ is negative (i.e. $\sigma_{33} < \bar{\bar{\sigma}}_{33}$), $d\varepsilon_{33}$ will be negative indicating expansion in a radial direction. This is consistent with positive energy dispersion internal to the soil specimen in both axial and radial directions as the stress differences are of the same sign as the strain-increments. Equation [9.35] does not give the total work input. At equilibrium, the stresses axially and radially must balance, and the work input $\delta W'$ under the applied stresses is given by:

$$\delta W' = \int \sigma_{11} d\varepsilon_{11} + 2 \int \sigma_{33} d\varepsilon_{33} \qquad [9.36]$$

While again the axial energy dispersion is positive in the test, the radial dispersion of energy is negative as the radial strain is in the opposite direction to the cell pressure. As for previous analyses, for infinitesimal changes the integration symbols in Equations 9.35 and 9.36 can be omitted.

Similar conclusions can be drawn for the straining of the aggregates and between the aggregates in an unsaturated soil, and indicate that while the strain-increments under the stress differences between equilibrium states are consistent with positive energy dispersion, when considering macro-mechanical imposed stresses at equilibrium, apparent negative energy dispersion may be indicated. Accordingly, care must be taken in interpretation of experimental data for $d\varepsilon_{w,11}$, $d\varepsilon_{w,33}$, $d\varepsilon_{a,11}$ and $d\varepsilon_{a,33}$, recognising that internal to a soil specimen, expansive strains must be associated with expansive forces and compressive strains must be associated with compressive forces.

9.9 Analysis of triaxial experimental data on kaolin

The results of a number of triaxial tests on specimens of kaolin are discussed in the following to highlight the use of the analysis and to show how the interpretation of the data sheds light on the behaviour during shearing. It is important to note that the analysis shows the change in conditions from those at the beginning of the shearing process. The conditions of the specimens at the start are important and it is necessary to recognise that the distribution and form of air voids and aggregates will be different for different preparation techniques, leading to differences in stress–strain behaviour.

9.9.1 Specimens A and B (constant suction shearing tests on initially one-dimensionally compressed specimens. Specimen A sheared under fully drained conditions and Specimen B sheared under constant pressure conditions)

First the results of these two distinctly different triaxial shearing tests on normally consolidated unsaturated specimens of kaolin (100 mm in length and 50 mm in diameter) will be examined and compared. The test procedures for the two specimens are outlined below:

Specimen A: The specimen was initially prepared by one-dimensional compression in a mould and then consolidated isotropically in a triaxial cell. The cell pressure was incrementally raised to 500 kPa and held constant during shearing. Prior to shearing, $v = 2.140$ and $v_w = 1.729$. The suction was held constant at 300 kPa during both

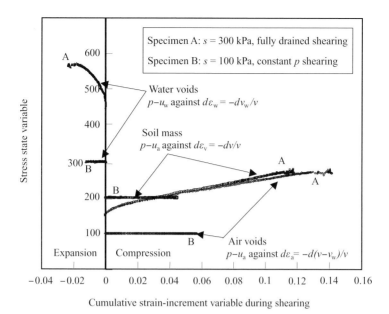

Figure 9.4   Specimens A and B – stress state variables against cumulative strain-increment variables.

consolidation and shearing using the axis translation technique. The specimen was sheared under fully drained conditions with increasing axial loading.

Specimen B: As for specimen A, specimen B was initially prepared by one-dimensional compression and then consolidated isotropically in a triaxial cell. However, the cell pressure for specimen B was incrementally raised to only 350 kPa and held constant during shearing. Prior to shearing, $v = 1.998$ and $v_w = 1.845$. The suction was held constant at 100 kPa during consolidation and shearing. The specimen was sheared by increasing the deviator stress but with $\bar{p}$ and p' maintained constant by adjusting u_a and u_w using the axis translation technique. This is referred to as a constant pressure test.

In Figure 9.4 the plots of $\bar{p} = (p - u_a)$ against cumulative volumetric strain-increment $d\varepsilon_v$ (condition 1 of Table 9.1) illustrate a decrease in the volume of both specimens during shearing, as indicated by an increase in cumulative $d\varepsilon_v$. Other plots in the figure indicate that the net volume decrease of both specimens is made up of a reduction in the volume of the air voids (increasing cumulative $d\varepsilon_a$) and an increase in the water content (decrease in cumulative $d\varepsilon_w$). For specimen A, in accordance with condition 2, an increase in stress within the aggregates is indicated by the plot of increasing $p' = (p - u_w)$ against cumulative $d\varepsilon_w$, and an increase in the stress between the aggregates is given by the plot of increasing $\bar{p}$ against cumulative $d\varepsilon_a$. For specimen B, in accordance with the test procedure, $\bar{p}$ and p' remain constant.

Though the specimens were deemed normally consolidated, in that they had not previously experienced externally imposed stresses greater than those at the commencement of shearing, the dual stress regime entailed stress conditions within the aggregates greater than those between the aggregates. The plot of p' against cumulative $d\varepsilon_w$ for specimen A in Figure 9.4 indicates expansive intra-aggregate dilation on uptake of water by the

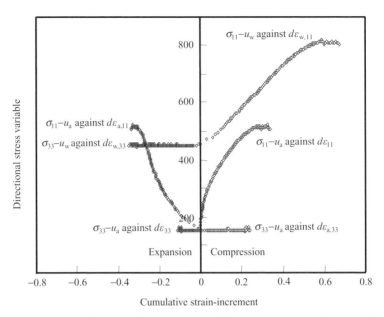

Figure 9.5   Specimen A – directional stress state variables against cumulative strain-increments.

aggregates during shearing. The plot is markedly curved, consistent with expansive plastic softening of the aggregates, characteristic of 'overconsolidated' behaviour. As specimen A experienced overall volume reduction, the aggregates must have expanded into the surrounding relatively large inter-aggregate air void spaces. Sivakumar *et al.* (2006b, 2010b) deduced similar behaviour from measurements on wetting of kaolin specimens.

Figure 9.5 presents the directional stress state variables against directional cumulative strain-increments for specimen A based on conditions 1 and 2 of Table 9.1. Following the discussion of Section 9.8, the directional strain-increments for the aggregates and the air voids cannot be directly related to linear straining of the aggregates and air voids as they account for the energy dispersion as a result of the various strain components identifiable on a micro-mechanical level. They include the elastic stored energy and plastic dissipated energy components resulting from the work input. Thus, while $d\varepsilon_{11} = d\varepsilon_{w,11} + d\varepsilon_{a,11}$ and $d\varepsilon_{33} = d\varepsilon_{w,33} + d\varepsilon_{a,33}$ and, accordingly, the overall continuity conditions for $d\varepsilon_v$ and $d\varepsilon_q$ are satisfied, the strains $d\varepsilon_{w,11}, d\varepsilon_{w,33}, d\varepsilon_{a,11}$ and $d\varepsilon_{a,33}$ have magnitudes in excess of those that might be expected from external measurements, and the strain-increments $d\varepsilon_{w,11}$ and $d\varepsilon_{w,33}$ are of opposite sign, as are the strain-increments $d\varepsilon_{a,11}$ and $d\varepsilon_{a,33}$. The strain-increments $d\varepsilon_{w,11}$ and $d\varepsilon_{a,11}$ are also of opposite sign as are the strain-increments $d\varepsilon_{w,33}$ and $d\varepsilon_{a,33}$.

While the strain-increments include the numerous energy dispersion components, positive increments indicate compression and the associated dispersion of energy, while negative increments indicate expansion and the associated dispersion of energy. The figures included in Section 9.9 identify compressive and expansive behaviour as an aid to clarity.

The areas below the plots give a measure of the summed mechanical stored and dissipated energy components, which for convenience is defined as the dispersed energy. It is apparent from the plot of σ'_{11} against $d\varepsilon_{w,11}$ in Figure 9.5 that most of the dispersed energy, as a result of the application of the external deviator stress, is positive compressive energy dispersion in an axial direction within the aggregates. While care must be exercised in interpreting the plots because of the uncertainty of how the strain-increments relate to the components of energy dispersion, the plot of σ'_{11} against $d\varepsilon_{w,11}$ in Figure 9.5 is consistent with compression of the aggregates axially, compliant with net axial compression of the specimen. The plot of σ'_{33} against $d\varepsilon_{w,33}$ indicates negative dispersed energy, suggesting expansive straining of the aggregates in a radial direction, again compatible with radial expansion of the specimen and the uptake of water by the specimen during shearing.

The plots of $\overline{\sigma}_{11}$ against $d\varepsilon_{a,11}$ and $\overline{\sigma}_{33}$ against $d\varepsilon_{a,33}$ in Figure 9.5 are indicative of energy dispersion between the aggregates. The behaviour is opposite to that of the aggregates, i.e. between the aggregates there is negative dispersive energy in a vertical direction and positive dispersive energy in a radial direction. The magnitudes of the energy dispersion components for the air voids are less than those of the water voids. The reverse effect is a necessary consequence of the need to balance the work input requirements on a macro-mechanical level. The energy dispersion between the aggregates includes the influence on the inter-aggregate void spaces of the movement and compression of air as a result of displacement and deformation of the aggregates into the air spaces. The individual energy dispersion effects are notional rather than quantifiable.

As both specimens A and B were normally consolidated and sheared to the critical state, they can be expected to have experienced plastic dissipation deformations under the increasing triaxial loading. For soils, yielding and subsequent plastic deformation are generally appraised from the overall response of a specimen based on plots, often to a log scale, of $\overline{p}$ against cumulative volumetric deformation $d\varepsilon_v$ (e.g. Mitchell, 1970; Tavenas and Leroueil, 1977; Graham *et al.*, 1983). This approach is not always valid as shown for specimen B in Figure 9.4 for which $\overline{p}$ remained constant during the test and the plot is a straight horizontal line. However, it is possible using the conjugate stress and strain-increment variables of Table 9.1 to gain some insight into the influence of the bi-modal structure and anisotropic behaviour of unsaturated soils even for the constant pressure test on specimen B. Figure 9.6, which shows the directional stresses and cumulative directional strain-increments, indicates significantly curved relationships for the dispersion of energy

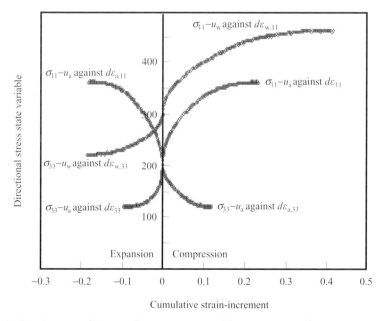

Figure 9.6   Specimen B – directional stress state variables against cumulative strain-increments.

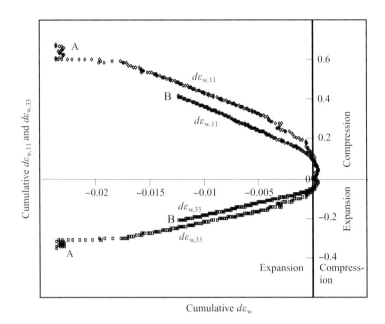

Figure 9.7   Specimens A and B – cumulative $d\varepsilon_{w,11}$ and $d\varepsilon_{w,33}$ against cumulative $d\varepsilon_w$.

within and between the aggregates, suggesting plastic dissipation of energy from the very early stages of the test. As for specimen A, the plots indicate the greater part of the energy dispersion as taken up by the aggregates in an axial direction. The plots suggest flattening of the aggregates as the specimen approaches the critical state where straining occurs under relatively constant loading.

Figure 9.7 compares the relationships between the directional cumulative strain-increments $d\varepsilon_{w,11}$ and $d\varepsilon_{w,33}$ and the cumulative volumetric strain-increment $d\varepsilon_w$ of the aggregates for specimens A and B. The curved relationships illustrate the distortion of the aggregates under increasing volumetric strain. The plots for both specimens A and B exhibit a similar form though the test conditions differ significantly.

Figures 9.5 and 9.6 are interpreted as indicating that the decreases in the air voids for specimens A and B in Figure 9.4 are a result of axial expansion (negative cumulative $d\varepsilon_{a,11}$) and radial compression (positive cumulative $d\varepsilon_{a,33}$) of the air voids, even though the total strains were compressive vertically and expansive radially. This suggests displacement straining of the air voids in a reverse direction to the behaviour of the aggregates. The curved directional relationships for the air voids are consistent with plastic deformations between the aggregates with the increase in axial air voids tentatively suggested as indicating progress towards vertical fissuring. Following testing, dried specimens of kaolin have been observed to exhibit vertical fissures. This suggests the predisposition towards such features during shearing. Fissuring is reminiscent of brittle materials such as rock and concrete tested under triaxial test conditions (Pine *et al.*, 2007). Using X-ray computed tomography, Sun *et al.* (2004) analysed cross sections of silty clay during shearing in a triaxial cell and the images clearly indicated the tendency towards the development of vertical, axially orientated fissures. Halverson *et al.* (2005) used X-ray computed tomography to examine the deformation characteristics of a one-dimensionally compressed silt specimen. The specimen was tested to failure under increasing axial loading under triaxial stress

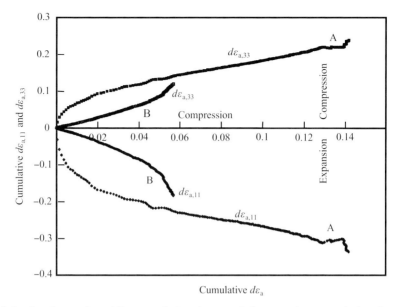

Figure 9.8 Specimens A and B – cumulative $d\varepsilon_{a,11}$ and $d\varepsilon_{a,33}$ against cumulative $d\varepsilon_{a}$.

conditions, and subsequently allowed to dry. The specimen exhibited typical shear banding and radial bulging, but also exhibited vertical cracking in the centre of the specimen.

Figure 9.8 compares the relationships between the cumulative incremental-strains $d\varepsilon_{a,11}$ and $d\varepsilon_{a,33}$ and volumetric cumulative-strain $d\varepsilon_{a}$ for the air voids for specimens A and B. The curves suggest distortion of the air void spaces under increasing strain, but unlike the plots for the aggregates in Figure 9.7, which indicate comparable trends for specimens A and B, the plots for the air void spaces for specimens A and B exhibit distinctly different curved forms.

The overall water volume increase for both specimens A and B is interpreted from Figure 9.7 as the net effect of 'flattening' of the aggregates under increasing axial loading, consistent with axial compression and lateral expansion, the reverse of the air void response to increased axial loading. The aggregates can be viewed as deformable large particles experiencing volume change and yielding with reorientation and redistribution of the soil particles influencing the soil structure. The anisotropic plastic stress–strain behaviour within unsaturated aggregated soils is complex and the analysis sheds some light on the behavioural trends that are masked in assessing behaviour from overall stress–volume change measurements.

The deviator stress q can be viewed in accordance with condition 2 of Table 9.1 as comprising the deviator stress within the aggregates q_d with a conjugate deviator strain-increment $d\varepsilon_{qw}$, and the deviator stress between the aggregates q_c with a conjugate deviator strain-increment $d\varepsilon_{qa}$. Plots of these deviator stresses against the conjugate cumulative strain-increments for specimens A and B are presented in Figure 9.9. For both specimens the greater part of the deviator stress is shown as generated within the aggregates, where the stress level given by p' is greater than between the aggregates, where the stress level is given by $\bar{p}$. The positive values of cumulative $d\varepsilon_{qw}$ are consistent with the interpretation of flattening axially of the aggregates, and the negative values of cumulative $d\varepsilon_{qa}$ of extension

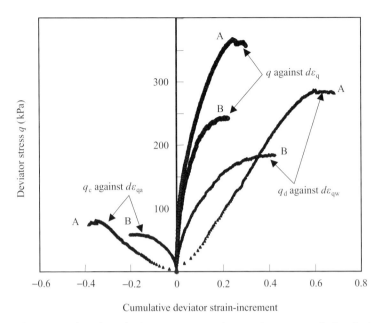

Figure 9.9   Specimens A and B – deviator stresses against conjugate cumulative deviator strain-increments.

axially of the air voids, though it is stressed that the actual values of the strain-increments are a function of dispersion of energy rather than solely a measure of pure distortion.

For positive $d\varepsilon_q$, the cumulative deviator strain of the aggregates $d\varepsilon_{qw}$ and the cumulative deviator strain of the air voids $d\varepsilon_{qa}$ will have opposite signs. In accordance with Equations 9.25 and 9.26, the cumulative strain-increment $d\varepsilon_{qw}$ will be positive and $d\varepsilon_{qa}$ will be negative, as observed in the tests, when $\eta_a > \eta_b > 0.5\eta_a$. Figure 9.10 illustrates that this inequality existed throughout the tests. Equation 9.18 indicates that the inequality necessitates $v_a/v < (\Omega/M_a) - 0.5$. For $\Omega = 0.6$ and $M_a = 0.86$, as observed in the tests undertaken (Figures 8.4 and 9.2), this necessitates $v_a/v < 0.198$ and accordingly $v_w/v > 0.802$. For specimen A the values of v_a/v decreased from 0.192 at the beginning of the shearing stage to 0.057 at the end of the test and the corresponding values for specimen B were 0.076 and 0.021. Thus, at all times the values of v_a/v were less than the critical value for both specimens.

9.9.2 Specimen C (constant suction shearing test on initially one-dimensionally compressed specimen sheared under fully drained conditions)

The investigation into the behaviour of kaolin tested under triaxial stress conditions is continued by examination of the data for a specimen prepared and sheared under the following conditions:

Specimen C was initially prepared by one-dimensional compression in a mould and then consolidated isotropically in a triaxial cell. The consolidation cell pressure was incrementally raised to 450 kPa and held constant during shearing. Prior to shearing,

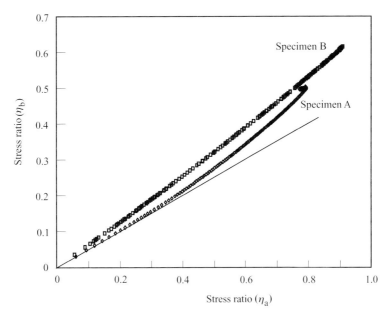

Figure 9.10 Specimens A and B – mobilised stress ratio η_b against η_a.

$v = 2.188$ and $v_w = 1.726$. The suction was held constant at 300 kPa during both consolidation and shearing. The specimen was sheared under fully drained conditions with increasing axial loading.

Comparison of the procedures in the preparation and shearing of specimens A and C indicates little difference other than that the consolidation pressure and cell pressure during shearing for specimen C (i.e. 450 kPa) was slightly less than that used for specimen A (500 kPa). Specimen C had a larger specific volume than specimen A at the commencement of shearing, indicating slightly greater air voids content as the specific water volumes v_w were very similar. Accordingly, specimen C was less compact, with a more open inter-aggregate structure than specimen A, and less confining pressure was applied during shearing. While Figures 9.4 and 9.11 indicate decrease in the volume of air voids and expansion of the water voids during shearing for both specimens, when the data are analysed in greater detail, significant differences in behaviour are uncovered. The differences are a function of the consolidation of the specimens prior to shearing.

Figure 9.5 for specimen A can be compared to Figure 9.12 for specimen C. The net water content increase for specimen A was interpreted as made up of radial expansion and axial compression of the aggregates, while the net air void decrease was interpreted as made up of radial compression and axial expansion. However, Figure 9.12 indicates a singularity in behaviour for specimen C (abrupt changes in energy as discussed in Section 9.6).

Figure 9.12 indicates that while the plot of axial net stress $\overline{\sigma}_{11}$ against cumulative axial strain-increment $d\varepsilon_{11}$ shows nothing untoward, there are dramatic changes in the dispersion of energy within and between the aggregates in an axial direction during shearing. Initially, as the axial stresses increase, the plots show positive energy dispersion between the aggregates indicated by positive, compressive values of the cumulative axial

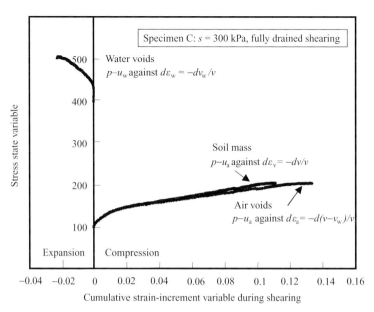

Figure 9.11   Specimen C – stress state variables against cumulative strain-increment variables.

strain-increment $d\varepsilon_{a,11}$, but as straining continues this changes dramatically to negative energy dispersion. The reverse is true of the aggregates which change from negative, expansive energy dispersion to positive, compressive energy dispersion as the axial loading increases as indicated by the change from negative to positive values of the axial strain-increment $d\varepsilon_{w,11}$. This is interpreted as axial compression of the air voids initially

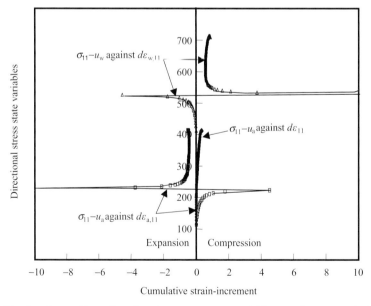

Figure 9.12   Specimen C – directional stress state variables against cumulative strain-increments.

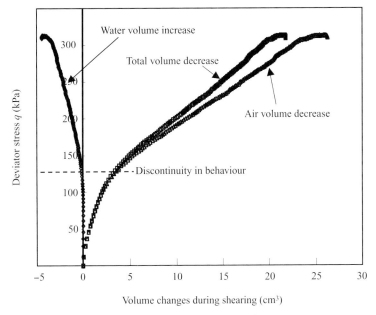

Figure 9.13   Specimen C – deviator stress against volume changes.

as a consequence of the relatively large inter-aggregate air voids within the loose soil structure, followed by compression of the aggregates in an axial direction as the air voids close and the aggregates begin to interact to a greater degree. After the perturbation in behaviour, specimen C behaves in a manner similar to specimens A and B with the net increase in the air voids axially again suggestive of the predilection to axial fissuring during shearing in the triaxial cell, and is consistent with the barrelling effect observed as a result of lateral expansion under axial compression.

Figure 9.13 sheds further insight into the behaviour of specimen C which was subject to increasing deviator stress with u_a constant at 350 kPa and u_w constant at 50 kPa (i.e. $s = 300$ kPa). Plots of deviator stress against the total voids, air voids and water voids are shown. Before the discontinuity in behaviour, the distortion of the aggregates was at constant volume as indicated by the plot for water volume. Up to this point, all volume reduction was due to reduction in the inter-aggregate void spaces consistent with the foregoing interpretation of Figure 9.12. There is thus a marked difference in behaviour before and after the discontinuity in behaviour. After the discontinuity, besides the obvious increase in the volume of the water and thus increase in volume of the aggregates, the rate of reduction in the total volume and air voids increased with increasing deviator stress. As for specimens A and B, increase in the volume of water after the discontinuity, along with reduction in total specimen volume, indicates expansion of the aggregates into the larger inter-aggregate air void spaces. The singularity in behaviour exhibited in Figure 9.12 is not just a mathematical phenomenon but a reflection of what the specimen has experienced.

Figure 9.14 presents the results for the behaviour radially of the water and air voids for specimen C. Again the singularity in behaviour is indicated. As discussed in Section 9.6, this occurs when $2\eta_b = \eta_a$ and in accordance with Equation 9.18 when $v_a/v = (\Omega/M_a) - 0.5$. For $\Omega = 0.6$ and $M_a = 0.86$, this gives $v_a/v = 0.198$. For specimen A, v_a/v was always less

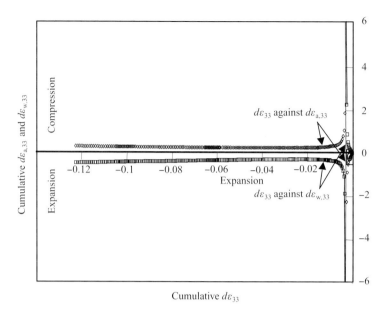

Figure 9.14 Specimen C – cumulative $d\varepsilon_{a,33}$ and $d\varepsilon_{w,33}$ against cumulative $d\varepsilon_{33}$.

than this critical value, being more compact than that for specimen C. However, specimen C had a value of $v_a/v = 0.211$ at the commencement of the test, which decreased to 0.187 by the end, thus passing through the critical value.

Figures 9.15 and 9.16 illustrate the energy dispersion changes of the water and air voids during the drained shearing for specimen C. The behaviour is distinctly different to that of specimen A and indicates a dramatic change in energy dispersion and, consequently,

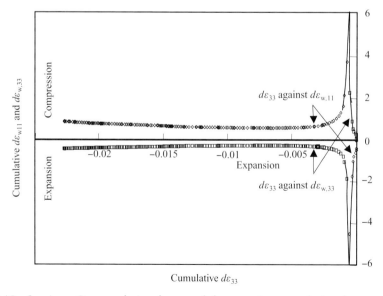

Figure 9.15 Specimen C – cumulative $d\varepsilon_{w,11}$ and $d\varepsilon_{w,33}$ against cumulative $d\varepsilon_w$.

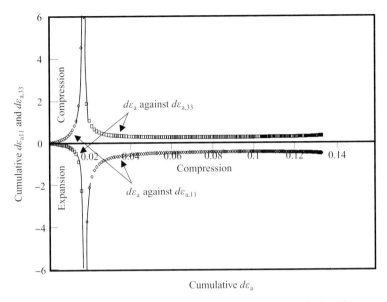

Figure 9.16   Specimen C – cumulative $d\varepsilon_{a,11}$ and $d\varepsilon_{a,33}$ against cumulative $d\varepsilon_a$.

straining during shearing. The analysis demonstrates that small differences in preparation and testing procedure can result in significant differences in behaviour characteristics depending on a critical inter-aggregate void spacing and thus on the degree of compaction.

As for specimens A and B, the deviator stresses can be viewed in accordance with condition 2 of Table 9.1 as comprising the deviator stress within the aggregates q_d with a conjugate deviator strain-increment $d\varepsilon_{qw}$, and the deviator stress between the aggregates q_c with a conjugate deviator strain-increment $d\varepsilon_{qa}$. Plots of the deviator stresses against the conjugate cumulative strain-increments for specimen C are presented in Figure 9.17. As the deformation of the specimen increases, the greater part of the deviator stress is generated within the aggregates, where stress levels given by p' are greater than between the saturated regions, where stress levels are given by $\bar{p}$. However, the condition $q = q_c + q_d$ given by Equation 9.12b is always satisfied.

The discontinuity in cumulative strain measurements is reflected in the deviator strain measurements presented in Figure 9.17. In accordance with Equations 9.25 and 9.26, the cumulative deviator strain of the aggregates $d\varepsilon_{qw}$ and the cumulative deviator strain between the aggregates $d\varepsilon_{qa}$ have opposite signs though the condition $d\varepsilon_q = d\varepsilon_{qw} + d\varepsilon_{qa}$ given by Equation 9.24 is always satisfied. However, unlike for specimens A and B, $d\varepsilon_{qw}$ is negative initially and $d\varepsilon_{qa}$ positive, but there is a change of signs as the deviator stress increases. The change of sign occurs either side of the condition $2\eta_b = \eta_a$. Figure 9.18 illustrates that $\eta_b < 0.5\eta_a$ exists in the early part of the test but $\eta_b > 0.5\eta_a$ exists after the discontinuity in behaviour. As previously, the change occurs at $v_a/v = 0.198$.

Under increasing axial stress it is only necessary that axial compression of the specimen is balanced by the net changes of the aggregates and air voids axially. This is also true of lateral expansion of the specimen. Indeed the data suggest that for specimens A, B and C there is a net increase in the air voids axially during shearing, and flattening of the aggregates, thus aggregate compression in an axial direction. Specimen C is shown to exhibit dramatic changes internally during the test with the initial relatively loose

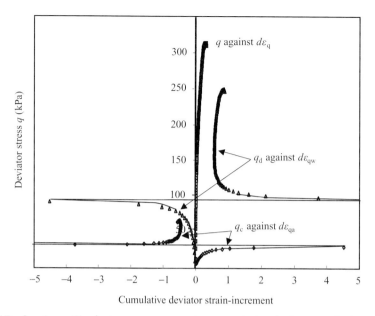

Figure 9.17   Specimen C – deviator stresses against cumulative deviator strain-increments.

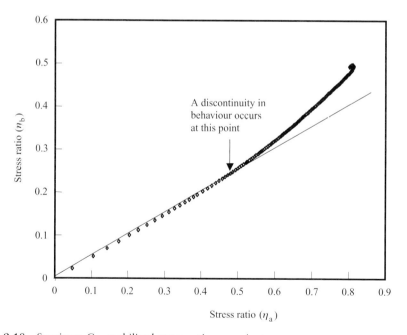

Figure 9.18   Specimen C – mobilised stress ratio η_b against η_a.

compaction of the specimen important and suggestive of an initial unstable fabric which adjusted to a more stable state following a discontinuity in energy dispersion.

9.9.3 Specimens D and E (constant water mass shearing tests. Specimen D prepared under truly isotropic conditions and specimen E by initial one-dimensional compression)

Sivakumar *et al.* (2010a) reported the results of tests on isotropically and one-dimensionally compressed specimens of kaolin. The research was aimed primarily at assessing the influence of stress-induced anisotropy on the performance of unsaturated soils. Figure 9.19 presents five paired plots of deviator stress q against suction s during the shearing stages of the isotropically (IS) and one-dimensionally (1D) prepared specimens sheared under constant water mass conditions. The paired specimens were prepared to similar initial volumetric conditions. It is interesting to note that though the tests were at constant water mass, the suction in the IS tests remained essentially constant while those in the 1D tests tended to decrease from the condition at the start of the shearing process. The results from two of the tests reported by Sivakumar *et al.* (2010a), and identified as specimens D and E, are discussed in detail.

A thermocouple psychrometer measured changes in the soil suction. The psychrometer, attached to the top-loading cap in the triaxial cell, was inserted in a small hole drilled in the specimen. When in the cell, the IS- and 1D-prepared specimens were isotropically compressed under pre-selected values of $\bar{p}$ before shearing. The results at the critical state in the constant water mass tests are included in Figures 8.7, 8.8 and 8.21, but are not the tests where p'_c/s was close to 1.0 for which the volumetric terms are considered influenced by fissuring.

The differences in specimen preparation procedure between specimens D and E are important. The methodology used to prepare the truly IS-prepared specimen D is described in detail in Chapter 3. Briefly, wet kaolin at a targeted water content of 25% was passed through a sieve with an aperture size of 1.18 mm and sealed in a plastic bag for 48 hours. The material was then poured into a 100-mm-diameter rubber membrane in a triaxial cell and pressurised isotropically. The sample was left in the compression system for 3 days though consolidation took place rapidly. At the end of this time the confining pressure was reduced to zero and a thin wall sample tube was used to extract a sub-sample of 50 mm in diameter and 100 mm in height. Specimen D was initially prepared by isotropic compression to a target specific volume of 2.19 and a water content of 25%. The specimen was then consolidated isotropically under a mean net stress of 300 kPa and sheared under constant cell pressure and water mass.

In contrast to the preparation of specimen D, the material of specimen E was initially conditioned to a water content of 25% and sieved as for specimen D, but was compressed one-dimensionally in layers to a target specific volume of 2.19. Specimen E was then consolidated isotropically in a triaxial cell under a mean net stress of 300 kPa and sheared under constant cell pressure and water mass. Specimens D and E were thus prepared under similar conditions other than that specimen D was isotropically compressed initially and specimen E one-dimensionally compressed initially.

The fourth plot of q against s in Figure 9.19 for $\bar{p} = 300$ kPa is for specimens D and E. The IS-prepared specimen D had an initial suction of around 820 kPa, which remained essentially constant during shearing, and the 1D-prepared specimen E had an initial suction of 560 kPa, which decreased to a value of around 430 kPa at the critical state.

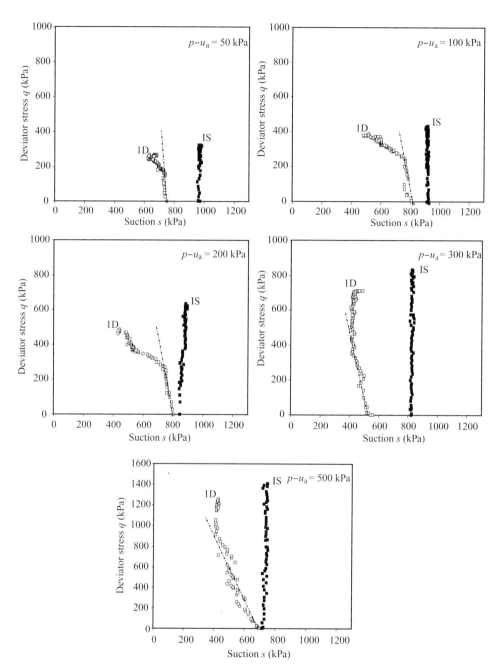

Figure 9.19 Effect of stress-induced anisotropy on the stress path under constant water mass conditions (Sivakumar *et al.*, 2010a).

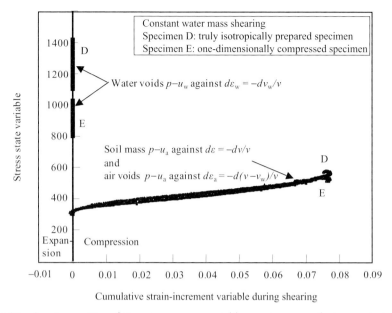

Figure 9.20   Specimens D and E – stress state variables against cumulative strain-increment variables, constant water mass tests.

Figure 9.20 confirms that there was no change in the specific water volume for either specimen during shearing (v_w remained constant and cumulative $d\varepsilon_w = 0$), but for the IS-prepared specimen D the value of p' was greater than for the 1D-prepared specimen E, suggesting greater stress within the aggregates as a result of the greater suction (lower values of u_w). The coincidence of the plots of $\bar{p}$ against both $d\varepsilon_v$ and $d\varepsilon_a$ for both specimens indicates similar average stress conditions between the aggregates $\bar{p}$, and comparable volumetric compression of the air voids. The coincidence of these plots conforms with the principle that the stress state variable controlling the volumetric behaviour between the aggregates is $\bar{p}$.

Figures 9.21 and 9.22 present the directional stress state variables and directional strain-increment variables for specimens D and E respectively. The plots indicate overall compression axially (cumulative $d\varepsilon_{11}$ increasing) and expansion radially (cumulative $d\varepsilon_{33}$ decreasing), as for the earlier reported tests on specimens A, B and C. However, the plots for the directional stresses σ'_{11}, σ'_{33}, $\bar{\sigma}_{11}$ and $\bar{\sigma}_{33}$ against the conjugate cumulative strain-increments indicate behaviour somewhat different to that for specimens A, B and C. The magnitudes of the cumulative strain-increments associated with the air voids $d\varepsilon_{a,11}$ and $d\varepsilon_{a,33}$ are greater both axially and radially than those of the aggregates $d\varepsilon_{w,11}$ and $d\varepsilon_{w,33}$, which is the reverse of the behaviour exhibited by specimens A, B and C. The cumulative strain-increments are also of opposite sign to those of specimens A, B and C other than for the early stages of the test on specimen C before the singularity in behaviour.

Unsaturated fine-grained soils develop aggregations of particles that behave in many respects as a granular material, and the greater the suction the stiffer the aggregates and the less the influence of inter-aggregate bonding (Wheeler *et al.*, 2003). The suctions in specimens D and E of 430–820 kPa were greater than those of specimens A, B and C of 100–300 kPa and the aggregates can be expected to present a greater resistance to deformation.

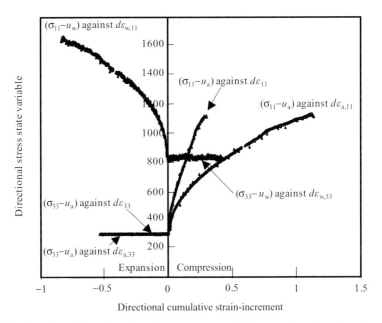

Figure 9.21 Specimen D – directional stress state variables against directional cumulative strain-increments, constant water mass test.

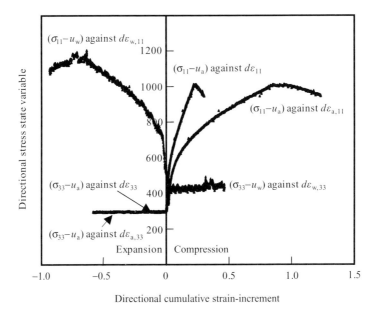

Figure 9.22 Specimen E – directional stress state variables against directional strain-increments, constant water mass tests.

Specimens A, B and C were tested under constant, relatively low suctions where inter-aggregate suction can be expected to have maintained a degree of continuity between the aggregates. These tests are interpreted as indicating flattening of the aggregates under axially increased loading. Following a similar line of argument for specimens D and E, the elevated suctions will have allowed more freedom of movement or displacement straining of the relatively rigid aggregates because of a reduced influence from inter-aggregate suction. This is thought to go a long way to explaining the test results for specimens D and E.

In Figures 9.21 and 9.22, the axial compressive (positive) strain-increments $d\varepsilon_{11}$ are the difference between the positive strain-increments between the aggregates $d\varepsilon_{a,11}$ and the negative strain-increments of the aggregates $d\varepsilon_{w,11}$. The radial expansive (negative) strain-increments $d\varepsilon_{33}$ are the difference between the negative strain-increments between the aggregates $d\varepsilon_{a,33}$ and the positive strain-increments of the aggregates $d\varepsilon_{w,33}$. The plots of Figures 9.21 and 9.22 exhibit behaviour similar to the early stages of the plots for specimen C, when, before the discontinuity in behaviour, the straining was also at constant water mass. Consistent with the conclusions for specimen C, the form of the plots in Figures 9.21 and 9.22 for specimens D and E can best be explained as net closure of the air voids in an axial direction as the deviator stress increases, along with opening up of the air voids radially. The cumulative strain-increments $d\varepsilon_{a,11}$ and $d\varepsilon_{a,33}$ portray the energy dispersion in this process. The opposing straining of the aggregates, $d\varepsilon_{w,11}$ and $d\varepsilon_{w,33}$, suggests energy dispersion due to movement of the aggregates to displace the air voids in an axial direction, with the opposite effect in a radial direction. This suggests inter-aggregate dilation in sensitive regions of the soil specimens giving rise to apparent negative energy dispersion. As noted in Section 9.8.5, dilation is a phenomenon observed in testing of granular soils in the triaxial cell (Alramahi and Alshibli, 2006) and gives rise to negative energy components as the soil expands against the imposed stress. It is important to realise, however, that such behaviour is likely to be restricted to localised regions of a specimen and that overall specimen volume change may be compressive. In places the aggregates will move closer together while in others there will be inter-aggregate dilation. It is the net overall effect of energy dispersion that is indicated by the negative values of the cumulative strain-increments $d\varepsilon_{w,11}$. It is argued that there is a tendency for localised regions to experience dilation, which is likely to give rise to significant energy dispersion, weighting the tests data. Some justification for such an interpretation is provided by Caner and Bažant (2002). They carried out axial loading tests and finite-element analysis of the lateral confinement needed to suppress 'softening' of concrete in compression. They presented evidence of softening, which would be consistent with dilation in soils, notably towards the outer edges of specimens.

While the plots indicate the directions of the cumulative strain-increments are similar for specimens D and E, there are obvious differences as the specimen structure altered under the increasing axial loading and the critical state was approached. In particular, the greater suction throughout the IS test on specimen D has led to greater directional stresses within the aggregates given by greater σ'_{11} and σ'_{33} values.

Figure 9.23 indicates that in addition to the overall volumetric changes given by the change in the air voids (i.e. $d\varepsilon_v = d\varepsilon_a$) for both specimens D and E in Figure 9.20, the directional cumulative strain-increments of the air voids $d\varepsilon_{a,11}$ and $d\varepsilon_{a,33}$ in the constant water mass tests are also comparable for both specimens.

Figure 9.24 presents the deviator stresses for specimens D and E in accordance with condition 2 of Table 9.1. In accordance with this condition, the total deviator stress is made up of the deviator stress within the aggregates q_d with a conjugate deviator strain-increment $d\varepsilon_{qw}$, and the deviator stress between the aggregates q_c with a conjugate

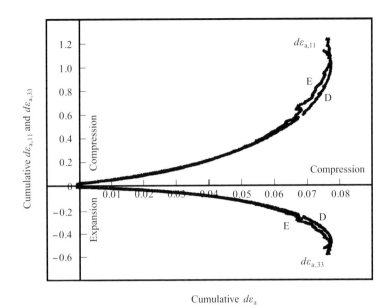

Figure 9.23   Specimens D and E – cumulative $d\varepsilon_{a,11}$ and $d\varepsilon_{a,33}$ against cumulative $d\varepsilon_a$.

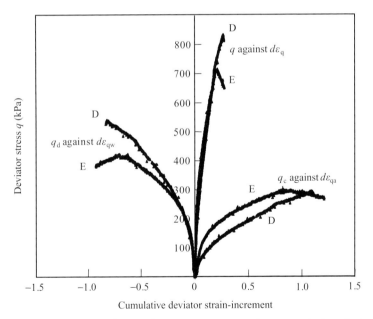

Figure 9.24   Specimens D and E – deviator stresses against cumulative deviator strain-increments.

deviator strain-increment $d\varepsilon_{qa}$. The total deviator stress q for the IS-prepared specimen D is greater than that for the 1D-prepared specimen E.

As for specimens A, B and C tested under different test conditions, the greater part of the deviator stress for specimens D and E is generated within the aggregates, where stress levels given by p' are greater than between the aggregates, where stress levels are given by $\bar{p}$. However, it is important to note that the relative proportion of the total deviator stress contributed by the deviator stress within the aggregates is less for specimens D and E than for specimens A, B and C. Conversely, the proportion of the total deviator stress contributed by the deviator stress between the aggregates is greater. This suggests greater shearing between the aggregates in the tests on specimens D and E than in the tests on specimens A, B and C. This is again consistent with the aggregates being more rigid and displacing to allow shearing between the aggregates. Specimens A, B and C exhibited a plastic failure mechanism while specimens D and E exhibited a more brittle failure mechanism though further testing would be warranted before firm conclusions can be drawn on the significance of the strain-increments and the type of failure. However, it is not surprising, based on the foregoing arguments, that as discussed in Section 8.5 and illustrated in Figure 8.32, differences in specimen preparation and test procedure lead to different critical state relationships.

As previously, in accordance with Equations 9.23 and 9.24, for positive $d\varepsilon_q$, the shear strain of the aggregates $d\varepsilon_{qw}$ and the shear strain between the aggregates $d\varepsilon_{qa}$ will have opposite signs. However, while for specimens A and B tested under constant suction, $d\varepsilon_{qw}$ is always positive and $d\varepsilon_{qa}$ always negative, for specimens D and E, $d\varepsilon_{qw}$ is always negative and $d\varepsilon_{qa}$ always positive, as shown in Figure 9.24. Specimen E exhibits a change between the two conditions of shearing.

For the test on specimens D and E, Figure 9.25 indicates $\eta_b < 0.5\eta_a$ at all stages of the tests. This is consistent with $d\varepsilon_{qw}$ being negative and $d\varepsilon_{qa}$ being positive. Equation 9.18 indicates that the inequality necessitates $v_a/v > (\Omega/M_a) - 0.5$. For $\Omega = 0.6$ and $M_a = 1.00$ as observed in the tests undertaken (Figure 8.7), $v_a/v > 0.100$. For specimen D the values of v_a/v decreased from 0.171 at the beginning of the shearing stage to 0.165 at the end of the test and the corresponding values for specimen B were 0.229 and 0.165; thus at all times they were greater than the critical value of 0.100.

9.10 Conclusions

In unsaturated soils, a bi-modal structure was shown in Chapter 7 to lead to three alternative equations describing the stress regime. The equations highlighted the thermodynamic importance of assigning volumetric variables to conjugate stress state variables. The current chapter has extended the analysis to develop three alternative equations for the work input into an unsaturated soil under triaxial test conditions. The analysis has led to the following major conclusions:

- From the work input equations, work conjugate stress and strain-increment variables have been determined and are presented in terms of two mobilised stress ratios.
- Equations have been developed for the axial strain-increments, the radial strain-increments and the deviator strain-increments that can be used to appraise the anisotropic behaviour both of the aggregates and between the aggregates in unsaturated fine-grained soils.

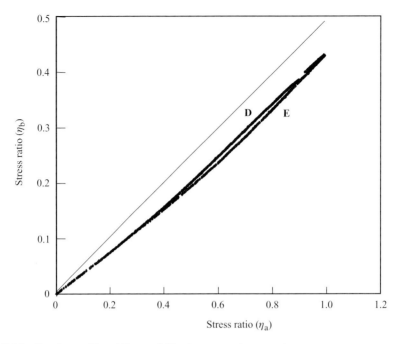

Figure 9.25   Specimens D and E – mobilised stress ratio η_b against η_a.

- The externally measured strains whether directional, volumetric or deviatoric must be the net effects of the displacement, rotation and distortional straining, plus phase mass changes and other energy dispersion effects, internal to the specimen. The directional strain-increments for the aggregates and between the aggregates are shown to reflect the energy dispersion due to the complex internal phase straining. On the micro-mechanical level, simple linear changes in dimensions do not satisfy the work input as there are identifiable mechanical changes internal to a soil specimen that give rise to energy dispersion but not to linear strain-increments. Nevertheless, the overall dimensional changes of a specimen must equate to the net internal changes as a result of energy dispersion. The analysis complies with these compatibility and work analysis requirements.
- The necessity of carefully assigning the stress state variables and deviator stresses with the conjugate volumetric and strain-increment variables in constitutive modelling, in order to obtain meaningful relationships for unsaturated soil, has been demonstrated.
- The shear behaviour of specimens of unsaturated kaolin in the triaxial cell has been examined using the stress state variables and deviator stresses along with their cumulative conjugate strain-increment variables. The analysis is shown to provide a logical interpretation of experimental data. In particular, the anisotropic behaviour of the aggregates and between the aggregates of kaolin is illustrated and takes account of the bi-modal structure of unsaturated soils.
- In the shearing tests on those specimens of kaolin where suction was held constant by the axis translation technique, the increase in air voids in an axial direction was consistent with the predisposition towards the development of axial fissures as the

triaxial specimen expanded laterally. This is reminiscent of the behaviour of brittle materials.

■ A specimen of kaolin with initially relatively large inter-aggregate voids and tested under triaxial compression with suction maintained constant using the axis translation technique experienced a discontinuity in behaviour prior to exhibiting behavioural characteristics similar to those described above. The specimen had a v_a/v ratio less than a critical value of 0.198 determined from the mobilised stress ratios taking $\Omega = 0.6$ and $M_a = 0.86$. The initial relatively loose compaction of the specimen suggested a fabric that was unstable during shearing and adjusted to a more stable state.

■ Tests on a truly isotropically prepared specimen and a one-dimensionally compressed specimen of kaolin sheared under constant water mass conditions in the triaxial cell have been analysed and compared. The overall volume reduction of the specimens comprised compression of the air voids. The conjectured behaviour based on considerations of the energy dispersion was one of localised inter-aggregate dilation and compression of the air voids as shearing proceeded.

■ Tests on soils are generally directed towards evaluating material properties that can be used in the design of engineering structures. This requires the development and validation of stress-deformation constitutive models and their implementation in numerical predictive analysis. The stress-deformation constitutive model approach described lends itself to further insightful research into the stress conditions and mechanisms governing the behaviour of unsaturated soils and to the determination of fundamental material parameters that may be incorporated in predictive numerical analysis.

■ While discussion has been presented on the significance and observations of the elasto-plastic behaviour of unsaturated soils, the formulation of a framework for yielding, the identification of plastic potential, a flow rule and hardening (or softening) is beyond the scope of this book. However, the anisotropic plastic stress–strain behaviour within unsaturated soils is shown to be complex and it is essential to consider yielding at a micro-mechanical level for a soil with a bi-modal structure. The analysis sheds light on the behavioural trends at a micro-mechanical level. These trends are masked in assessing behaviour from overall stress–volume change measurements. The evidence supports the assertion that the dispersion of energy at a micro-mechanical level, which includes the various frictional dissipative strains and free energy recoverable strains, must be considered in the development of a comprehensive elasto-plastic model.

Notes

1. This is consistent with Chapters 4 and 5. In Section 5.6.2 the work done to a specimen is given by $\delta W = -p\,dV$, where p is pressure and V is volume. Accordingly, $dV < 0$ for compression of a soil specimen gives $\delta W > 0$.
2. Free energy describes the elastic energy that is recoverable and can be converted to do work on removal of applied loading.

References

Abu-Hassanein, Z.S., Benson, C.H. and Blatz, L.R. (1996). Electrical conductivity of compacted clays. *Journal of Geotechnical Engineering, ASCE*, **122**(5), 397–406.

Ahmed, S., Lovell, C.W. and Diamond, S. (1974). Pore sizes and strength of compacted clay. *Journal of Geotechnical Engineering, ASCE*, **100**(4), 407–425.

Aitchison, G.D. (1961). Relationship of moisture and effective stress functions in unsaturated soils. In: *Pore Pressure and Suction in Soils*. Butterworths, London, pp. 47–52.

Aitchison, G.D. (1965). *Moisture Equilibria and Moisture Changes in Soils Beneath Covered Areas, a Symposium in Print*. Butterworths, Sydney, Australia.

Aitchison, G.D. (1967). Separate roles of site investigation, quantification of soil properties, and selection of operational environment in the determination of foundation design on expansive clays. In: *Proceedings of the Third Asian Regional Conference on Soil Mechanics and Foundation Engineering*, Haifa, Israel, Vol. 3, pp. 72–77.

Aitchison, G.D. (1973). The quantitative description of the stress-deformation behaviour of expansive soils – preface to set of papers. In: *Proceedings of the Third International Conference on Expansive Soils*, Haifa, Israel, Vol. 2, pp. 79–82.

Aitchison, G.D. and Martin, R. (1973). A membrane oedometer for complex stress-path studies in expansive clays. In: *Proceedings of the Third International Conference on Expansive Soils*, Haifa, Israel, Vol. 2, pp. 83–88.

Aitchison, G. D. and Woodburn, J. A. (1969). Soil suction in foundation design. In: *Proceedings of the Seventh International Conference on Soil Mechanics and Foundation Engineering*, Mexico, Vol. 2, pp. 1–8.

Al-Homoud, A.S. Basma, A.A. Hussein Malkawi, A.I. and Al-Bashabshah, M.A. (1995). Cyclic swelling behaviour of clays. *Journal of Geotechnical Engineering, ASCE*, **121**(7), 562–565.

Al-Mukhtar, M. (1995). Macroscopic behaviour and microstructural properties of a kaolinite clay under controlled mechanical and hydraulic state. In: *Proceedings of the First International Conference on Unsaturated Soils, UNSAT'95*, Paris, France, pp. 3–9.

Al-Mukhtar, M., Belanteur, N., Tessier, D. and Vanapalli, S.K. (1996). The fabric of a clay soil under controlled mechanical and hydraulic stress states. *Applied Clay Science*, **11**, 99–115.

Al-Mukhtar, M., Qi, Y., Alcover, J.F. and Bergaya, F. (1999). Oedometric and water-retention behaviour of highly compacted unsaturated smectites. *Canadian Geotechnical Journal*, **36**, 675–684.

Alonso, E.E., Gens, A. and Hight, D.W. (1987). Special problem soils. General report. In: *Proceedings of the Ninth European Conference on Soil Mechanics and Foundation Engineering*, Dublin, Ireland, Vol. 3, pp. 1087–1146.

Alonso, E.E. (1988). Keynote lecture: Modelling expansive soil behaviour. In: *Proceedings of the Second International Conference on Unsaturated Soils*, Beijing, China, Vol. 1, pp. 37–70.

Alonso, E.E., Gens, A. and Josa, A. (1990). A constitutive model for partly saturated soils. *Geotechnique*, **40**(3), 405–430.

Alonso, E.E., Josa, A. and Gens, A. (1992). *Modelling the Behaviour of Compacted Soils Upon Wetting*, Raul Marsal (ed.). Soc. Mexicana de Mecánica de Suelos, A.C. Mexico, pp. 207–223.

Alonso, E.E., Lloret, A., Gens, A. and Yang, D.Q. (1995). Experimental behaviour of highly expansive double-structure clay. In: *Proceedings of the First International Conference on Unsaturated Soil, UNSAT'95*, Paris, France, Vol. 1, pp. 11–16.

Alonso, E.E., Vaunat, J. and Gens, A. (1999). Modelling the mechanical behaviour of expansive clays. *Engineering Geology*, **54**, 173–183.

Alonso, E.E. and Olivella, S. (2006). Unsaturated soil mechanics applied to geotechnical problems. In: *Unsaturated Soils*, Vol. **1**. Geotechnical Special Publication No. 147. ASCE, Reston, VA, pp. 1–35. Proceedings of the Fourth International Conference on Unsaturated Soils, UNSAT'06, Carefree, AZ.

Alonso, E.E., Romero, E., Hoffmann, C. and Garcia-Escudero, E. (2005). Expansive bentonite-sand mixtures in cyclic controlled-suction drying and wetting. *Engineering Geology*, **81**(3), 213–226.

Alramahi, B.A. and Alshibli, K.A. (2006). Application of computer tomography (CT) to characterize the internal structure of geomaterials. *Structure of Geomaterials: Limitations and Challenges, Site and Geomaterial Characterisation*. ASCE, pp. 88–95.

Amenu, G.G., Kumar, P. and Liang, X-Z. (2005). Interannual variability of deep-layer hydrologic memory and mechanisms of its influence on surface energy fluxes. *Journal of Climate, American Meteorological Society*, **18**, 5024–5045.

Arch, J., Stehenson, E. and Maltman, A.J. (1993). Factors affecting the containment properties of natural clays. In: *The Engineering Geology of Waste Disposal and Storage, 29th Annual Conference of the Engineering Group of the Geological Society*, Cardiff, UK, pp. 263–272.

ASTM (1997). *D5298–94: Standard Test Method for the Measurement of Soil Potential (Suction) Using Filter Paper*, Vol. 04.09.

Atkinson, J. (1993). *An Introduction to the Mechanics of Soils and Foundations*. McGraw-Hill, London.

Atkinson, J. and Bransby, P.L. (1978). *The Mechanics of Soils*. McGraw-Hill, London.

Aversa, S. and Nicotera, M.V. (2002). A triaxial and oedometer apparatus for testing unsaturated soils. *Geotechnical Testing Journal, ASTM*, **25**(1), 3–15.

Badonis, M.E. and Kavvadas, M.J. (2008). Modifying the Barcelona Basic Model to account for residual void ratio and subsequent decrease of shear strength relative to suction. In: *Unsaturated Soils: Advances in Geo-Engineering*, Toll, D.G. *et al.* (eds). Routledge, London, pp. 589–595. Proceedings of the First European Conference on Unsaturated Soils, E-UNSAT, Durham, UK.

Barbour, S.L. (1999). The soil water characteristic curve: a historical perspective. *Canadian Geotechnical Journal*, **35**, 873–894.

Barden L., Madedor, A.O. and Sides, G.R. (1969). Volume change characteristics of unsaturated clay. *Journal of the Soil Mechanics and Foundations Division, ASCE*, **95**(1), 33–51.

Barden, L., and Sides, G.R. (1967). The diffusion of air through the pore water of soils. In: *Proceedings of the Third Asian Regional Conference on Soil Mechanics and Foundation Engineering*, Haifa, Israel, Vol. 1, pp. 135–138.

Barden, L. and Sides, G.R. (1970). Engineering behaviour and structure of compacted clay. *Journal of the Soil Mechanics and Foundations Division, ASCE*, **96**(SM4), 1171–1197.

Bell, F.G. and Culshaw, M.G. (2001). Problem soils: a review from a British perspective. In: *Problematic Soils: Proceedings of the Symposium Held at the Nottingham Trent University on 8 November 2001*, Jefferson, I., Murray, E.J., Faragher, E. and Fleming, P.R. (eds). Thomas Telford, London, pp. 1–35.

Bishop, A.W. (1959). The principle of effective stress. *Teknisk Ukeblad*, **106**(39), 859–863.

Bishop, A.W., Alpan, I., Blight, G.E. and Donald, I.B. (1960). Factors controlling the strength of partly saturated soils. In: *Proceedings of the ASCE Research Conference on Shear Strength of Cohesive Soils*, Boulder, CO, pp. 503–532.

Bishop, A.W. and Blight, G.E. (1963). Some aspects of effective stress in saturated and unsaturated soils. *Geotechnique*, **13**(3), 177–197.

Bishop, A.W. and Donald, I.B. (1961). The experimental study of partly saturated soil in the triaxial apparatus. In: *Proceedings of the Fifth International Conference on Soil Mechanics and Foundation Engineering*, Paris, Vol. 1, pp. 13–21.

Black, J. (1775). The supposed effect of boiling upon water, in disposing it to freeze more readily, ascertained by experiments. Letter to Sir John Pringle, Bart. P. R. S. *Philosophical Transactions (1683–1775) of the Royal Society*, **65**(1775), 124–128.

Blatz, J.A., Cui, Y.-J., Oldecop, L. (2008). Vapour equilibrium and osmotic suction technique for suction control. *Journal of Geotechnical and Geological Engineering*, **26**(6), 661–673.

Blatz, J.A. and Graham, J. (2000). A system for controlled suction in triaxial tests. *Geotechnique*, **50**(4), 465–469.

Blatz, J.A. and Graham, J. (2003). Elastic-plastic modelling of unsaturated soil using results from a new triaxial tests with controlled suction. *Geotechnique*, **53**(1), 113–122.

Blight, G.E. (1965). A study of effective stress for volume change. In: *Moisture Equilibria and Moisture Changes in Soils Beneath Covered Areas, a Symposium in Print*, Aitchison, G.D. (ed.). Butterworths, Sydney, Australia, pp. 259–269.

Bloodworth, M.E. and Page, J.B. (1957). Use of thermistor for the measurement of soil moisture and temperature. *Soil Science Society of America, Proceedings*, **21**, 11–15.

Bolton, M.D., Nakata, Y. and Cheng, Y.P. (2008). Micro- and macro-mechanical behaviour of DEM crushable materials. *Geotechnique*, **58**(6), 471–480.

Borsic, A., Comina, C., Foti, S. Lancellotta, R. and Musso, G. (2005). Imaging heterogeneities with electrical impedance tomography: laboratory results. *Geotechnique*, **55**(7), 539–547.

Boso, M., Tarantino, A. and Mongiovi, L. (2005). A direct shear box improved with the osmotic technique. In: *International Symposium on Advanced Experimental Unsaturated Soil Mechanics*, Trento, Italy, Tarantino, A, Romero, E. and Cui, Y. (eds), pp. 85–91.

Brackley, I.J.A. (1971). Partial collapse in unsaturated expansive clays. In: *Proceedings of the Fifth Regional Conference for Africa on Soil Mechanics and Foundation Engineering*, South Africa, pp. 23–30.

Brackley, I.J.A. (1973). Swell pressure and free swell in a compacted clay. In: *Proceedings of the Third International Conference on Expansive Soils*, Haifa, Israel, Vol. 1. Academic Press, Jerusalem, pp. 169–176.

Brackley, I.J.A. (1975). A model of unsaturated clay structure and its application to swell behaviour. In: *Proceedings of the Sixth Regional Conference for Africa on Soil Mechanics and Foundation Engineering*, Durban, Vol. 1, Balkema, Rotterdam, the Netherlands, pp. 65–70.

Brady, K.C. (1988). Soil suction and the critical state. *Geotechnique*, **38**(1), 117–120.

Brennen, C.E. (1995). *Cavitation and Bubble Dynamics*. Oxford Engineering Science Series 44. Oxford University Press, New York.

Brown, J.L. (2009). *The Stress Regime in Unsaturated Compacted Clays When Laterally Confined*. PhD thesis, Queen's University of Belfast, Belfast, Northern Ireland.

Brown, R.W. (1970). *Measurement of Water Potential with Thermocouple Psychrometers: Construction and Applications*. U.S.D.A. Forest Service Research Paper INT-80.

B.S.1377 (1990). *Methods of Test for Soils for Civil Engineering Purposes*. British Standards Institutions.

Bulut, R. and Leong, E.C. (2008). Indirect measurement of suction. *Journal of Geotechnical and Geological Engineering*, **26**(6), 633–644.

Bulut, R., Lytton, R.L. and Wray, W.K. (2001). Soil suction measurements by filter paper. In: *Expansive Clays Soils and Vegetative Influence on Shallow Foundations*, Vipulanandan, C. Addison, M.B. and Hasen, M. (eds). Geotechnical Special Publication, No. 115. ASCE, Houston, TX, pp. 243–261,

Bulut, R. and Wray, W.K. (2005). Free energy of water suction in filter papers. *Geotechnical Testing Journal*, **28**(4), 355–364.

Burland, J.B. (1964). Effective stresses in partly saturated soils. Discussion on 'some aspects of effective stress in saturated and partly saturated soils', by G.E. Blight and A.W. Bishop. *Geotechnique*, **14**, 65–68.

Burland, J.B. (1965). Some aspects of the mechanical behaviour of partly saturated soils. In: *Moisture Equilibria and Moisture Changes in Soils Beneath Covered Areas, a Symposium in Print*, Aitchison, G.D. (ed.). Butterworths, Sydney, Australia, pp. 270–278.

Callen, H.B. (1965). *Thermodynamics and an Introduction to Thermostatistics*, John Wiley & Sons, New York.

Campbell, G.S. (1979). Improved thermocouple psychrometers for measurement of soil water potential in a temperature gradient. *Journal of Physics E: Scientific Instruments*, **12**, 739–743.

Campbell, G.S. and Gardner, W.H. (1971). Psychometric measurement of soil potential: temperature and bulk density effects. *Soil Science Society of America, Proceedings*, **35**, 8–12.

Caner, F.C. and Bažant, Z. (2002). Lateral confinement needed to suppress softening of concrete in compression. *Journal of Engineering Mechanics, ASCE*, **128**(12), 1304–1313.

Caruso, M. and Tarantino, A. (2004). A shear box for testing unsaturated soils at medium to high degrees of saturation. *Geotechnique*, **54**(4), 281–284.

Casagrande, A. (1932). The structure of clay and its importance in foundation engineering. In: *Contributions to Soil Mechanics, 1925–1940*. Boston Society of Civil Engineers, pp. 72–113.

Chandler, R.J. (1974). Lias clay: the long-term stability of cutting slopes. *Geotechnique*, **24**(1), 21–38.

Chandler, R.J. and Davis, A.G. (1973). Further work on the engineering properties of Keuper Marl. CIRIA Report 47.

Chandler, R.J. and Gutierrez, C.I. (1986). The filter-paper method of suction measurement. *Geotechnique*, **36**(2), 265–268.

Chen, J. and Kumar, P. (2004). A modeling study of the ENSO influence on the terrestrial energy profile in North America. *Journal of Climate, American Meteorological Society*, **17**, 1657–1670.

Childs, E.C. (1969). *An Introduction to the Physical Basis of Soil Water Phenomena*. Wiley Intersciences, London.

Chiu, C.F. and Ng, C.W.W. (2003). A state-dependent elasto-plastic model for saturated and unsaturated soils. *Geotechnique*, **53**(9), 809–829.

Clifton, A.W., Barbour, S.L. and Wilson, G.W. (1999). *The Emergence of Unsaturated Soil Mechanics – Fredlund Volume*. NRC Research Press, Ottawa, Canada.

Collins, I.F. (2005). The concept of stored plastic work or frozen elastic energy in soil mechanics. *Geotechnique*, 55(5), 373–382.

Collins, I.F. and Muhunthan, B. (2003). On the relationship between stress-dilatancy, anisotropy, and plastic dissipation for granular materials. *Geotechnique*, 53(7), 611–618.

Collins, I.F, Muhunthan, B, Tai, A.T. and Pender, M. (2007). The concept of a 'Reynolds-Taylor' state and the mechanics of sand. *Geotechnique*, 57(5), 437–447.

Comstock, J.P. (2000). Correction of thermocouple psychrometer readings for the interaction of temperature and actual water potential. *Crop Science*, 40, 709–712.

Craig, R.F. (1997). *Soil Mechanics*, 6th edn. E & FN Spon, London.

Crilley, M.S., Schreiner, H.D. and Gourley, C.S. (1991). A simple field suction probe. In: *Proceedings of the Tenth Regional Conference for Africa on Soil Mechanics and Foundation Engineering*, Maseru, Lesotho, pp. 291–298.

Croney, D. and Coleman, J.D. (1948). Soil thermodynamics applied to the movement of moisture in road foundations. In: *Proceedings of the Seventh International Congress of Applied Mechanics*, London, Vol. 3, pp. 163–177.

Croney, D. and Coleman, J.D. (1954). Soil structure in relation to soil suction (pF). *Journal of Soil Science*, 5(1), 75–84.

Croney, D. and Coleman, J.D. (1960). *Proceedings of Conference on Pore Pressure and Suction in Soils*. Institution of Civil Engineers. Butterworths, London, pp. 31–37.

Croney, D., Coleman, J.D. and Black, W.P.M. (1958). *The Movement and Distribution of Water in Soil in Relation to Highway Design and Performance*. Highway Research Board Special Report No. 40, Washington, DC.

Crooks, J.H.A. and Graham, J. (1976). Geotechnical properties of the Belfast estuarine deposits. *Geotechnique*, 26(2), 293–315.

Cui, L. and O'Sullivan, C. (2006). Exploring the micro- and macro-scale response of an idealised granular material in the direct shear box. *Geotechnique*, 56(7), 455–468.

Cui, Y.J. and Delage, P. (1996). Yielding and plastic behaviour of an unsaturated compacted silt. *Geotechnique*, 46(2), 291–311.

Cuisinier, O. and Deneele, D. (2008). Long-term behaviour of lime-treated expansive soil submitted to cyclic wetting and drying. In: *Unsaturated Soils: Advances in Geo-Engineering*, Toll, D.G. *et al.* (eds). Routledge, London, pp. 327–333. Proceedings of the First European Conference on Unsaturated Soils, E-UNSAT, Durham, UK.

Cuisinier, O. and Laloui, L. (2004). Fabric evolution during hydromechanical loading of a compacted silt. *International Journal for Numerical and Analytical Methods in Geomechanics*, 28(6), 483–499.

Cuisinier, O. and Masrouri, F. (2004). Testing the hydromechanical behaviour of a compacted swelling soil. *Geotechnical Testing Journal*, 72(6), 598–606.

Cunningham, M.R., Ridley, A.M., Dineen, K. and Burland, J.B. (2003). The mechanical behaviour of a reconstituted unsaturated silty clay. *Geotechnique*, 53(2), 183–194.

De Campos, T.M.P. and Carrillo, C.W. (1995). Direct shear test on an unsaturated soil from Rio de Janeiro. In: *Unsaturated Soils*, Vol. 1, Alonso, E.E. and Delage, P. (eds). A.A. Balkema/Presses des Ponts et Chaussées, Paris, France, pp. 31–38. Proceedings of the First International Conference on Unsaturated Soils, Paris.

Degueldre, C., Pleinert, H., Lehmann, P., Missimer, J. Hammer, J., Leenders, K., Böck, H. and Townsend, D. (1996). Porosity and pathway determination in crystalline rock by positron emission tomography and neutron radiography. *Earth and Planetary Science Letters*, 140, 213–225.

Dela, B.F. (2001). *Measurement of Soil Moisture Using Gypsum Blocks*. Danish Building and Urban Research, By og Byg Document 004.

Delage, P. (2002). Experimental unsaturated soil mechanics: state-of-the-art report. In: *Proceedings of the Third International Conference on Unsaturated Soils, UNSAT'02*, Recife, Brazil, Vol. 3, Juca, J.F.T., deCampos, T.M.P. and Marinho, F.A.M. (eds), pp. 973–996.

Delage, P., Audiguier, M. Cui, Y-J. and Howat, M.D. (1996). Microstructure of compacted silt. *Canadian Geotechnical Journal*, **33**, 150–158.

Delage, P. and Graham, J. (1995). Mechanical behaviour of unsaturated soils: understanding the behaviour of unsaturated soils requires reliable conceptual models. In: *Proceedings of the First International Conference on Unsaturated Soils*, Paris, France, pp. 1223–1256.

Delage, P, Howat, M.D. and Cui, Y.J. (1998). The relationship between suction and swelling properties in a heavily compacted unsaturated clay. *Engineering Geology*, **50**, 31–48.

Delage, P., Romero, E. and Tarantino, A. (2008). Recent developments in the techniques of controlling and measuring suction in unsaturated soils. In: *Unsaturated Soils: Advances in Geo-Engineering*, Toll, D.G. *et al.* (eds). Routledge, London, pp. 33–52. Proceedings of the First European Conference on Unsaturated Soils, E-UNSAT, Durham, UK.

Delage, P., Suraj, De Silva, G.P.R. and De Laure, E. (1987). Un nouvel appareil triaxial pour les soils non saturés. In: *Proceedings of the Ninth European Conference on Soil Mechanics and Foundation Engineering*, Dublin, Ireland, Vol. 1, pp. 26–28.

Delage, P. Suraj De Silva, G.P.R. and Vicol, T (1992). Suction controlled testing on non saturated soils with an osmotic consolidometer. In: *Proceedings of the Seventh International Conference on Expansive Soils*, Dallas, TX, pp. 206–211.

Diamond, S. (1970). Pore size distribution in clays. *Clays and Clay Minerals*, **18**, 7–23.

Dimos, A. (1991). Measurement of soil suction using transistor psychrometer. Internal Report (IR/3−19), Special Research Section, Materials Tech. Dept., Vic Roads.

Dineen, K and Burland, J.B. (1995). A new approach to osmotically controlled oedometer testing. In: *Unsaturated Soils*, Vol. **2**, Alonso, E.E. and Delage, P. (eds). A.A. Balkema/Presses des Ponts et Chaussées, Paris, France, pp. 459–465. Proceedings of the First International Conference on Unsaturated Soils, Paris.

Donald, I.B. (1956). Shear strength measurements in unsaturated non-cohesive soils with negative pore pressures. In: *Proceedings of the Second Australian–New Zealand Conference on Soil Mechanics and Foundation Engineering*, Christchurch, New Zealand, pp. 200–205.

Dorsey, N.E. (1940). *Properties of Ordinary Water-Substances*. American Chemical Society, Monograph Series. Reinhold, New York.

Dueck, A. (2004). *Hydro-Mechanical Properties of a Water Unsaturated Sodium Bentonite – Laboratory Study and Theoretical Interpretation*. PhD. thesis, Lund University, Sweden.

Edlefsen, N.E. and Anderson, A.B.C. (1943). Thermodynamics of soil moisture. *Hilgardia*, **15**(2), 31–298.

Erné, B.H. (2000). Thermodynamics of water superheated in the microwave oven. *Journal of Chemical Education*, 77(10), 1309–1310.

Escario, V. (1980). Suction controlled penetration and shear tests. In: *Proceedings of the Fourth International Conference on Expansive Soils*, Denver, CO, Vol. 2, pp. 781–797.

Escario, V. and Juca, J. (1989). Strength and deformation of partly saturated soils. In: *Proceedings of the 12th International Conference Soil Mechanics and Foundation Engineering*, Rio de Janeiro, Brazil, Vol. 2, pp. 46–63.

Escario, V. and Saez, J. (1973). Measurement of the properties of swelling and collapsing soil under controlled suction. In: *Proceedings of the Third International Conference on Expansive Soils*, Haifa, Israel, Vol. 2, pp. 195–200.

Escario, V. and Saez, J. (1986). The shear strength of partly saturated soils. *Geotechnique*, 36(3), 453–456.

Esterban, V and Saez, J. (1988). A device to measure the swelling characteristics of rock samples with control of the suction up to very high values. In: *ISRM Symposium on Rock Mechanics and Power Plants*, Madrid, Spain, pp. 2.

Fawcett, R.G. and Collis-George, N. (1967). A filter-paper method for determining the moisture characteristics of soil. *Australian Journal of Experimental Agriculture and Animal Husbandry*, 7, 162–167.

Ferreira, P.M.V. and Bica, A.V.D. (2006). Problems in identifying the effects of structure and critical state in a soil with transitional behaviour. *Geotechnique*, 56(7), 445–454.

Fityus, S. and Buzzi, O. (2009). The place of expansive clays in the framework of unsaturated soil mechanics. *Applied Clay Science*, 43, 150–155.

Fredlund, D.G. (1975). A diffused air volume indicator for unsaturated soils. *Canadian Geotechnical Journal*, 12, 533–539.

Fredlund, D.G. (1979). Second Canadian Geotechnical Colloquium: appropriate concepts and technology for unsaturated soils. *Canadian Geotechnical Journal*, 16, 121–139.

Fredlund, D.G. and Morgenstern, N.R. (1976). Constitutive relations for volume change in unsaturated soils. *Canadian Geotechnical Journal*, 13(3), 261–276.

Fredlund, D.G. and Morgenstern, N.R. (1977). Stress state variables for unsaturated soils. *Journal of the Geotechnical Engineering Division*, ASCE, 103, 447–446.

Fredlund, D.G., Morgenstern, N.R. and Widger, R.A. (1978). Shear strength of unsaturated soils. *Canadian Geotechnical Journal*, 15, 313–321.

Fredlund, D.G. and Rahardjo, H. (1993). *Soil Mechanics for Unsaturated Soils*. John Wiley, New York.

Fredlund, D.G., Rahardjo, H. and Gan, J. (1987). Non-linearity of strength envelope for unsaturated soils. In: *Proceedings of the Sixth International Conference on Expansive Soils*, New Delhi, India, Vol. 1, pp. 49–54.

Fredlund, D.G., Shuai, F. and Feng, M. (2000). Increased accuracy in suction measurements using an improved thermal conductivity sensor. In: *Proceedings of the Seventh International Conference on Tailing and Mine Waste*, Fort Collins, CO, pp. 443–450.

Fredlund, D.G. Vanapalli, S.K., Xing, A. and Pufahl, D.E. (1995). Predicting the shear strength function for unsaturated soils using the soil-water characteristic curve. In: *Unsaturated Soils*, Vol. 1, Alonso, E.E. and Delage, P. (eds). A.A. Balkema/Presses des Ponts et Chaussées, Paris, France, pp. 63–69. Proceedings of the First International Conference on Unsaturated Soils, Paris.

Fredlund, D.G., Xing, A., Fredlund, M.D. and Barbour, S.L. (1995). The relationship of the unsaturated soil shear strength to the soil-water characteristic curve. *Canadian Geotechnical Journal*, 32(3), 440–448.

Futai, M.M. and Almeida, M.S.S. (2005). An experimental investigation of the mechanical behaviour of an unsaturated gneiss residual soil. *Geotechnique*, 55(3), 201–213.

Gallipoli, D. Gens, A. Sharma, R. and Vaunat, J. (2003). An elasto-plastic model for unsaturated soil incorporating the effects of suction and degree of saturation on mechanical behaviour. *Geotechnique*, 53(1), 123–135.

Gan, J.K.-M. and Fredlund, D.G. (1996). Shear strength characteristics of two saprolitic soils. *Canadian Geotechnical Journal*, 33, 595–609.

Gan, J.K.-M., Fredlund, D.G. and Rahardjo, H. (1988). Determination of the shear strength parameters of an unsaturated soil using the direct shear tests. *Canadian Geotechnical Journal*, 25(3), 500–510.

Gardner, W.R. (1956). Calculation of capillary conductivity from pressure plate outflow data. *Soil Science Society of America, Proceedings*, **20**, 317–320.

Gasmo, J.M., Hritzuk, K.J., Rahardjo, H. and Leong, E.C. (2000). Instrumentation of an unsaturated residual soil slope. *Geotechnical Testing Journal, ASTM*, **23**(2), 128–137.

Gasparre, A., Nishimura, S., Coop, M.R. and Jardine, R.J. (2007). The influence of structure on the behaviour of London clay. Symposium in print, Stiff Sedimentary Clays – Genesis and Engineering Behaviour: Part 1. *Geotechnique*, **57**(1), 49–62.

Gee, G., Campell, M., Campbell, G. and Campbell, J. (1992). Rapid measurement of low soil potentials using a water activity meter. *Soil Science Society of America Journal*, **56**, 1068–1070.

Gens, A. (2010). Soil-environment interactions in geotechnical engineering. *Geotechnique*, **60**(1), 3–74.

Gens, A. and Alonso, E.E. (1992). A framework for the behaviour of unsaturated expansive clays. *Canadian Geotechnical Journal*, **29**, 1013–1032.

Gens, A., Alonso, E.E., Suriol, J. and Lloret, A. (1995). Effect of structure on the volumetric behaviour of a compacted soil. In: *Proceedings of the First International Conference on Unsaturated Soils, UNSAT'95*, Paris, France, pp. 83–88.

Georgiadis, K., Potts, D.M. and Zdravkovic, L. (2008). An improved constitutive model for unsaturated and saturated soils. In: *Unsaturated Soils: Advances in Geo-Engineering*, Toll, D.G. *et al.* (eds). Routledge, London, pp. 581–588. Proceedings of the First European Conference on Unsaturated Soils, E-UNSAT, Durham, UK.

Graham, J., Crooks, J.H.A. and Lau, S.L.K. (1988). Yield envelope: identification and geometric properties. *Geotechnique*, **38**(1), 125–134.

Graham J. and Houlsby, G.T. (1983). Anisotropic elasticity in a natural plastic clay. *Geotechnique*, **33**, 165–180.

Graham, J., Noonan, M.L. and Lew, K.V. (1983). Yield states and stress-strain relationships in a natural plastic clay. *Canadian Geotechnical Journal*, **20**, 502–516.

Graham, J., Wood, D.M., Yin, J.-H., Azizi, F. (1989). Prediction of triaxial stress-strain behaviour of Winnipeg clay using an anisotropic elastic-plastic model. In: *Proceedings of 42nd Canadian Geotechnical Conference*, Winnipeg, MB, Canada, pp. 280–288.

Griffiths, F.J. and Joshi, R.C. (1989). Change in pore size distribution due to consolidation of clays. *Geotechnique*, **39**(1), 159–167.

Grim, R.E. (1962). *Applied Clay Mineralogy*. McGraw-Hill, New York.

Guan, Y. (1996). *The Measurement of Soil Suction*. PhD thesis, University of Saskatchewan, Canada.

Habib, S.A. (1995). Lateral pressure of unsaturated expansive clay in looped stress path. In: *Unsaturated Soils*, Vol. **1**, Alonso, E.E. and Delage, P. (eds). A.A. Balkema/Presses des Ponts et Chaussées, Paris, France, pp. 95–100. Proceedings of the First International Conference on Unsaturated Soils, Paris, France.

Habib, S.A., Kato, T. and Karube, D. (1992). One dimensional swell behaviour of unsaturated soil. In: *Proceedings of the Seventh International Conference on Expansive Soils*, Dallas, TX, Vol. 2, pp. 222–226.

Habib, S.A., Kato, T. and Karube, D. (1995). Suction controlled one dimensional swelling and consolidation behaviour of expansive soil. In: *Unsaturated Soils*, Vol. 1, Alonso, E.E. and Delage, P. (eds). A.A. Balkema/Presses des Ponts et Chaussées, Paris, France, pp. 101–107. Proceedings of the First International Conference on Unsaturated Soils, Paris, France.

Halverson, C., White, D.J. and Gray, J. (2005). Application of X-ray CT scanning to characterize geomaterials used in transportation construction. In: *Proceedings of 2005 Mid-Continent Transportation Research Symposium*, Ames, IA.

Hamblin, A.P. (1981). Filter paper method for routine measurement of field water potential. *Journal of Hydrology*, **53**, 355–360.

Han, K.K., Rhardjo, H. and Broms, B.B. (1995). Effects of hysteresis on a residual soil. In: *Unsaturated Soils*, Vol. **2**, Alonso, E.E. and Delage, P. (eds). A.A. Balkema/Presses des Ponts et Chaussées, Paris, France, pp. 499–504. Proceedings of the First International Conference on Unsaturated Soils, Paris, France.

He, L-C. (1999). *Evaluation of Instruments for Measurement of Suction in Unsaturated Soils*. MSc thesis, School of Civil and Structural Engineering, Nanyang Technological University, Singapore.

Hilf, J.W. (1956). *An Investigation of Pore-Water Pressure in Compacted Cohesive Soils*. PhD thesis, Technical Memo No. 654, United States Bureau of Reclamation, Denver, CO.

Ho, D.Y.F. and Fredlund, D.G. (1982). The multi-stage triaxial tests for unsaturated soils. *Geotechnical Testing Journal, ASTM*, **5**, 18–25.

Ho, D.Y.F. and Fredlund, D.G. (1995). Determination of the volume change modulii and respective inter-relationships for two unsaturated soils. In: *Unsaturated Soils*, Vol. **1**, Alonso, E.E. and Delage, P. (eds). A.A. Balkema/Presses des Ponts et Chaussées, Paris, France, pp. 117–122. Proceedings of the First International Conference on Unsaturated Soils, Paris, France.

Ho, D.Y.F. and Fredlund, D.G. and Rahardjo, H. (1992). Volume change indices during loading and unloading of an unsaturated soil. *Canadian Geotechnical Journal*, **29**(2), 195–207.

Hossain, D. (1992). Prediction of permeability of fissured tills. *Quarterly Journal of Engineering Geology*, **25**(4), 331–342.

Houlsby, G.T. (1979). The work input to a granular material. *Gotechnique*, **29**(3), 354–358.

Houlsby, G.T. (1997). The work input to an unsaturated granular material. *Gotechnique*, **47**(1), 193–196.

Houlsby, G.T., Amorosi, A. and Rojas, E. (2005). Elastic moduli of soils dependent on pressure: a hyperelastic formulation. *Gotechnique*, **55**(5), 383–392.

Houston, S.L., Houston, W.N. and Wagner, A. (1994). Laboratory filter paper suction measurements. *Geotechnical Testing Journal, ASTM*, **17**(2), 185–194.

Hoyos, L.R. (1998). *Experimental and Computational Modelling of Unsaturated Soil Behaviour Under True Triaxial Stress States*. PhD dissertation, Georgia Institute of Technology, Atlanta, GA.

Hoyos, L.R., Laikram, A. and Puppala, A.J. (2008). A novel suction-controlled true triaxial apparatus for unsaturated soils. In: *Unsaturated Soils: Advances in Geo-Engineering*, Toll, D.G. *et al.* (eds). Routledge, London, pp. 83–95. Proceedings of the First European Conference on Unsaturated Soils, E-UNSAT, Durham, UK.

Hoyos, L.R. and Macari, E.J. (2001). Development of a stress/suction-controlled true triaxial testing device for unsaturated soils. *Geotechnical Testing Journal*, **24**(1), 5–13.

Huang, J.F. and Bartell, L.S. (1995). Kinetics of homogeneous nucleation in the freezing of large water clusters. *Journal of Physical Chemistry*, **99**, 3924–3931.

Hutchison, J.N., Somerville, S.H. and Petley, D.J. (1973). A landslide in periglacial disturbed Etruria Marl at Bury Hill, Staffordshire. *Quarterly Journal of Engineering Geology*, **6**, 377–404.

Jennings, J.E. (1961). A revised effective stress law for use in the prediction of the behaviour of unsaturated soils. In: *Pore Pressure and Suction in Soils*. Butterworths, London, pp. 26–30.

Jennings, J.E. and Burland, J.B. (1962). Limitations of the use of effective stresses in partly saturated soils. *Geotechnique*, **12**(2), 125–144.

Johnston, L.N. (1942). Water permeability jacketed thermal radiators as indicators of field capacity and permanent wilting percentage in soils. *Soil Science*, **54**, 123–126.

Jommi, C. (2000). Remarks on the constitutive modelling of unsaturated soils. In: *Experimental Evidence and Theoretical Approaches in Unsaturated Soils*. Tarantino, A. and Mancuso, C. (eds), Balkema, Rotterdam, the Netherlands, pp. 139–153.

Josa, A., Balmaceda, A, Gens, A. and Alonso, E.E. (1992). An elasto-plastic model for partially saturated soils exhibiting a maximum collapse. In: *Proceedings of the Third International Conference on Computational Plasticity*, Barcelona, Spain, pp. 815–826.

Kassiff, G. and Ben Shalom, A. (1971). Experimental relationship between swell pressure and suction. *Geotechnique*, **21**, 245–255.

Keller, J.B. (1964). Growth and decay of gas bubbles in liquids. In: *Proceedings of the Symposium on Cavitation in Real Liquids*, General Motors Research Laboratories, Warren, Michigan, 1962, Davies R. (ed.). Elsevier, Amsterdam, pp. 19–29.

Kelvin, L. (1852). On a universal tendency in nature to the dissipation of mechanical energy. *Proceedings of the Royal Society of Edinburgh*, **3**, 139.

Koliji, A., Vulliet, L., Carminati, A., Kaestner, A., Flühler, H., Lehmann, P. Hassanien, R., Lehmann, E. and Vontobel, P. (2006). Structure degradation of dry aggregated soils: experimental evidence and model formulation. In: *Unsaturated Soils*, Vol. **2**. Geotechnical Special Publication No. 147. ASCE, Reston, VA, pp. 2174–2185. Proceedings of the Fourth International Conference on Unsaturated Soils, UNSAT'06, Carefree, AZ.

Komine, H. and Ogata, N. (1999). Experimental study on swelling characteristics of bentonites. *Journal of Geotechnical and Geoenvironmental Engineering, ASCE*, **130**(8), 818–829.

Komornik, A., Livneh, M. and Smucha, S. (1980). Shear strength and swelling of clays under suction. In: *Proceedings of the Fourth International Conference on Expansive Soils*, Denver, CO.

Koorevaar, P., Menelik, G. and Dirksen, C. (1983). *Elements of Soil Physics*. Elsevier, Amsterdam, the Netherlands.

Krahn, J. and Fredlund, D.G. (1972). On total matric and osmotic suction. *Journal of Soil Science*, **114**(5), 339–348.

Kunhel, R.A. and Van Der Gaast, S.J. (1993). Humidity-controlled diffractometry and its applications. *Advances in X-Ray Analysis*, **36**, 439–449.

Lambe, T.W. (1951). The structure of compacted clay. *Journal of the Soil Mechanics and Foundations Division, ASCE*, **84**(SM2), 1–34.

Lambe, T.W. (1955). The permeability of fine-grained soils. ASTM Special Technical Publication No. 163, pp. 56–67.

Lambe, T.W. and Whitman, R.V. (1969). *Soil Mechanics*. Wiley, New York.

Landau, L. D. and Lifshitz, E. M. (1986). *Theory of Elasticity (Course of Theoretical Physics*, Vol. 7), 3rd edn. Translated from Russian by J.B. Sykes and W.H. Reid. Butterworths-Heinemann, Boston, MA.

Lang, A.R.G. (1967). Osmotic coefficients and water potentials of sodium chloride solutions from 0–40°C. *Australian Journal of Chemistry*, **20**, 2017–2023.

Lee, H.C. and Wray, W.K. (1992). Evaluation of soil suction instruments. In: *Proceedings of the Seventh International Conference on Expansive Soils*, Dallas, TX, Vol. 1, pp. 307–312.

Lee, I.K. and Coop, M.R. (1995). The intrinsic behaviour of a decomposed granite soil. *Geotechnique*, **45**(1), 117–130.

Leong, E-C. He, L. and Rahardjo, H. (2002). Factors affecting the filter paper method for total and matric suction measurements. *Geotechnical Testing Journal, ASTM*, **25**(3), 322–333.

Leong, E-C., Tripathy, S. and Rahardjo, H. (2003). Total suction measurement of unsaturated soils with a device using the chilled-mirror dew-point technique. *Geotechnique*, **53**(2), 173–182.

Leong, E-C., Widiastuti, S., Lee, C.C. and Rahardjo, H. (2007). Accuracy of suction measurements. *Geotechnique*, **57**(6), 547–556.

Leroueil, S. (1997). Critical state soil mechanics and the behaviour of real soils. In: *Recent Developments in Soil and Pavement Mechanics*, Almeda, R.A. (ed.). Balkema, Rotterdam, the Netherlands, pp. 41–78.

Leroueil, S., Lerout, P., Hight, D.W. and Powell, J.J.M. (1992). Hydraulic conductivity of a recent estuarine silty clay at Bothkennar, *Geotechnique*, **42**(2), 275–288.

Leroueil, S. and Vaughan, P.R. (1990). The general and congruent effects of structure in natural soils and weak rocks. *Geotechnique*, **40**(3), 467–488.

Lewis, G.N. and Randall, M. (1961). *Thermodynamics*, 2nd edn. (Revised by Pitzer, K.S. and Brewer, L.). McGraw-Hill, New York.

Li, X.S. (2003). Effective stress in unsaturated soil: a microstructural analysis. *Gotechnique*, **53**(2), 273–277.

Li, X.S. (2007a). Thermodynamics-based constitutive frameworks for unsaturated soils. 1: theory. *Geotechnique* **57**(5), 411–422.

Li, X.S. (2007b). Thermodynamics-based constitutive frameworks for unsaturated soils. 2: a basic triaxial model. *Geotechnique* **57**(5), 423–435.

Little, J.A., Muir-Wood, D., Paul, M.A. and Bouaz a, A. (1992). Some laboratory measurements of permeability of Bothkennar clay in relation to soil fabric. *Geotechnique*, **42**(2), 355–361.

Lloret, A. and Alonso, E.E. (1985). State surfaces for partially saturated soils. In: *Proceedings of the 11th International Conference on Soil Mechanics and Foundation Engineering*, San Francisco, CA, Vol. 2, pp. 557–562.

Lloret, A. Villar, M.V., Sanchez, M., Gens, A., Pintado, X. and Alonso, E.E. (2003). Mechanical behaviour of heavily compacted bentonite under high suction changes. *Geotechnique*, **53**(1), 27–40.

Lloret, M, Sanchez, M, Karstunen, M and Wheeler, S. (2008). Generalised elasto-plastic stress-strain relations of a fully coupled hydro-mechanical model. In: *Unsaturated Soils: Advances in Geo-Engineering*, Toll, D.G. *et al.* (eds). Routledge, London, pp. 567–573. Proceedings of the First European Conference on Unsaturated Soils, E-UNSAT, Durham, UK.

Lourenço, S.D.N., Gallipoli, D. Toll, D.G. and Evans, F.D. (2006). Development of a commercial tensiometer for triaxial testing of unsaturated soils. In: *Unsaturated Soils*, Vol. **2**. Geotechnical Special Publication No. 147. ASCE, Reston, VA, pp. 1875–1886. Proceedings of the Fourth International Conference on Unsaturated Soils, UNSAT'06, Carefree, AZ.

Lu, N and Likos, W.J. (2004). *Unsaturated Soil Mechanics*. John Wiley, Hoboken, NJ.

Lupini, J.P., Skinner, A.E. and Vaughan, P.R. (1981). The drained residual strength of cohesive soils. *Geotechnique* **31**(2), 181–213.

Maâtouk, A., Leroueil, S. and La Rochelle, P. (1995). Yielding and critical state of a collapsible unsaturated silty soil. *Geotechnique*, **45**(3), 465–477.

Manheim, F.T. (1996). *A Hydraulic Squeezer for Obtaining Interstitial Water from Consolidated and Unconsolidated Sediment*. U.S. Geological Survey Professional Paper 550-C, pp. 256–261.

Masson, S. and Martinez, J. (2001). Micromechanical analysis of the shear behaviour of a granular material. *Journal of Engineering Mechanics, ASCE*, **127**(10), 1007–1016.

Matsuoka, H, Sun, D.A., Kogane, A., Fukuzawa, N. and Ichihara, W. (2002). Stress-strain behaviour of unsaturated soil in true triaxial tests. *Canadian Geotechnical Journal*, **39**, 608–619.

Matyas, E.L. and Radhakrishna, H.S. (1968). Volume change characteristics of partially saturated soils. *Geotechnique*, **18**(4), 432–448.

Maxwell, J. (1872). *A to Z of Thermodynamics*. Oxford University Press, New York.

McLean, A.C. and Gribble, C.D. (1988). *Geology for Civil Engineers*. E & FN Spon, London.

Meeuwig, R.O. (1972). A low-cost thermocouple psychrometer recording system. In: *Psychrometry in Water Relations Research*, Brown, R.W. and Van Haveren, B.P. (eds). Utah Agricultural Experiment Station, Utah State University, Logan.

Meilani, I., Rahardjo, H., Leong, E.C. and Fredlund, D.G. (2002). Mini suction probe for matric suction measurement. *Canadian Geotechnical Journal*, 39(6), 1427–1432.

Merchán, V., Vaunat, J. Romero, E. and Meca, T. (2008). Experimental study of the influence of suction on the residual friction angles of clay. In: *Unsaturated Soils: Advances in Geo-Engineering*, Toll, D.G. *et al.* (eds). Routledge, London, pp. 423–428. Proceedings of the First European Conference on Unsaturated Soils, E-UNSAT, Durham, UK.

Mitchell, J.K. (1993). *Fundamentals of Soil Behaviour*. John Wiley, New York.

Mitchell, R.J. (1970). On the yielding and mechanical strength of Leda Clay. *Canadian Geotechnical Journal*, 7, 297–312.

Modell, M. and Reid, R.C. (1983). *Thermodynamics and Its Applications*, 2nd edn. Prentice-Hall, Englewood Cliffs, NJ.

Mollins, L.H., Stewart, D.I. and Cousens, T.W. (1999). Drained strength of bentonite-enhanced sand. *Geotechnique*, 49(4), 523–528.

Monroy, R., Zdravkovic, L. and Ridley, A. (2010). Volumetric behaviour of compacted London Clay during wetting and loading. In: *Unsaturated Soils: Advances in Geo-Engineering*, Toll, D.G. *et al.* (eds). Routledge, London, pp. 315–320. Proceedings of the First European Conference on Unsaturated Soils, E-UNSAT, Durham, UK.

Monroy, R., Zdravkovic, L. and Ridley, A. (2008b). Evolution of microstructure in compacted London Clay during wetting and loading. *Geotechnique*, 60(2), 105–199.

Morgenstern, N.R. (1979). Properties of compacted clays. In: Contribution to Panel Discussion, Session IV, *Proceedings of the Sixth Panamerican Conference on Soil Mechanics and Foundation Engineering*, Lima, Peru, Vol. 3, pp. 349–354.

Mottes, Y. (1975). Tensiometer, U.S. Patent 3,884,067. See Guan (1996).

Mullen, J.W. (2001). *Crystallization*, 4th edn. Butterworths-Heinemann, Oxford, UK.

Murray, B.J. and Bertram, A.K. (2006). Formation and stability of cubic ice in water droplets. *Physical Chemistry Chemical Physics*, 8, 186–192.

Murray, B.J., Knopf, D.A. and Bertram, A.K. (2005). The formation of cubic ice under conditions relevant to Earth's atmosphere. *Nature*, 434, 202– 205.

Murray, E.J. (2002). An equation of state for unsaturated soils. *Canadian Geotechnical Journal*, 39(1), 125–140.

Murray, E.J. and Brown, J. (2006). Assumptions in equilibrium analysis and experimentation in unsaturated soils. In: *Unsaturated Soils*, Vol. 2. Geotechnical Special Publication No. 147. ASCE, Reston, VA, pp. 2401–2407. Proceedings of the Fourth International Conference on Unsaturated Soils, UNSAT'06, Carefree, AZ.

Murray, E.J. and Geddes, J.D. (1995). A conceptual model for clay soils subjected to negative pore water pressures. *Canadian Geotechnical Journal*, 32(5), 905–912.

Murray, E.J., Murray, B.J. and Sivakumar, V. (2008). Discussion on meta-stable equilibrium in unsaturated soils. In: *Unsaturated Soils: Advances in Geo-Engineering*, Toll, D.G. *et al.* (eds). Routledge, London, pp. 553–558. Proceedings of the First European Conference on Unsaturated Soils, E-UNSAT, Durham, UK.

Murray, E.J. and Sivakumar, V. (2004). Discussion on 'critical state parameters for an unsaturated residual sandy soil', Toll, D.G. and Ong. *Geotechnique*, 54(1), 69–71.

Murray, E.J. and Sivakumar, V. (2005). Stresses and conjugate strain-increments in plotting experimental data for unsaturated soils. In: *International Symposium on Advanced*

Experimental Unsaturated Soil Mechanics, Tarantino, A., Romero, E. and Cui, Y.J. (eds). Trento, Italy.

Murray, E.J. and Sivakumar, V. (2006). Equilibrium stress conditions in unsaturated soils. In: *Unsaturated Soils*, Vol. **2**, Miller, G.A., Zapata, C.E., Houston, S.L. and Fredlund, D.G. (eds). Geotechnical Special Publication No. 147. ASCE, Reston, VA, pp. 2392–2400. Proceedings of the Fourth International Conference on Unsaturated Soils, UNSAT'06, Carefree, AZ.

Murray, E.J., Sivakumar, V. and Tan, W.C. (2002). Use of a coupling stress for unsaturated soils. In: *Proceedings of the Third International Conference on Unsaturated Soils, UN-SAT'02*, Recife, Brazil, Vol. 1, Juca, J.F.T., deCampos, T.M.P. and Marinho, F.A.M. (eds), pp. 121–124.

Navaneethan, T., Sivakumar, V., Wheeler, S.J. and Doran, I.G. (2005). Assessment of suction measurements in saturated clays. In: *Geotechnical Engineering Proceedings*, Institution of Civil Engineers, Vol. 158, pp. 15–24.

Ng, C.W.W. and Chiu, C.F. (2003). Laboratory study of loose saturated and unsaturated decomposed granitic soil. *Journal of Geotechnical and Geoenvironmental Engineering, ASCE*, **129**(6), 550–559.

Ng, C.W.W., Cui, Y., Chen, R. and Delage, P. (2007). The axis-translation and osmotic techniques in shear testing of unsaturated soils: a comparison. *Soils and Foundations*, **47**(4), 678–684.

Ng, C.W.W. and Menzies, B. (2007). *Advanced Unsaturated Soil Mechanics and Engineering*. Taylor & Francis, London.

Ng, C.W.W., Zhan, L.T. and Cui, Y.J. (2002). A new simple system for measuring volume changes in unsaturated soils. *Canadian Geotechnical Journal*, **39**, 757–764.

Nishimura, T. and Fredlund, D.G. (2003). A new triaxial apparatus for high suctions using relative humidity. In: *Proceedings of the 12th Asian Regional Conference on Soil Mechanics and Geotechnical Engineering*, Singapore, Vol. 1, pp. 65–68.

Okochi, Y. and Tatsuoka, F. (1984). Some factors affecting K_o values of sand measured in the triaxial cell. *Soils and Foundations*, **24**(3), 52–68.

Oldecop, L. and Alonso, E.E. (2004). Testing rock under relative humidity control. *Geotechnical Testing Journal*, **27**(3), 1–10.

Oloo, S.Y. and Fredlund, D.G. (1996). A method for determination of ϕ^b for statically compacted soils. *Canadian Geotechnical Journal*, **33**, 272–280.

Ostwald, W. (1897). Studien über die Bildung und Umwandlung fester Körper. *Zeitschrift für Physikalische Chemie*, **22**, 289–330.

Oteo-Mazo, C., Saez-Aunon, J. and Esteban, F. (1995). Laboratory tests and equipment with suction control. In: *Unsaturated Soils*, Vol. **3**, Alonso, E.E. and Delage, P. (eds). A.A. Balkema/Presses des Ponts et Chaussées, Paris, France, pp. 1509–1515. Proceedings of the First International Conference on Unsaturated Soils, Paris.

Padilla, J.M., Perera, Y.Y. and Fredlund, D.G. (2004). Performance of Fredlund thermal conductivity sensor. In: *Tailings and Mine Waste'04*, Fort Collins, CO, pp. 125–133.

Painter, L.I. (1966). Method of subjecting growing plants to a continuous soil moisture stress. *Agronomy Journal*, **58**, 459–460.

Papa, R., Evangelista, A., Nicoreta, N.V. and Urcioli, G. (2008). Mechanical properties of unsaturated pyroclastic soils affected by fast landslide phenomena. In: *Unsaturated Soils: Advances in Geo-Engineering*, Toll, D.G. *et al.* (eds). Routledge, London, pp. 917–923. Proceedings of the First European Conference on Unsaturated Soils, E-UNSAT, Durham, UK.

Peck, A.J. and Rebbidge, R.M. (1966). Direct measurement of moisture potential: a new technique. In: *Proceedings of UNESCO–Netherlands Goverment Symposium on Water in the Unsaturated Zone*, Paris, France, pp. 165–170.

Peck, A.J. and Rebbidge, R.M. (1969). Design and performance of an osmotic tensiometer for measuring capillary potential. *Soil Science Society of America, Proceedings*, **33**(2), 196–202.

Penrose, R. (2005). *The Road to Reality – A Complete Guide to the Laws of the Universe*. Knopf (Random House), New York.

Penumandu, D. and Dean, J. (2000). Compressibility effect in evaluating the pore-size distribution of kaolin clay using mercury intrusion porosimetry. *Canadian Geotechnical Journal*, 37, 393–405.

Phene, C.J., Hoffman, G.J. and Rawlings, S.L. (1971). Measuring soil matric potential in situ by sensing heat dissipation with a porous body: theory and sensor construction. *Soil Science Society of America, Proceedings*, **35**, 27–32.

Pine, R.J., Owen, D.R.J., Coggan, J.S. and Rance, J.M. (2007). A new discrete fracture modelling approach for rock masses. *Geotechnique*, **57**(9), 757–766.

Rahardjo, H. and Fredlund, D.G. (1995). Pore pressure and volume change behaviour during undrained and drained loadings of an unsaturated soil. In: *Unsaturated Soils*, Vol. **1**, Alonso, E.E. and Delage, P. (eds). A.A. Balkema/Presses des Ponts et Chaussées, Paris, France, pp. 177–187. Proceedings of the First International Conference on Unsaturated Soils, Paris.

Rahardjo, H. and Leong, E.C. (2006). Suction measurements. In: *Unsaturated Soils*, Vol. **1**. Geotechnical Special Publication No. 147. ASCE, Reston, VA, pp. 81–104. Proceedings of the Fourth International Conference on Unsaturated Soils, UNSAT 2006, Carefree, AZ.

Rahardjo, H., Lim, T.T., Chang, M.F. and Fredlund, D.G. (1994). Shear strength characteristics of a residual soil. *Canadian Geotechnical Journal*, **32**, 60–77.

Rampino, C. Mancuso, C. and Vinale, F. (1999). Laboratory testing on an unsaturated soil: equipment, procedures, and first experimental results. *Canadian Geotechnical Journal*, **36**, 1–12.

Richards, B.G. (1965). Measurement of the free energy of soil moisture by the psychrometric technique using thermistors. In: *Moisture Equilibria and Moisture Changes in Soils Below Covered Areas, A Symposium in Print*, Aitchison, G.D. (ed.). Butterworths, Sydney, Australia, pp. 39–46.

Richards, B.G. (1966). The significance of moisture flow and equilibria in unsaturated soils in relation to the design of engineering structures built on shallow foundations in Australia. In: *Symposium on Permeability and Capillarity*. ASTM, Atlantic City, NJ.

Richards, L.A. (1941). A pressure membrane extraction apparatus for soil suction. *Soil Science* 51(5), 377–386.

Richards, L.A. and Ogata, G. (1958). Thermocouple for vapour pressure measurements in biological and soil systems at high humidity. *Science*, **128**, 1089–1090.

Richards, L.A. and Ogata, G. (1961). Psychrometric measurements of soil samples equilibrated on pressure membranes. *Soil Science Society of America, Proceedings*, **25**(6), 456–459.

Richards, L.A., Russell, M.B. and Neal, O.R. (1937). Further developments on apparatus for field moisture studies. *Soil Science Society of America, Proceedings*, **2**, 55–64.

Ridley, A.M. (1995). Discussion on 'Laboratory filter paper suction measurements' by Houston, S.L., Houston, W.N. and Wagner A-M. *Geotechnical Testing Journal*, **18**(3), 391–396.

Ridley, A.M. and Burland, J.B. (1993). A new instrument for the measurement of soil moisture suction. *Gotechnique*, **43**(2), 321–324.

Ridley, A.M. and Burland, J.B. (1995). Measurement of suction in materials which swell. *Applied Mechanics Reviews*, **48**(9), 727–732.

Ridley, A.M. and Burland, J.B. (1996). A pore pressure probe for the in situ measurement of a wide range of soil suctions. In: *Advances in Site Investigation Practice*, Craig. C. (ed.). Thomas Telford, London, pp. 510–520.

Ridley, A.M. and Burland, J.B. (1999). Use of the tensile strength of water for the direct measurement of high soil suction. Discussion. *Canadian Geotechnical Journal*, **36**, 178–180.

Ridley, A.M., Dineen, K., Burland, J.B. and Vaughan, P.R. (2003). Soil matric suction: some examples of its measurement and application in geotechnical engineering. *Geotechnique*, **53**(2), 241–253.

Ridley, A.M. and Wray, W.K. (1995). Suction measurement: a review of current theory and practices. In: *Unsaturated Soils*, Vol. **3**, Alonso, E.E. and Delage, P. (eds). A.A. Balkema/Presses des Ponts et Chaussées, Paris, France, pp. 1293–1322. Proceedings of the First International Conference on Unsaturated Soils, Paris.

Rojas, J.C., Mancuso, C. and Vinale, F. (2008). A modified triaxial apparatus to reduce testing time: equipment and preliminary results. In: *Unsaturated Soils: Advances in Geo-Engineering*, Toll, D.G. *et al.* (eds). Routledge, London, pp. 103–109. Proceedings of the First European Conference on Unsaturated Soils, E-UNSAT, Durham, UK.

Romero, E., Gens, A. and Lloret, A. (2003). Suction effects on a compacted clay under non-isothermal conditions. *Geotechnique*, **53**(1), 65–81.

Romero, E., Lloret, A. and Gens, A. (1995). Development of a new suction and temperature controlled oedometer. In: *Unsaturated Soils*, Vol. **2**, Alonso, E.E. and Delage, P. (eds). A.A. Balkema/Presses des Ponts et Chaussées, Paris, France, pp. 553–559. Proceedings of the First International Conference on Unsaturated Soils, Paris, France.

Romero, E. and Simms, P.H. (2008). Microstructure investigation in unsaturated soils: a review with special attention to contribution of mercury intrusion porosimetry and environmental scanning electron microscopy. *Geotechnical and Geological Engineering*, **26**(6), 705–727.

Roscoe, K.H. and Burland, J.B. (1968). On the generalised stress-strain behaviour of 'wet' clay. In: *Engineering Plasticity*, Heyman, J. and Leckie, F.A. (eds). Cambridge University Press, Cambridge, UK, pp. 535–609.

Roscoe, K.H., Schofield, A.N. and Thurairajah, A. (1963). Yielding of clays in states wetter than critical. *Geotechnique*, **13**(3), 211–240.

Saiyouri, N., Hicher, P.T. and Tessier, D. (2000). Microstructural approach and transfer water modelling in highly compacted unsaturated swelling clays. *Mechanics of Cohesive Frictional Materials*, **5**(1), 41–60.

Sánchez, M., Gens, A., Gimaraes, L.N. and Olivella, S. (2001). Generalized plasticity model for THM simulations involving expansive clays. In: *Proceedings of the Sixth International Workshop Key Issues on Waste Isolation Research*, Paris, France, pp. 397–415.

Sharma, R.S. (1998). *Mechanical behaviour of unsaturated highly expansive clays*. PhD thesis, University of Oxford, UK.

Schofield, A.N. (2005). *Disturbed Soil Properties and Geotechnical Design*. Thomas Telford, London.

Schofield, A.N. (2006). Interlocking, and peak and design strengths. *Geotechnique*, **56**(5), 357–358.

Schofield, A.N. and Wroth, C.P. (1968). *Critical State Soil Mechanics*. McGraw-Hill, London.

Schofield, R.K. (1935). The pF of water in soil. *Transactions of the Third International Congress on Soil Science*, Oxford, UK, Vol. **2**, pp. 37–48.

Schuurman, E. (1966). The compressibility of an air–water mixture and a theoretical relation between the air and water pressures. *Geotechnique*, **16**(4), 269–281.

Shaw, B and Baver, L.D. (1939a). Heat conductivity as an index of soil moisture. *Journal of the American Society of Agronomy*, **31**, 886–891.

Shaw, B and Baver, L.D. (1939b). An electrothermal method for following moisture changes of the soil in situ. *Soil Science Society of America, Proceedings*, **4**, 78–83.

Shuai, F and Fredlund, D.G. (2000). Use of a new thermal conductivity sensor to measure soil suction. In: *Proceedings of GeoDenver Conference*, Denver, CO, pp. 1–12.

Sivakumar, R. (2005). *Effects of Anisotropy on the Behaviour of Unsaturated Compacted Clay*. PhD thesis submitted to the Queen's University of Belfast.

Sivakumar, R., Sivakumar, V., Blatz, J. and Vimalan, J. (2006a). Twin-cell stress path apparatus for testing unsaturated soils. *Geotechnical Testing Journal, ASTM*, **29**(2), 1–5.

Sivakumar, V. (1993). *A Critical State Framework for Unsaturated Soils*. PhD thesis submitted to the University of Sheffield, UK.

Sivakumar, V., Doran, I.G. and Graham, J. (2002). Particle orientation and its influence on the mechanical behaviour of isotropically consolidated reconstituted clay. *Engineering Geology*, **66**, 197–209.

Sivakumar, V., Doran, I.G., Graham, J. and Johnson, A.S. (2001). The effect of anisotropic elasticity on the yielding characteristics of overconsolidated natural clay. *Canadian Geotechnical Journal*, **38**, 125–137.

Sivakumar, V. and Ng, P. (1998). Yielding of unsaturated soils. In: *Proceedings of the Second International Conference on Unsaturated Soils*, Beijing, China, pp. 131–136.

Sivakumar, V., Sivakumar, R., Boyd, J.L. and MacKinnon, P. (2010a). Mechanical behaviour of unsaturated kaolin with isotropic and anisotropic stress history. Part 2: performance under shear loading. *Geotechnique* [doi: 10.1680/geot.8.P.008].

Sivakumar, V., Sivakumar, R., Murray, E.J., MacKinnon, P. and Boyd J.L. (2010b). Mechanical behaviour of unsaturated kaolin with isotropic and anisotropic stress history. Part 1: wetting and compression behaviour. *Geotechnique* [doi: 10.1680/geot.8.P.007].

Sivakumar, V., Tan, W.C., Murray, E.J. and McKinley, J.D. (2006b). Wetting, drying and compression characteristics of compacted clay, *Geotechnique*, **56**(1), 57–62.

Sivakumar, V. and Wheeler, S.J (2000). Influence of compaction procedure on the mechanical behaviour of an unsaturated compacted clay. Part 1: wetting and isotropic compression. *Geotechnique*, **50**(4), 359–368.

Skempton, A.W. (1953). The colloidal activity of clay. In: *Proceedings of the Third International Conference on Soil Mechanics and Foundation Engineering*, Zurich, Switzerland, Vol. 1, pp. 57–61.

Smith, P.R., Jardine, R.T. and Hight, D.W (1992). The yielding of Bothkennar clay. *Geotechnique*, **42**(2), 275–274.

Songyu, L., Heyuan, L., Peng, J. and Yanjun, D. (1998). Approach to cyclic swelling behaviour of compacted clays. In: *Proceedings of the Second International Conference on Unsaturated Soils*, Beijing, China, Vol. 2, pp. 219–225.

Spanner, D.C. (1951). The Peltier effect and its use in the measurement of suction pressure. *Journal of Experimental Botany*, **11**, 145–168.

Sposito, G. (1981). *The Thermodynamics of Soil Solutions*. Clarendon Press, Oxford, UK.

Sposito, K. (1989). *The Chemistry of Soils*. Oxford University Press, New York.

Sridharan, A., Altschaeffl, A.G. and Diamond, S. (1971). Pore size distribution studies. *Journal of the Soil Mechanics and Foundation Engineering Division, ASCE*, **97**(5), 771–787.

Stannard, D.I. (1986). Theory, construction and operation of simple tensiometers. *Ground Water Monitoring Review*, **6**(3), 70–78.

Stannard, D.I. (1990). Tensiometers – theory, construction, and use. In: *Groundwater and Vodose Zone Monitoring, ASTM STP*, 1053, Neilson, D.M. and Johnson, A.I. (eds). ASTM, Philadelphia, PA, pp. 34–51.

Stewart, D.I., Tay, Y.Y. and Cousens (2001). The strength of unsaturated bentonite-enriched sand. *Geotechnique*, **51**(9), 767–776.

Subba Rao, K.S. and Satyadas, G.C. (1987). Swelling potentials with cycles of swelling and partial shrinkage. In: *Proceedings of the Sixth International Conference on Expansive Soils*, New Delhi, India, Vol. 1, pp. 137–147.

Sun, H, Chen, J.F. and Ge, X.R. (2004). Deformation characteristics of silty clay subjected to triaxial loading, by computerised tomography. *Geotechnique*, **54**(5), 307–314.

Swarbrick, G.E. (1995). Measurement of soil suction using the filter paper method. In: *Unsaturated Soils*, Vol. 2, Alonso, E.E. and Delage, P. (eds). A.A. Balkema/Presses des Ponts et Chaussées, Paris, France, pp. 653–658. Proceedings of the First International Conference on Unsaturated Soils, Paris.

Tamagnini, R. (2004). An extended Cam-clay model for unsaturated soils with hydraulic hysteresis. *Geotechnique*, **54**(3), 223–228.

Tan, W.C. (2004). *Wetting, drying and shear strength characteristics of compacted clay*. PhD thesis, Queen's University of Belfast, Northern Ireland.

Tang, A.-M. and Cui, Y.-J. (2009). Modelling the thermomechanical volume change behaviour of compacted expansive clays. *Geotechnique*, **59**(3), 185–195.

Tang, G.X. and Graham, J. (2002). A possible elasto-plastic framework for unsaturated soils with high plasticity. *Canadian Geotechnical Journal*, **39**, 894–907.

Tang, G.X., Graham, J., Blatz, J. Gray, M. and Rajapakse, R.K.N.D. (2002). Suctions, stresses and strengths in unsaturated sand-bentonite. *Engineering Geology*, **64**(2–3), 147–156.

Tarantino, A. (2009). A water retention model for deformable soils. *Geotechnique*, **59**(9), 751–762.

Tarantino, A. and De Col, E. (2008). Compaction behaviour of clay. *Geotechnique*, **58**(3), 199–213.

Tarantino, A. and Mongiovi, L. (2000). A study of the efficiency of semipermeable membranes in controlling soil matrix suction using the osmotic technique. In: *Proceedings of the Asian Conference on Unsaturated Soils*, Singapore. Balkema, Rotterdam, the Netherlands, pp. 309–314.

Tarantino, A. and Mongiovi, L. (2002). Design and construction of a tensiometer for direct measurement of matric suction. In: *Proceedings of the Third International Conference on Unsaturated Soils, UNSAT'02*, Recife, Brazil, Vol. 1, Juca, J.F.T., deCampos, T.M.P. and Marinho, F.A.M. (eds), pp. 319–324.

Tarantino, A., Mongiovi, L. and Bosco, G. (2000). An experimental investigation on the independent isotropic stress variables for unsaturated soils. *Geotechnique*, **50**(3), 275–282.

Tarantino, A., Romero, E. and Cui, Y-J. (2009). *Laboratory and Field Testing of Unsaturated Soils*. Springer.

Tarantino, A. and Tombolato, S. (2005). Coupling of hydraulic and mechanical behaviour in unsaturated compacted clay. *Geotechnique*, **55**(4), 307–317.

Tatsuoka, F. (1988). Some recent developments in triaxial testing for cohesionless soils. ASTM STP 977, Philadelphia, PA.

Tavenas, F. and Leroueil, S. (1977). Effects of stress and time on yield of clays. In: *Proceedings of the Ninth International Conference on Soil Mechanic and Foundation Engineering*, Tokyo, Japan, Vol. 1, pp. 319–326.

Taylor, D.W. (1948). *Fundamentals of Soil Mechanics*. John Wiley, New York.

Terzaghi, K. (1936). The shear resistance of saturated soils. In: *Proceedings of the First International Conference on Soil Mechanics and Foundation Engineering*, Cambridge, MA, Vol. 1, pp. 54–56.

Terzaghi, K. (1943). *Theoretical Soil Mechanics*. Wiley, New York.

Thom, R., Sivakumar, R. Sivakumar, V. Murray, E.J. and Mackinnon, P. (2007). Pore size distribution of unsaturated compacted kaolin: the initial states and final states following saturation. *Geotechnique*, 57(5), 469–474.

Thom, R., Sivakumar, V., Brown, J.L. and Hughes, D.A. (2008). A simple triaxial system for evaluating the performance of unsaturated soils under repeated loading. *Geotechnical Testing Journal, ASTM*, 31(2), 107–114.

Toker, N.K., Germaine, J.T., Sjoblom, K.J. and Culligan, P.J. (2004). A new technique for rapid measurement of continuous soil moisture characteristic curves. *Geotechnique*, 54(3), 179–186.

Toll, D.G. (1990). A framework for unsaturated soil behaviour. *Geotechnique*, 40(1), 31–44.

Toll, D.G. (2003). On the shear strength of unsaturated soils. In: *International Conference on Problematic Soil*, Nottingham, UK, Vol. 1, pp. 127–136.

Toll, D.G., Ali Rahman, Z. and Gallipoli, D. (2008). Critical state conditions for an artificially bonded soil. In: *Unsaturated Soils: Advances in Geo-Engineering*, Toll, D.G. *et al.* (eds). Routledge, London, pp. 435–440. Proceedings of the First European Conference on Unsaturated Soils, E-UNSAT, Durham, UK.

Toll, D.G. and Ong, B.H. (2003). Critical state parameters for an unsaturated residual sandy soil. *Geotechnique*, 53(1), 93–103.

Tombolato, S., Tarantino, A. and Mongiovi, L (2008). A simple shear apparatus for testing unsaturated soils. In: *Unsaturated Soils: Advances in Geo-Engineering*, Toll, D.G. *et al.* (eds). Routledge, London, pp. 89–95. Proceedings of the First European Conference on Unsaturated Soils, E-UNSAT, Durham, UK.

Truong, H.V.P. and Holdon, J.C. (1995). Soil suction measurement with transistor psychrometers. In: *Unsaturated Soils*, Vol. 2, Alonso, E.E. and Delage, P. (eds). A.A. Balkema/Presses des Ponts et Chaussées, Paris, France, pp. 659–665. Proceedings of the First International Conference on Unsaturated Soils, Paris.

Vanapalli, S.K., Fredlund, D.G., Pufahl, D.E. and Clifton, A.W. (1996). Model for the prediction of shear strength with respect to soil suction. *Canadian Geotechnical Journal*, 33, 379–392.

Vanapalli, S.K., Nicoreta, M.V. and Sharma, S. (2008). Axis translation and negative water column techniques for suction control. *Geotechnical and Geological Engineering*, 26(6), 645–660.

Vasallo, R., Mancuso, C. and Vinale, F. (2007). Effects of net stress and suction on the small strain stiffness of a compacted clayey silt. *Canadian Geotechnical Journal*, 44, 447–462.

Villar, M.V. and Lloret, A. (2001). Variation of the intrinsic permeability of expansive clay upon saturation. In: *Clay Science for Engineering*, Adachi, K. and Fukue, M. (eds). Balkema, Rotterdam, the Netherlands, pp. 259–898.

Waldron, L.J. and Manbeian, T. (1970). Soil moisture characteristics by osmosis with polyethylene glycol: a simple system for osmotic pressure data and some results. *Soil Science*, 110(6), 401–404.

Wan, A.W., Graham, J. and Gray, N.N. (1990). Influence of soil structure on the stress-strain behaviour of sand-bentonite mixtures. *Geotechnical Testing Journal*, 13(3), 179–187.

Wang, Q., Pufahl, D.E. and Fredlund, D.G. (2002). A study of critical state on an unsaturated silty soil. *Canadian Geotechnical Journal*, 39, 213–218.

Ward, W.H., Samuels, S.G. and Butler M.E. (1959). Further studies of the properties of London clay. *Geotechnique*, 9, 33–58.

Watson, K.K. (1967). A recording field tensiometer with rapid response characteristics. *Journal of Hydrology*, 5, 33–39.

Washburn, E.W. (1921). Note on a method of determining the distribution of pore sizes in a porous material. *Proceedings of the National Academy of Sciences of the USA*, 7, 115–116.

Wheeler, S.J. (1986). *The Stress-Strain Behaviour of Soils Containing Large Gas Bubbles*. Model. DPhil thesis, University of Oxford, Oxford, UK.

Wheeler, S.J. (1991). An alternative framework for unsaturated soil behaviour. *Geotechnique*, **41**(2), 257–261.

Wheeler, S.J., Sharma, R.S. and Buisson, M.S.R. (2003). Coupling of hydraulic hysteresis and stress–strain behaviour in unsaturated soils. *Geotechnique*, **53**(1), 41–54.

Wheeler, S.J. and Sivakumar, V. (1992). Critical state concepts for unsaturated soils. In: *Proceedings of the Seventh International Conference on Expansive Soils*, Dallas, TX, Vol. 1, pp. 167–172.

Wheeler, S.J. and Sivakumar, V. (1993). Development and application of a critical state model for unsaturated soils. In: *Predictive Soil Mechanics*, Houlsby, G.T. and Schofield, A.N. (eds). Thomas Telford, London, pp. 709–728.

Wheeler, S.J. and Sivakumar, V. (1995). An elasto-plastic critical state framework for unsaturated soils. *Geotechnique*, **45**(1), 35–53.

Wheeler, S.J. and Sivakumar, V. (2000). Influence of compaction procedure on the mechanical behaviour of an unsaturated compacted clay. Part 2: shearing and constitutive modelling. *Geotechnique*, **50**(4), 369–376.

Wiebe, H.H., Brown, R.W., Daniel, T.W. and Campbell, E. (1970). Water potential measurement in trees. *Biosciences*, **20**(4), 225–226.

Witt, K.J. and Brauns, J. (1983). Permeability-anisotropy due to particle shape. *Journal of Geotechnical Engineering*, ASCE, **109**(9), 1181–1187.

Wood, D.M. (1990). *Soil Behaviour and Critical State Soil Mechanics*. Cambridge University Press, Cambridge, UK.

Woodburn, J.A., Holden, J. and Peter, P. (1993). The transistor psychrometer: a new instrument for measuring soil suction. In: *Unsaturated Soils*, Houston, S.L. and Wray, W.K. (eds). Geotechnical Special Publications No. 39. ASCE, Dallas, TX, pp. 91–102.

Woodburn, J.A. and Lucas, B. (1995). New approaches to the laboratory and field measurement of soil suction. In: *Unsaturated Soils*, Vol. 2, Alonso, E.E. and Delage, P. (eds). A.A. Balkema/Presses des Ponts et Chaussées, Paris, France, pp. 667–671. Proceedings of the First International Conference on Unsaturated Soils, Paris.

Wroth, C.P. (1973). A brief review of the applicability of plasticity theory to soil mechanics. In: *Proceedings of the Symposium on the Role of Plasticity in Soil Mechanics*, Cambridge, UK, pp. 1–11.

Zakaria, I., Wheeler, S.J. and Anderson, W.F. (1995). Yielding of unsaturated compacted kaolin. In: *Proceedings of the First International Conference on Unsaturated Soils*, Paris, France, pp. 223–228.

Zur, B. (1966). Osmotic control of the matrix soil water potential. *Soil Science*, **102**, 394–398.

Index